AF602082

Power Systems

André Veltman, Duco W.J. Pulle, Rik W. De Doncker
Fundamentals of Electrical Drives

More information about this series at http://www.springer.com/series/4622

André Veltman • Duco W.J. Pulle
R.W. De Doncker

Fundamentals of Electrical Drives

Second Edition

André Veltman
TU Eindhoven
Piak Electronic Design
Culemborg, Gelderland
The Netherlands

Duco W.J. Pulle
EMSynergy
Milperra, NSW
Australia

R.W. De Doncker
RWTH Aachen FB
FB 6 Elektrotechnik
Aachen, Germany

ISSN 1612-1287 ISSN 1860-4676 (electronic)
Power Systems
ISBN 978-3-319-29408-7 ISBN 978-3-319-29409-4 (eBook)
DOI 10.1007/978-3-319-29409-4

Library of Congress Control Number: 2016931895

© Springer International Publishing Switzerland 2007, 2016
This work is subject to copyright. All rights are reserved by the Publisher, whether the whole or part of the material is concerned, specifically the rights of translation, reprinting, reuse of illustrations, recitation, broadcasting, reproduction on microfilms or in any other physical way, and transmission or information storage and retrieval, electronic adaptation, computer software, or by similar or dissimilar methodology now known or hereafter developed.
The use of general descriptive names, registered names, trademarks, service marks, etc. in this publication does not imply, even in the absence of a specific statement, that such names are exempt from the relevant protective laws and regulations and therefore free for general use.
The publisher, the authors and the editors are safe to assume that the advice and information in this book are believed to be true and accurate at the date of publication. Neither the publisher nor the authors or the editors give a warranty, express or implied, with respect to the material contained herein or for any errors or omissions that may have been made.

Printed on acid-free paper

This Springer imprint is published by Springer Nature
The registered company is Springer International Publishing AG Switzerland

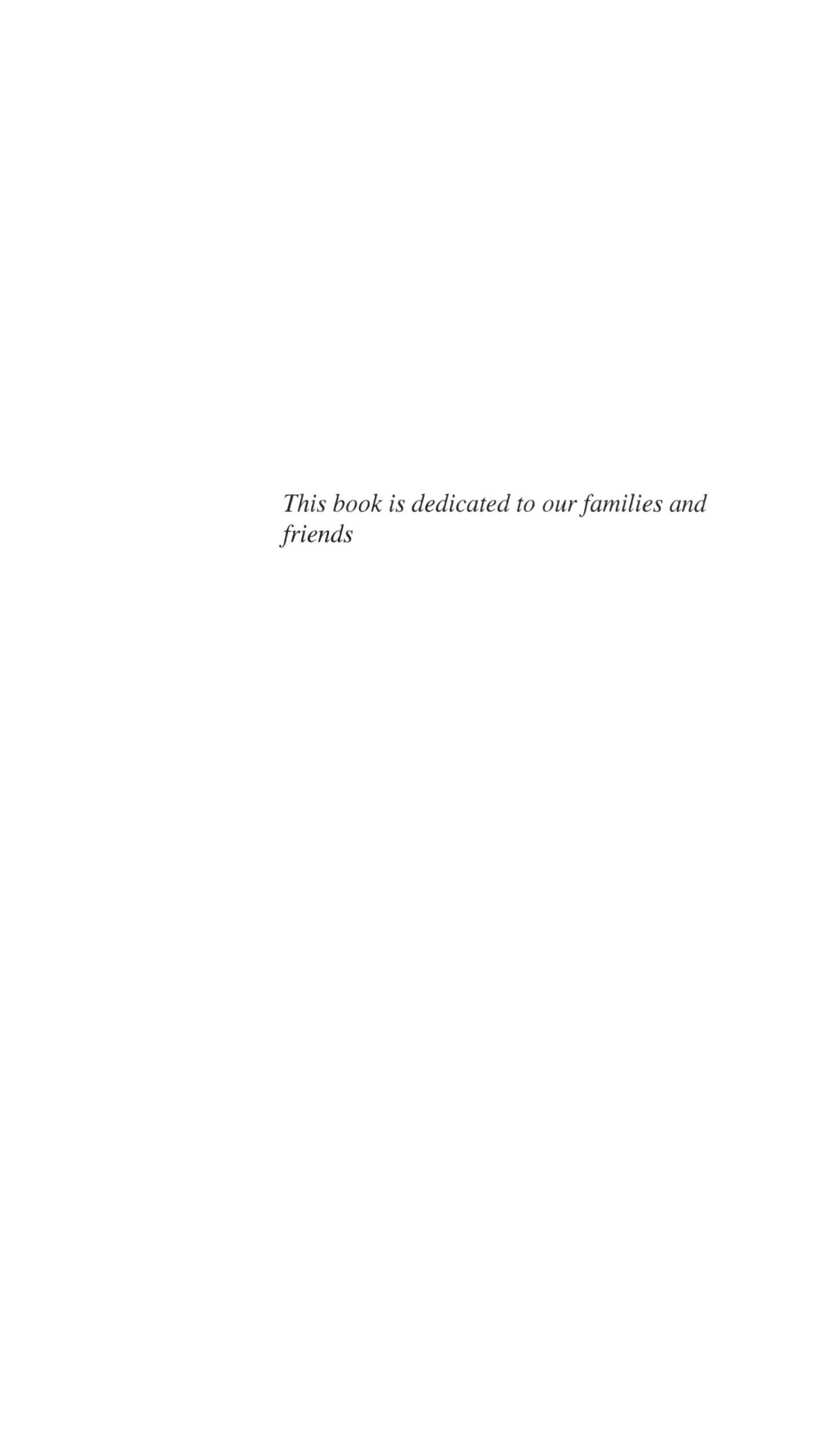

This book is dedicated to our families and friends

Foreword

Within one academic lifetime, the electric drive has progressed from the three-machine DC drive called the Ward-Leonard system to today's sophisticated AC drives utilizing PWM inverter power electronics and field orientation or direct torque control. Roughly around the same period, machine theory progressed from the classical "one machine at a time" approach to the generalized or unified approach emphasizing similarities between machine types. This unified theory also utilized much more sophisticated mathematical tools to obtain models applicable to transients as well as steady state. This enabled theoretical modeling a host of important machine problems, but almost always required computer solutions as opposed to more general analytic solutions. This often left one with a feeling of detachment from the physical reality of inrush currents, the whine of spinning rotors, and the smell of over-warm electrical insulation.

Partway through my academic lifetime, I was introduced to the next phase of unified theory; the use of complex notation to model the effective spatial orientation of quantities within a machine. This concept, often called space vector theory, provides a much clearer mathematical picture of what is happening in a machine, but at the expense of another level of abstraction in the model. However, the insights provided to one initiated in the method are so significant that today essentially all work in drive control is presented in this format. And therein lies a problem. To the uninitiated these presentations appear quite unintelligible, and a route to becoming initiated is generally hard to find and often harder to follow once found.

This purpose of this book is to show the theory and notation used, in modern electric drive analysis and design at an introductory level. The authors, bring an exceptional breadth of knowledge to this book making it stand out from other books that only providing mathematical foundation for advanced work. This strong effort is made to present the physical basis for all of the major steps in development, and to give the space vector physical and mathematical meaning. Readers using the book for self-study will find the sets of simulation tutorials at the end of each chapter of special value in mastering the implications and fine points of the material covered in the chapter.

Electric machine theory with its interacting temporal, spatial variations and multi-winding topologies can appear to be a very complicated and difficult subject. The approach followed in this book is, I believe, one that will help eliminate this perception by providing a fundamental, coherent, and user-friendly introduction to electric machines for those beginning a serious study of electric drive systems.

Madison, WI, USA Donald W. Novotny

Preface

Our motivation and purpose for writing this book stems from our belief that there is a practical need for a learning platform which will allow the motivated reader to gain a basic understanding of the modern multidisciplinary principles which govern electrical drives. The book in question should appeal to those readers who have an elementary understanding of electrical circuits and magnetics and who have an interest or need to comprehend advanced textbooks in the field of electrical drives. Consideration has also been given to those interested in using this book as a basis for teaching this subject matter. In this context, a Springer website *Extra Materials* has been set up which contains the simulation examples and tutorials discussed in this book. Furthermore, all the figures in this book are available on the Springer website, in order to assist lecturers with the preparation of electronic "power point" type lectures.

Electrical drives consist of a number of components: the electrical machine, converter, and controller, all of which are discussed at various levels. A brief résumé of magnetic and electrical circuit principles is given in Chap. 1 together with a set of generic building modules which are used throughout this book to represent dynamic models. Chapter 2 is designed to familiarize the reader with the process of building a dynamic model of a coil with the aid of generic modules. This part of the text contains an introduction on phasors as required for steady-state analysis. The approach taken in this and the following chapters is to present a physical model, which is then represented by a symbolic model with the relevant equation set. A generic model is then presented which forms the basis for a set of *build and play* simulations set out in various steps in the tutorial at the end of the chapter.

Chapter 3 introduces a single-phase *ideal transformer* (ITF) which forms the basis of a generic transformer model with leakage and magnetizing inductance. A phasor analysis is given to familiarize the reader with the steady-state model. The *build and play* tutorials at the end of the chapter give the reader the opportunity to build and analyze the transformer model under varying conditions. It is emphasized that the use of these *build and play* sets are essential components of the learning process throughout this book.

Chapter 4 deals with star and delta connected three-phase systems and introduces the generic modules required to model such systems. The space vector-type representation is also introduced in this part of the text. A set of *build and play* tutorials are given which reinforce the concepts introduced in this chapter.

Chapter 5 deals with the concepts of real and reactive power in single-as well as three-phase systems. Additional generic modules are introduced in this part of the text, and tutorial examples are given to familiarize the reader with this material.

Chapter 6 extends the ITF concept introduced earlier to a space vector-type model which is represented in a symbolic and generic form. In addition, a phasor-based model is also given in this part of the text. The *build and play* tutorials are self-contained step-by-step simulation exercises which are designed to show the reader the operating principles of the transformer under steady-state and dynamic conditions. At this stage of the text, the reader should be familiar with building and using simulation tools for space vector-type generic models which form the basis for a transition to rotating electrical machines.

Chapter 7 introduces a unique concept, namely, the *ideal rotating transformer* (IRTF), which is the fundamental building block that forms the basis of the dynamic electrical machine models discussed in this book. A generic space vector-based IRTF model is given in this part of the text which is instrumental in the process of familiarizing the reader with the torque production mechanism in electrical machines. This chapter also explores the conditions under which the IRTF module is able to produce a constant torque output. It is emphasized that the versatility of the IRTF module extends well beyond the electrical machine models discussed in this book. These advanced IRTF-based machine concepts are used in our second book *Advanced Electrical Drives* [2] and also in our third book *Applied Control of Electrical Drives* [10]. The latter-mentioned book has been recently introduced to facilitate the transition to experimental drives by the reader. The *build and play* tutorials at the end of this chapter serve to reinforce the IRTF concept and allow the reader to "play" with the conditions needed to produce a constant torque output from this module.

Chapters 8–9 deal with the implementation of the IRTF module for synchronous and asynchronous machines. In both cases, a simplified IRTF-based symbolic and generic model is given of the machine in question to demonstrate the operating principles. This model is then extended to a "full" dynamic model as required for modeling standard electrical machines. A steady-state analysis of the machines is also given in each chapter. In the sequel of each chapter, a series of *build and play* tutorials are introduced which take the reader through a set of simulation examples which steps up from a very basic model designed to show the operating principles, to a full dynamic model which can be used to represent the majority of modern AC electrical machines in use today.

Chapter 10 dealt with the DC machine, for which a dynamic model is introduced. In addition, the steady-state torque/speed characteristics of this machine with either PM or field excitation are discussed.

Chapter 11 deals with the converter, modulation, and control aspects of the electrical drive at a basic level. Both half- and full-bridge converter concepts are discussed together with the pulse width modulation (PWM) strategies that are in use in modern drives. A model-based current control algorithm is presented in combination with a DC machine. The *build and play* tutorials in the sequel of this chapter clearly show the operating principles of PWM-based current-controlled electrical drives.

The purpose, content, and approach of our book have been presented above. On the basis of this material, the following set of unique points are presented below in response to the question as to why prospective readers should purchase this book:

- The introduction of an *ideal rotating transformer* (IRTF) module concept is a basic didactic tool for introducing the elementary principles of torque production in electrical machines to the uninitiated reader. The apparent simplicity of this module provides the reader with a powerful tool which can be used for the understanding and modeling of a very wide range of electrical machines well beyond those considered in this book.
- The application of the IRTF module to AC machines provides a unique insight into their operation principles. The book shows the transitional steps needed to move from a very basic IRTF model to a full IRTF-based dynamic model usable for representing the dynamic and steady-state behavior of most machines in use today. In addition the IRTF based module can be readily extended to include more specific machine effects such as "skin effect" in asynchronous machines. Furthermore, the IRTF module can be extended to machine models outside the scope of this book. Examples which appear in the book *Advanced Electrical Drives* by the authors of this text are the salient pole PM machine and the single-phase IRTF-based induction machine.
- This text is designed to bridge the gap between advanced textbooks covering electrical drives and textbooks at either a fundamental electrical circuit level or more generalized mechatronic books. Our text is accompanied by a set of tutorials which are located in the *Extra Materials section* at the Springer website. This book should fit well into the undergraduate curriculum for students who have completed first or second year and who have an interest in seeking a career in the area of electrical drives. The book should also appeal to engineers with a non-drive background who have a need to acquire a better understanding of modern electrical drive principles.
- The use of *build and play*-type tutorials is of fundamental importance to understanding the theory presented in the text. The didactic role of modern simulation tools in engineering cannot be overestimated, and it is for this reason that extensive use is made of generic modules which are in turn used to build complete models of the drive. Such an approach allows the reader to visualize the complex equation set which is at the basis of these models. The simulation tool used in these tutorials is "PLECS®" which can be used with MATLAB/SIMULINK or (as is the case in this book) as "stand-alone" software. The said tutorials are linked directly to the generic modules discussed in the corresponding chapter and are included in the *Extra Material, Springer website:extras.springer.com* linked to this book.
- A series of "demonstration" laboratories are introduced which are used to experimentally verify key theoretical concepts/models introduced in this book.

Hence, it is hoped that the critical reader will be convinced that the material presented in this book is applicable to actual electrical drives.

The second edition of this book has been tailored to the text *Advanced Electrical Drives* by the same authors. Notably some changes have been made to ease the readers' transition to our textbooks *Advanced Electrical Drives* as well as *Applied Control of Electrical Drives*. Notably the new edition makes use of so-called "amplitude invariant" space vectors, which is in line with the approach used in *Applied Control of Electrical Drives*. Specifically, Chap. 3 has been extensively revised to introduce the so-called "universal-oriented model approach" at an early stage. Furthermore, Chap. 10 on DC machines has been simplified. Finally, in Chap. 11 the term "incremental flux" has been omitted and replaced by the variable *average voltage per sample* given its use in our other books. The said chapter has also been extended to cover "H-bridge" operation.

Culemborg, The Netherlands — André Veltman
Sydney, NSW, Australia — Duco W.J. Pulle
Aachen, Germany — R.W. De Doncker

Acknowledgments

The process of writing this book has not been without its difficulties. That this work has come to fruition stems from a deep belief that the material presented in this book will be of profound value to the engineering community as a whole and the educational institutions in particular.

The content of this second edition reflects upon the collective academic and industrial experience of the authors concerned. In this context, the input of students in general and other colleagues cannot be overestimated. In particular, the authors wish to thank the staff and students of the Institute for Power Electronics and Electrical Drives (ISEA), RWTH Aachen University, Germany.

In addition, the authors would like to thank the various industrial institutions (in alphabetical order) who have been involved with this work, namely, EMsynergy, Sydney, Australia; Piak Electronic Design, Culemborg, the Netherlands, and Plexim GmbH, Zurich, Switzerland.

A number of individuals have played a major role in terms of bringing the second edition of this book to fruition. In particular, we would like to thank Stefan Engel, Krassimir Gurov, Annegret Klein-Heßling, and Michael Schubert for their extensive work on editing graphs and diagrams for the PLECS®-based tutorials.

Contents

Chapter 1
Introduction

1.1 Why Use Electro-Mechanical Energy Conversion?

Electric motors are around us everywhere. Generators in power plants are connected to a three-phase power grid of alternating current (AC), pumps in your heating system, refrigerator, and vacuum cleaner are connected to a single phase AC grid and switched on or off by means of a simple contactor. In cars a direct current (DC) battery is used to provide power to the starter motor, windshield wiper motors, and other utilities. These motors run on direct current and in most cases they are activated by a relay switch without any control.

Many applications driven by electric motors require more or less advanced control. Lowering the speed of a fan or pump can be considered relatively simple. Perhaps one of the most difficult ones is the dynamic positioning of a tug in a wafer-stepper with nanometer accuracy while accelerating at several g's. Another challenging controlled drive is an electric crane in a harbor that needs to be able to move an empty hook at high speed, navigate heavy loads up and down at moderate velocities, and make a soft touchdown as close as possible to its intended final position. Other applications such as assembly robots, electric elevators, electric motor control in hybrid vehicles, trains, streetcars, or CD-players can, with regard to complexity, be situated somewhere in between.

Design and analysis of all electric drive systems requires not only knowledge of dynamic properties of different motor types, but also a good understanding of the way these motors interact with power electronic converters and their loads. These power converters are used to control motor currents or voltages in various manners.

Compared to other drive systems such as steam engines (still used for aircraft launch assist), hydraulic engines (famous for their extreme power per volume),

Electronic supplementary material The online version of this chapter (doi: 10.1007/978-3-319-29409-4_1) contains supplementary material, which is available to authorized users.

© Springer International Publishing Switzerland 2016

A. Veltman et al., *Fundamentals of Electrical Drives*, Power Systems,
DOI 10.1007/978-3-319-29409-4_1

pneumatic drives (famous for their simplicity, softness, and hissing sound), combustion engines in vehicles, or turbo-jet drives in helicopters or aircrafts, electric drive systems have a very wide field of applications thanks to some strong points:

- Large power range available: actuators and drives are used in a very wide range of applications from wrist watch micro-watt level to machines at the multi-megawatt level, e.g., as used in coal mines, steel industry, and ship propulsion systems.
- Electrical drives are capable of full torque at standstill, hence no clutches are required.
- Electrical drives can provide a very large speed range, usually gearboxes can be omitted.
- Clean operation, no oil-spills to be expected.
- Safe operation is possible in environments with explosive fumes (pumps in oil-refineries).
- Immediate use: electric drives can be switched on immediately.
- Low service requirement: electrical drives do not require regular service as there are very few components subject to wear, except the bearings. This means that electrical drives can have a long life expectancy, typically in excess of 20 years.
- Low no-load losses: when a drive is running idle, little power is dissipated since no oil needs to be pumped around to keep it lubricated. Typical efficiency levels for a drive are in the order of 85 %. In some cases this may be as high as 98 %. The higher the efficiency the more costly the drive technology, in terms of initial costs.
- Electric drives produce very little acoustic noise compared to combustion engines.
- Excellent control ability: electrical drives can be made to conform to precise user requirements. This may, for example, be in relation to realizing a certain shaft speed or torque level.
- "Four-quadrant operation": Motor and braking mode are both possible in forward or reverse direction, yielding four different quadrants: forward motoring, forward braking, reverse motoring, and reverse braking. Positive speed is called forward, reverse indicates negative speed. A machine is in motor mode when energy is transferred from the power source to the shaft, i.e., when both torque and speed have the same sign.

1.1.1 Modes of Operation

When a machine is in motoring mode, most of the energy is transferred from the electrical power source to the mechanical load. Motoring mode takes place in quadrants 1 and 3 (see Fig. 1.1b). If the shaft torque and shaft speed are in opposition, then the flow of energy is reversed, in which case the drive is in the so-called braking mode.

Braking comes in three "flavors." The first is referred to as "regenerative" braking operation, where most of the mechanical energy from the load is returned to

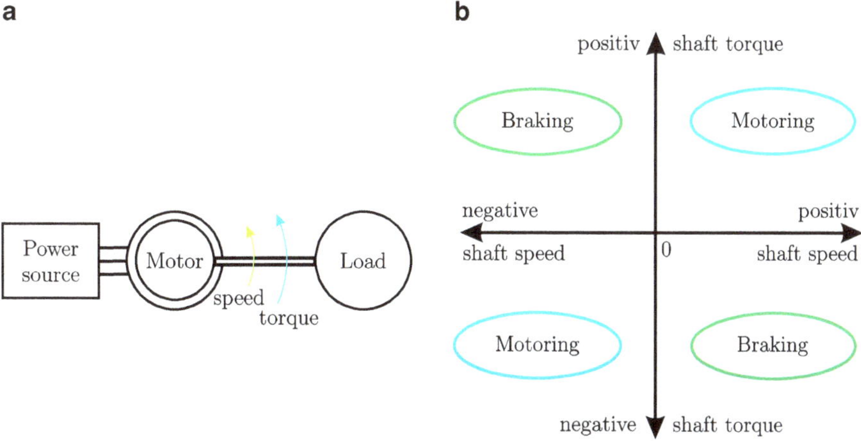

Fig. 1.1 Motoring and braking operation. (**a**) Motor with power supply. (**b**) Operating modes

the power source. Most drives which contain a converter (see Sect. 1.2) between motor and supply use a diode rectifier as a front end, hence power can only flow from the AC power grid to the DC-link in the drive and not the other way around. In such converters regenerative operation is only possible when the internal DC-link of the drive is shared with other drives that are able to use the regenerated power immediately. Sharing a common rectifier with many drives is economic and becoming standard practice. Furthermore, attention is drawn to the fact that some power sources are not able to accept any (or only a limited amount) of regenerated energy.

The second option is referred to as "dissipative" braking operation. Typically, this method is used to dissipate irreversibly the kinetic energy of the mechanical load system in an external brake-chopper-resistor. A brake-chopper can burn away a substantial part of the rated power for several seconds, designed to be sufficient to stop the mechanical system in a fast and safe fashion. One can regard such a brake-chopper as a big Zener-diode that prevents the DC-link voltage in the converter from rising too high. Brake-choppers come in all sizes, in off-shore cranes and locomotives power levels of several megawatts are common practice.

The third braking mode is the one where mechanical power is completely returned to the motor, while at the same time some electrical power may still be delivered, i.e., both mechanical and electrical input power are dissipated in the motor. Think of a permanent magnet motor being shorted, or an induction motor that carries a DC current in its stator, acting as an eddy-current-brake.

Of course there are also disadvantages when using electrical drive technology, a few of these are briefly outlined below.

- Low torque/force density compared to combustion engines or hydraulic systems. This is why aircraft control systems are still mostly hydraulic. However, there is an emerging trend in this industry to use electrical drives instead of hydraulic systems.

- High complexity: A modern electrical drive encompasses a range of technologies as will become apparent in this book. This means that it requires highly skilled personnel to repair or modify such systems.

1.2 Key Components of an Electrical Drive System

The "drive" shown in Fig. 1.1a is in fact only an electrical machine connected directly to a power supply. This configuration is widely in use but one cannot exert very much control in terms of controlling torque and/or speed. Such drives are either on or off with rather wild starting dynamics. The drive concept of primary interest in this book is capable of what is referred to as "adjustable speed" operation [8] which means that the machine can be made to operate over a wide speed range. A simplified structure of an adjustable speed drive is shown in Fig. 1.2. A brief description of the components is given below:

- Load: This component is central to the drive in that the purpose of the drive is to meet specific mechanical load requirements. It is emphasized that it is important to fully understand the nature of the load and the user requirements which must be satisfied by the drive. The load component may or may not have sensors to measure either speed, torque, or shaft angle. The sensors which can be used are largely determined by the application. The nature of the load may be translational or rotational and the drive designer must make a prudent choice whether to use a direct-drive with a large motor or geared drive with a smaller but faster one. Furthermore, the nature of the load in terms of the need for continuous or intermittent operation must be determined.
- Motor: A limited range of motor types is presently in use. Among these are the so-called classical machines, which have their origins at the turn of the nineteenth century. This classical machine set has displaced a large assortment

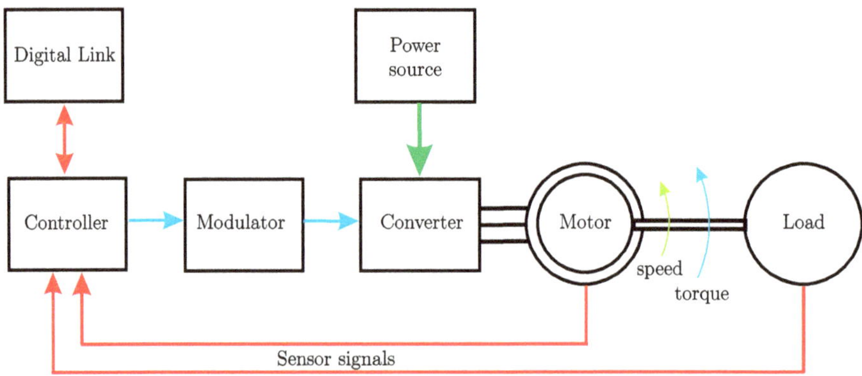

Fig. 1.2 Typical drive setup

of "specialized" machines used prior to the introduction of power electronic converters for speed control. This classical machine set contains the DC (Direct Current) machine, asynchronous (induction) machine, synchronous machine, and "variable reluctance" machine. Of these the "variable reluctance" machine will not be discussed in this book. A detailed discussion of this machine appears in the textbook *Advanced electrical drives* written by the authors of this book.

The term "motor" refers to a machine which operates as a motor, i.e., energy flows from the motor to the load. When the energy flows in the opposite direction a machine is said to operate as a generator.

- Converter: This unit contains a set of power electronic (semiconductor) switches which are used to manipulate the energy transfer between power supply and motor. The use of switches is important given that no power is dissipated (in the ideal case) when the switches are either open or closed. Hence, theoretically, the efficiency of such a converter is 100 %, which is important particularly for large converters given the fact that semiconductor devices cannot operate at high temperatures. Hence, it is not possible to absorb high losses which inevitably appear in the form of heat. A large range of power electronic switches is available to the designer to meet a wide range of applications.
- Modulator: The switches within the converter are controlled by the modulator which determines which switches should be on, and for what time interval, normally on a micro-second timescale. An example is the pulse width modulator that realizes a required pulse width at a given carrier-frequency of a few kHz.
- Controller: The controller, typically a digital signal processor (DSP), or micro-controller (MCU) in combination with programmable logic devices, contains a number of software based control loops which control and protect, for example, the currents in the converter and machine. In addition, torque, speed, and shaft angle control loops may be present within this module. Shown in the diagram are the various sensor signals which form the key inputs to the controller together with a number of user set-points (not shown in the diagram). The output of the controller is a set of control parameters which are used by the modulator.
- Digital link: This unit serves as the interface between the controller and an external computer. With the aid of this link drive set-points and diagnostic information can be exchanged with a remote user.
- Power supply: In most cases the converter requires a DC voltage source (DC voltage link). The power can be obtained directly from a DC power source, in case one is available, for example, batteries in electric vehicles. However, in most cases the DC power requirements are met via a rectification process, which makes use of the single or three-phase AC (referred to as the "grid") power supply as provided by the utility grid.

1.3 What Characterizes High Performance Drives?

Prior to moving to a detailed discussion of the various drive components it is important to understand the reasons behind the ongoing development of drives. Firstly, an observation of the drive structure (see Fig. 1.2) shows that the drive has components which cover a very wide field of knowledge. For example, moving from load to controller one needs to appreciate the nature of the load, have a thorough understanding of the motor, and comprehend the functioning of the converter and modulator. Finally, one needs to understand the control principles involved and how to implement (in software) the control algorithms into a micro-processor or DSP. Hence, there is a need to have a detailed understanding of a very wide range of topics which is perhaps one of the most challenging aspects of working in this field. The development of electrical machines occurred, as was mentioned earlier, more than a century ago. However, the step to a high performance adjustable speed drive took considerably longer and is in fact still ongoing. The main reasons as to why drive technology has improved over the last decades are briefly outlined below:

- Availability of fast and reliable power semiconductor switches for the converter: A range of switches is available to the user today to design and build a wide range of converter topologies. The most commonly used switching devices for motor drives are MOSFETs for low-voltage applications and IGBTs for medium (kW) and higher (MW) powers. In addition GCTs are available for medium-voltage and high-voltage applications.
- Availability of fast computers for (real time) embedded control: the controller needs to provide the control input to the modulator at a sampling rate which is typically in the order of 100 μs. Within that time frame the computer needs to acquire the input data from sensors and user set-points and apply the control algorithm in order to calculate the control outputs for the next cycle. The presence of low cost fast micro-processors or DSPs since the mid-eighties has been of key importance for drive development.
- Better sensors: A range of reliable and low cost sensors is available to the user which provides accurate inputs for the controller such as LEMs, incremental encoders, and Hall-effect sensors.
- Better simulation packages: The availability of sophisticated so-called finite-element computer aided design (CAD) packages for motor design has been instrumental in gaining a better understanding of machines. Furthermore, they have been and continue to be used for designing machines and for optimization purposes. In terms of simulating the entire drive structure there are simulators with graphical user interfaces, such as, among others, MATLAB/Simulink® and PLECS® which allow the user to analyze a detailed dynamic model of the entire system. This means that one can analyze the behavior of such a system under a range of conditions and explore new control techniques without the need of

actually building the entire system. This does not mean that implementing real life systems is no longer required. The proof of the pudding is in the eating, and only experimental validation can prove that the supposedly exact models are indeed valid for a real drive system, which was our motivation for writing our latest book *Applied Control of Electrical Drives* [10].

Simulation and experiment are never exactly the same. When the models are not able to describe the drive system under certain conditions, it might be useful to enhance the simulation model to incorporate some of the found differences. As engineers, we should be aware of the fact that drive systems are often closed-loop systems that are able to tolerate (to some extent) deviations in parameters and unknown load torques without any problem. To paraphrase Einstein, "A simulation model should be as simple as possible, but no simpler" is the key to a successful simulation. This means that essential dynamics or non-linearities, found in the real world system, need to be implemented in the (physics based) simulation model in order to study extreme situations with acceptable accuracy.

The simulation model used depends on what needs to be studied. Simulating pulse width modulated outputs requires a very short simulation time-step, in the order of sub-μs or so, while the overall mechanical system and the motor's response can be calculated at a hundred times larger time-step with negligible loss of accuracy, as long as the power converter is regarded as a non-switching controlled voltage source. Another extreme example is the study of thermal effects on the motor. In that case only the average power dissipation in terms of seconds or even minutes is of interest.

- Better materials: The availability of improved magnetic, electrical, and insulation materials has provided the basis for efficient machines capable of withstanding higher temperatures, thereby offering long application life and low life-cycle costs.

1.4 Notational Conventions

1.4.1 Voltage and Current Conventions

The conventions used in this book for the voltage and current variables are shown with the aid of Fig. 1.3. The diagram shows the variables: voltage u and current i, which are specifically given in "lower case" notation, because they represent instantaneous values, i.e., a function of time. The voltage and current "arrows" shown in Fig. 1.3 point to the negative terminal of the respective circuit, i.e., motoring arrow system, in which positive power $p = ui$ means power absorbed by the electrical circuit (load).

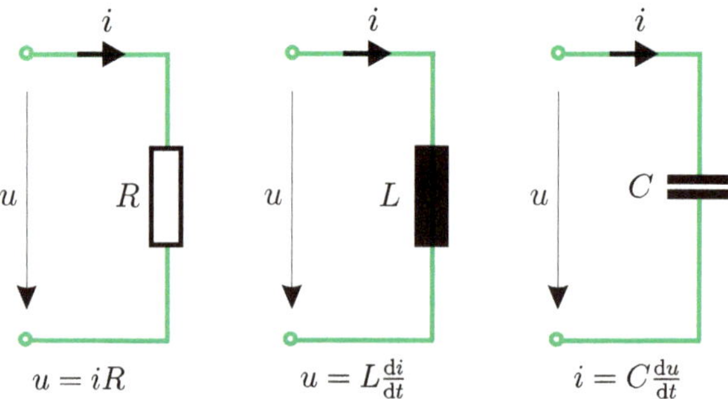

Fig. 1.3 Notation conventions used for electrical quantities

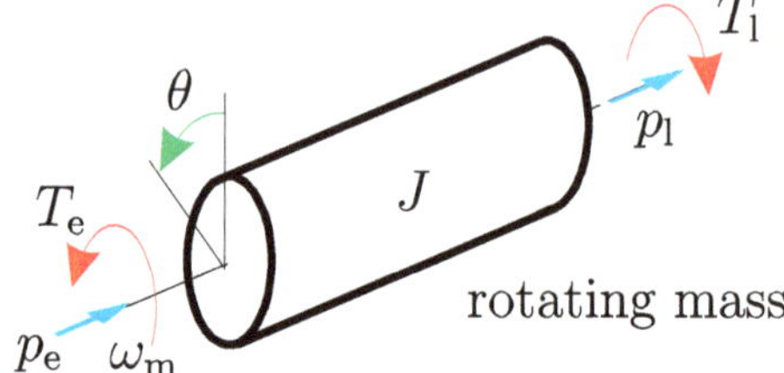

Fig. 1.4 Notation conventions used for mechanical quantities

1.4.2 Mechanical Conventions

The mechanical conventions used in this book are shown with the aid of Fig. 1.4. The electro-magnetic torque T_e produced by the machine corresponds with a power output $p_e = T_e\omega_m$, where ω_m represents the rotational speed, otherwise known as the angular frequency. The load torque T_l is linked to the power delivered to the load $p_l = T_l\,\omega_m$. Ignoring bearing and windage losses in the machine, the torque difference $T_e - T_l$ results in an acceleration $J{}^{d\omega_m}/_{dt}$ of the total rotating mass, which is characterized by its inertia J. This rotating structure is represented as a lumped mass formed by the rotor of the motor, motor shaft, and load. The corresponding mechanical equation which governs this system is of the form

$$J\frac{d\omega_m}{dt} = T_e - T_l \tag{1.1}$$

The angular frequency may also be written as $\omega_m = {}^{d\theta}/_{dt}$ where θ represents the rotor angle.

Figure 1.4 shows the machine operating as a motor, i.e., $T_e > 0$ and $\omega_m > 0$. These motor conventions are used throughout this book.

1.5 Use of Building Blocks to Represent Equations

Throughout this book the so-called generic models of drive components will be applied to build a useful simulation model of an electrical drive system [7]. Models of this type are directly derived from the so-called symbolic representation of a given drive component. The generic models are dynamic models which can be implemented as "control blocks" in a practical simulation environment such as PLECS [9]. Models in this form can then be analyzed by the reader in terms of the expected transient or steady-state response. Furthermore, changes can be made to a model to observe their effect. This interactive type of learning process is particularly useful to become familiar with the material.

An example of moving from symbolic to generic representation is given in Fig. 1.5. The symbolic model shown in Fig. 1.5 represents a resistance. As such, the resistance represents a relation between voltage and current as defined by Ohm's law: you can calculate current from voltage, voltage from current, or resistance from both voltage and current. The generic diagram assumes in this case that the voltage u is an input and the current i represents the output variable for this building block known as a gain module. The gain for this module must in this case be set to $^1/_R$. Also shown in this figure is the corresponding PLECS representation, as needed to implement this transfer function. Throughout this book additional building blocks will be introduced as they are required. At this point, a basic set will be given which will form the basis for the first set of generic models to be discussed in this book.

1.5.1 *Basic Generic Building Block Set*

The first set of building blocks given in Fig. 1.6 are linked to "example" transfer functions. Added to these building blocks are the equivalent PLECS "control blocks" which can be used to implement the transfer function under consideration. For example, the GAIN module has as input the current i and as output u, the gain is set to R. The INTEGRATOR example module has as input the variable ΔT and output ω_{m}. The gain of the integrator is $^1/_J$. *Note that the module shows the gain as J and not $^1/_J$.* The equivalent PLECS implementation requires two control blocks

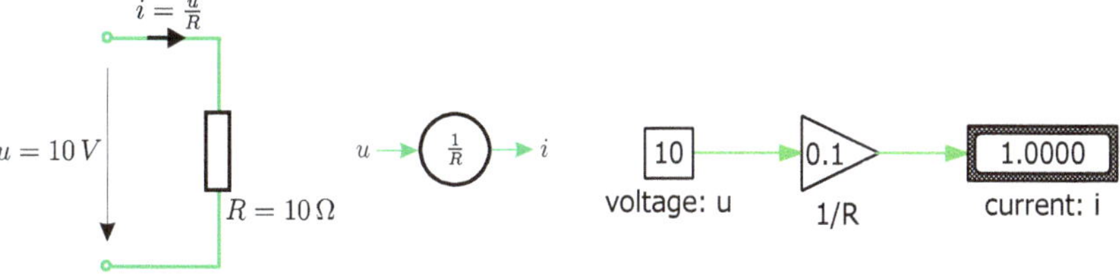

Fig. 1.5 Symbolic, generic, and PLECS representations

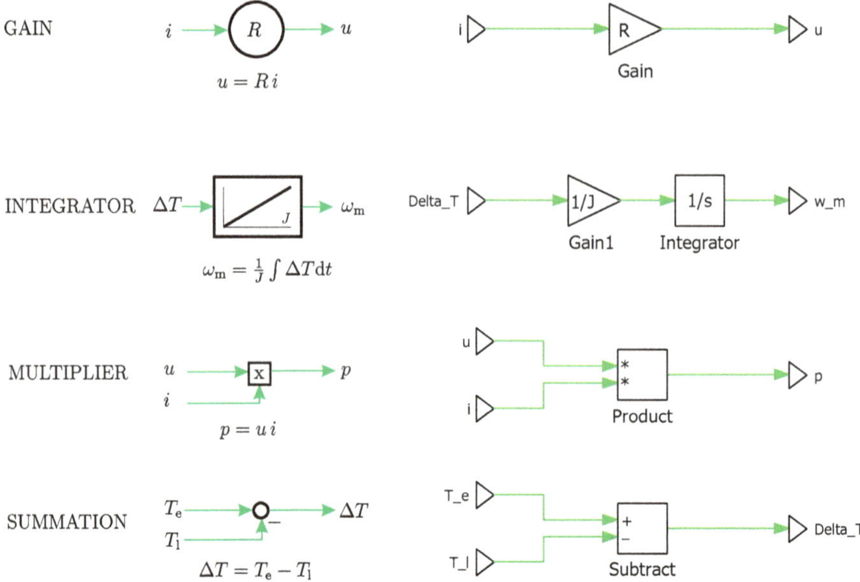

Fig. 1.6 Basic building block set

with gain `1/J` and an `integrator`. When multiplying two variables in the time domain, a MULTIPLIER module is used. This module differs from the given GAIN module in that the latter is used to multiply a variable with a constant. Finally, an example of a SUMMATION module is given. In this case the output is a variable ΔT and subtracts the input variable T_l from input variable T_e. *Note that in the case of adding two variables no "plus" symbol is placed. A "minus" sign is used when subtracting two variables.* In PLECS either an `Add` or `Subtract` control block is used, respectively, for adding or subtracting two variables.

An example of combining some of these modules is readily given by considering the following equation

$$u = iR + L\frac{di}{dt} \tag{1.2}$$

which represents the voltage across a series network in the form of an inductance L and resistance R. To build a generic representation with the voltage as input variable and current as output variable, it is helpful to rewrite the expression in its differential equation form

$$\frac{di}{dt} = \frac{1}{L}(u - iR) \tag{1.3}$$

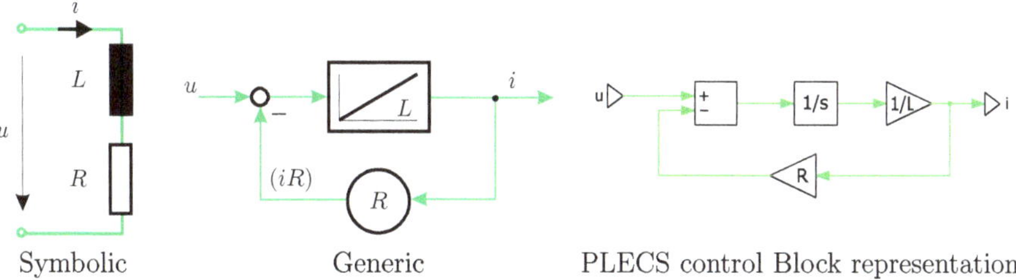

Fig. 1.7 Example of using basic building blocks

In this case the output of the integrator is the variable i and the input of the integrator is given as $(u - iR)$, hence

$$i = \frac{1}{L}\int (u - iR)\,\mathrm{d}t \tag{1.4}$$

The initial current is assumed to be zero, i.e., $i(0) = 0$. An observation of Eq. (1.4) shows that the integrator input is formed by the input variable u from which the term iR must be subtracted where use is made of a summation unit, as shown in the generic model presented in Fig. 1.7. The gain "$1/L$" present in Eq. (1.4) appears in the generic integrator module as "L" as discussed previously. The complete generic and symbolic diagrams for this example are given in Fig. 1.7. Also added to this figure is the PLECS implementation where use is made of "control blocks." Note that "symbolic" models can be implemented directly in PLECS as "electrical models" without further notable changes to appearance as will be shown in the tutorial section.

1.6 Magnetic Principles

Prior to looking at the various components of a drive it is important to revise the basic magnetic principles. On the basis of these principles we will examine the so-called *ideal transformer* (ITF) and the *ideal rotating transformer* (IRTF). The book by Hughes [5] is highly recommended as it provides an excellent primer in the area of magnetic principles and drives. We will follow a similar line of thinking for the magnetic principles section in this book.

1.6.1 Force Production

The production of electro-magnetic torque T_e in rotating electrical machines, such as those considered in this book, is directly linked to the question how forces are

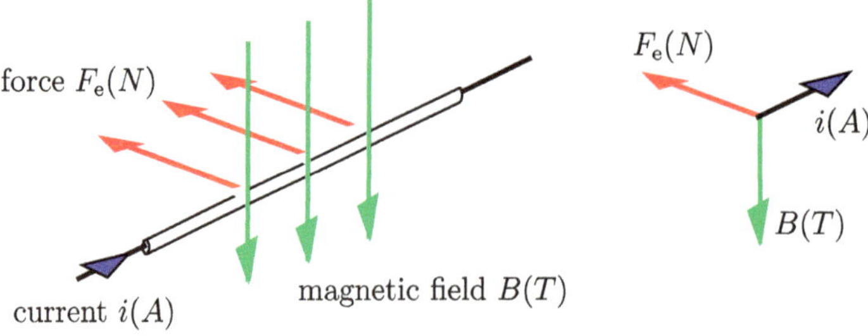

Fig. 1.8 Relationship between current, magnetic field, and force

produced. It is noted that other types of machines exist where torque production is based on either reluctance, electro-static, piezo-electric, or magneto-restrictive principles. Machines which abide with those principles are not considered in this book. The basic relationship between force, current in a conductor, and magnetic field has been discovered by Lorentz. The directions of the three variables are at right angles with respect to each other and under these circumstances the force magnitude acting on a conductor (exposed over a length l to a flux density B and carrying a current i) is given as

$$F_e = B i l \tag{1.5}$$

where l is the length (in meters) of the conductor section which is exposed to the field. Force is expressed in newtons (N) (see Fig.1.8).

1.6.2 *Magnetic Flux and Flux Density*

Prior to discussing the concept of flux density it is helpful to understand the meaning of flux lines. Consider a bar magnet as given in Fig. 1.9a, which also shows a set of so-called magnetic field lines. Between each pair of adjacent lines there is a fixed quantity of magnetic flux. This amount is represented as a "flux tube" and an example is given in Fig. 1.9a. The meaning of flux density B within such a tube is defined as the flux in the tube divided by the tube cross-section. For simplicity we will assume a unity length tube in the dimension perpendicular to the plane shown in Fig. 1.9a, hence the cross-section (of the tube) is directly proportional to the width of the tube shown in Fig. 1.9a. This means that the flux density in the tube increases as the tube becomes narrower. Within the magnet, the flux density is considerably higher than outside. A flux density plot of the same magnet is shown in Fig. 1.9b. The use of field line and flux density plots as presented in

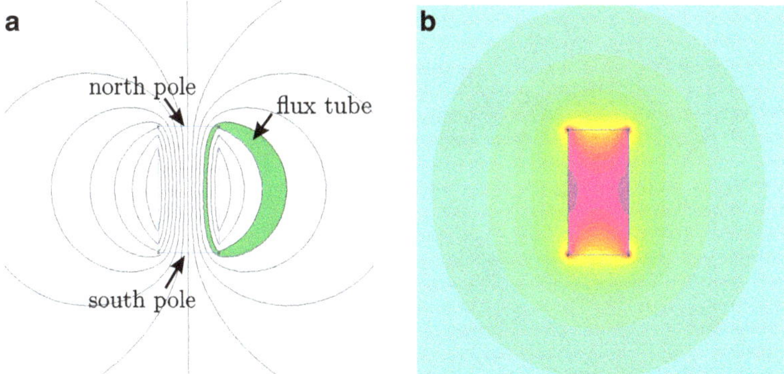

Fig. 1.9 Bar magnet flux and flux density plot. (**a**) Flux plot. (**b**) Flux density plot

Fig. 1.9a (and subsequent similar figures in this chapter) were obtained with the two-dimensional magnetic analysis package FEMM [3]. Such magnetic plots are extremely valuable to designers as they enable one to look at "hot spots," i.e., places where the flux density is very high. Colors indicate different flux density values, which 'red' being the highest. Clearly the bar magnet in its present form cannot be considered as a source with a uniform flux density.

1.6.3 Magnetic Circuits

It is interesting to see what can be achieved when magnetic steel is used to "shape" the field pattern. Furthermore, the permanent magnet will be replaced with an n turns circular coil, which carries a current i. The use of a coil has advantages in terms of being able to better control the flux. However, machines generally become more compact when permanent magnets are used. Furthermore, magnets provide flux without the use of an external power supply. An example of the field distribution produced by a coil *without* any magnetic material is shown in Fig. 1.10a. The coil is shown in cross-sectional form where the right section has the current "into" the diagram and the left side has the current coming towards the reader. The flux direction which corresponds to the current flowing "into" the winding half is clockwise. Hence, following the "right hand" rule, the "north" pole is on the top of the diagram which corresponds to the pole alignment shown for the bar magnet. Note that the field distribution is almost identical to that produced by the magnet. As with the bar magnet the flux density is highest in the coil, as may be observed from the flux density plot of the coil shown in Fig. 1.10b. The observant reader will note that there is also a "C" and "I" shaped outline shown in red in both figures. These are in fact the outlines of an iron core structure which in the case of Fig. 1.10 has been constructed of "air," i.e., the coil does not see this structure at this point of our discussion.

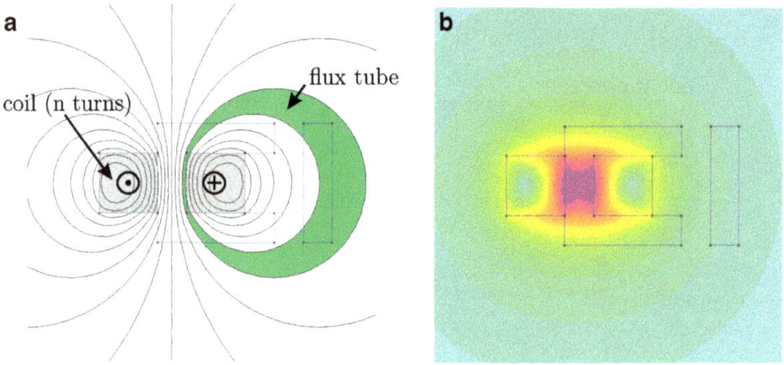

Fig. 1.10 Coil flux and flux density plot. (**a**) Flux plot. (**b**) Flux density plot

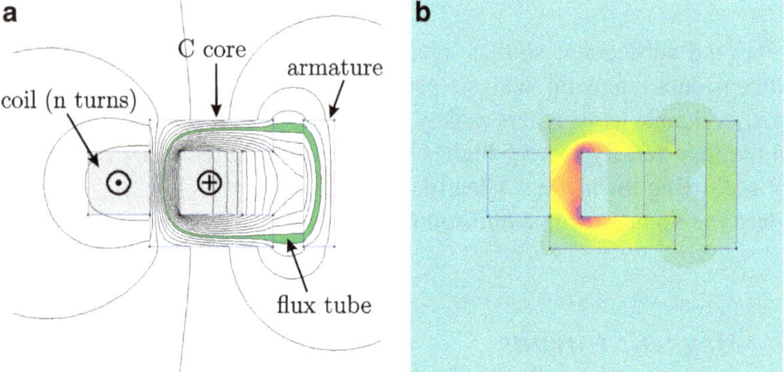

Fig. 1.11 Coil with "C" core and "I" shaped armature: (**a**) flux and (**b**) flux density plot

If we now introduce an iron "C" core and "I" section (known as the armature) with our coil, then we see a remarkable change to the field distribution, as may be observed from Fig. 1.11a. The flux lines are now mostly confined to the core. However, when the flux lines cross from the "C" core to the armature they tend to spread out, an effect referred to as *fringing*. If one looks to the "green" flux tube we see that it is very narrow in the coil and steel regions. The flux tube in question widens out when it crosses the airgaps located between the "C" core and armature. The airgap is large, to demonstrate clearly how the flux lines are affected when moving through air. However, in real induction machines the airgap is in the order of 0.3–0.7 mm which means that most of the flux tube area, when it passes through air, is not much wider than in the steel. In permanent magnet synchronous motors however, airgaps can be as high as several cms. The flux density in the structure of Fig. 1.11a is still relatively uneven, which means that the flux density is high within the core that has the coil wrapped around it. The flux density plot shown in Fig. 1.11b clearly shows this. The color red represents the highest flux density.

1.6.4 Electrical Circuit Analogy and Reluctance

The flux and flux density plots given in the previous section were derived with a two-dimensional magnetic analysis package [3], which enables the user to quickly observe flux patterns for a particular application. However, there is a need to make some "sanity checks" in every type of simulation. Hence, some way must be found to make a simple analytical calculation which will give us confidence in the results produced by a particular simulation. We can do this check by making use of Hopkinson's law, which for a magnetic circuit allows us to create, for example, an electric circuit of the structure given in Fig. 1.11. Hopkinson's law is in fact equivalent to Ohm's law for electrical circuits. Electrically Ohm's law tells us that the electric voltage u across a resistance is equal to the product of the current i and resistance R, i.e., $u = iR$. Hopkinson's law defines a 'magnetic potential u_{M}, which is the product of the flux ϕ in the magnetic circuit times the so-called magnetic reluctance R_{m}, i.e., $u_{\mathrm{M}} = \phi R_{\mathrm{m}}$. The method presented here is confined to the so-called linear magnetic circuits, which implies that the reluctance is neither a function of ϕ nor of u_{M}. In the equivalent circuit, the flux ϕ is in electrical terms equivalent to the current i. An approximate magnetic equivalent circuit of the structure given in Fig. 1.11 is of the form shown in Fig. 1.12. The approximation used is that all the flux lines cross the airgaps between the "C" core and the armature "I". Clearly, this is not the case here (see Fig. 1.11a) because the airgap is unrealistically large. The reluctance R_{m} is generally proportional to the length of the path and inversely proportional to the product of the cross-sectional area and the so-called permeability μ of the medium in which the flux travels. In the example given above two reluctances are given, namely, R_{airgap} and R_{airgap}. The "iron" reluctance represents the total reluctance of the steel sections ("C" core *and* armature). The term "iron" is commonly used to describe the magnetic steel sections. The "airgap" reluctance represents the total reluctance of *both* airgaps. Mathematically the reluctance may be written as

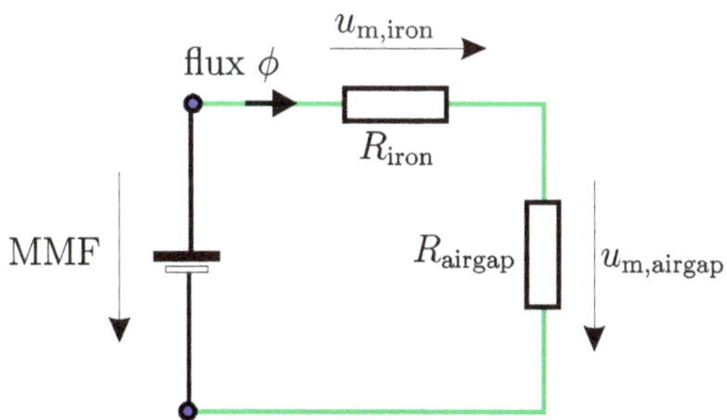

Fig. 1.12 Equivalent magnetic circuit representation

$$R_{\text{iron}} = \frac{l_{\text{iron}}}{A_{\text{iron}}\,\mu_{\text{iron}}} \tag{1.6a}$$

$$R_{\text{airgap}} = \frac{l_{\text{airgap}}}{A_{\text{airgap}}\,\mu_{\text{airgap}}} \tag{1.6b}$$

where l_{iron} and l_{airgap}, respectively, represent the total length the flux travels through iron and airgaps between "C" core and armature. Furthermore, A_{iron} and A_{airgap}, respectively, represent the cross-sectional areas of the steel sections and airgaps. The latter is not easily defined due to fringing effects. Hence, we will assume for this example that the airgap cross-section is equal to that in the iron sections. This means that we assume that the flux density in air and steel (iron) are equal, which they are not in this case. The permeability of the steel μ_{iron} and air μ_{airgap} differs considerably. Typically the permeability in steel (iron) is a factor 1000 higher than that of air. Consequently, the reluctance of the steel sections is considerably lower than that in air.

The magnetic potential across each reluctance is shown as $u_{\text{m,iron}}$ and $u_{\text{m,airgap}}$, respectively. Together they form the total magnetic potential of the circuit, which is equal to the magneto-motive force (MMF). The MMF is equal to the product of the number of coil turns n and current i as shown below

$$\text{MMF} = n\,i \tag{1.7}$$

The MMF can also be expressed in terms of the circuit magnetic potentials, namely,

$$\text{MMF} = u_{\text{m,iron}} + u_{\text{m,airgap}} \tag{1.8}$$

The magnetic field H (A/m) is directly linked to the MMF in the circuit by the expression:

$$\text{MMF} = \underbrace{H_{\text{iron}}\,l_{\text{iron}}}_{u_{\text{m,iron}}} + \underbrace{H_{\text{airgap}}\,l_{\text{airgap}}}_{u_{\text{m,airgap}}} \tag{1.9}$$

where $H = B/\mu$, with μ as the permeability. Consequently, a material with a high permeability will, for a given flux density (and geometry), yield a low magnetic field value and corresponding low magnetic potential.

The circuit flux ϕ in the circuit is of the form

$$\phi = \frac{\text{MMF}}{R_{\text{iron}} + R_{\text{airgap}}} \tag{1.10}$$

An interesting observation of Eqs. (1.8) and (1.10) is that in most cases the reluctance in iron can be ignored given that $R_{\text{iron}} \ll R_{\text{airgap}}$, which implies that under these circumstances MMF $\approx u_{\text{m,airgap}}$. Note that the airgap reluctance will become zero when the armature is placed against the "C" core. In that case the $u_{\text{m,airgap}}$ also goes to zero, and this also applies to the MMF. This in turn means that the current becomes zero. Flux *remains*, as its value (actually its derivative) is determined by the applied electrical voltage as will become apparent shortly. An alternative view of this problem is to consider the case where a current is forced into the coil, under these conditions a finite MMF would be present with zero reluctance, in which case the flux would theoretically become infinite.

Note that the flux ϕ is the same in each part of the circuit (see Fig. 1.12). Consequently, the product of flux density times cross-sectional area remains the same. Hence in a narrow part of the circuit the flux density will be higher than in a wider part. In the airgaps the effective cross-sectional area is increased due to fringing, hence the flux density in the airgap will be lower than in the adjacent iron circuit.

The magnetic example to be compared with the magnetic structure shown in Fig. 1.11 has a set of parameters as given in Table 1.1.

A convenient approach to simulate the structure according to Fig. 1.12 is possible by making use of the PLECS magnetic library which in this case leads to the model given in Fig. 1.13. Readily observable in this figure is the (different colored) magnetic circuit components which represent the reluctances of the magnetic model. Parameters for these are specified in Table 1.1. The electric part of this model is arbitrarily represented by a coil resistance of 1 Ω in series with a DC supply source

Table 1.1 Parameters for magnetic "C" core example

Parameters		Value
Total path length in iron	l_c	150 mm
Total path length in air	l_a	20 mm
Core cross-section	A_c	100 mm^2
Airgap cross-section	A_a	100 mm^2
Copper cross-section	A_{cu}	1600 mm^2
Permeability in iron	μ_c	0.008 H/m
Permeability in air	μ_0	$4\pi 10^{-7}$ H/m
Number of turns coil	n	1000 turns
Coil current	I	5 A

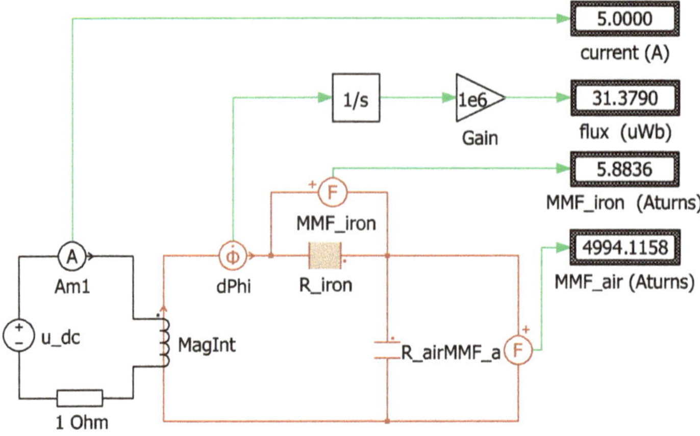

Fig. 1.13 PLECS magnetic circuit representation

chosen to generate the required 5 A current excitation. After running this simulation the steady-state value shown on the display appears. In PLECS the circuit flux is found by using an incremental flux sensor of which the output must be integrated as shown.

Some interesting observations can be made from the results shown in Fig. 1.13, namely

- The magnetic potential across the airgaps is an order of magnitude higher than in the iron. This confirms earlier comments with regard to this topic. Note that the sum of the magnetic potential in air and iron is equal to the MMF [$n{*}I = 5000$ At (ampère-turns)], which is the MMF provided by the winding.
- The corresponding flux density in, for example, the airgap is equal to $B_{\mathrm{a}} = {}^{\phi_{\mathrm{a}}}/_{A_{\mathrm{a}}}$ which in this case gives $B_{\mathrm{c}} = 0.31\,\mathrm{T}$, than the values found in Fig. 1.11b. According to the linear model the flux density in the core would also be $B = 0.31\,\mathrm{T}$ given that the same cross-sectional area is assumed. The core flux density value is however considerably lower than the value found in Fig. 1.11b. The flux density values, as indicated by the color, in the core and airgap were found to be 1.8 T and 0.3 T, respectively. The main reason for this is that the equivalent circuit model is a linear model which does not exhibit so-called saturation effects (which implies that the reluctance of the core is much higher than assumed here). This topic will be discussed shortly. Secondly, the airgap reluctance is not modeled well, i.e., the assumption that all the flux crosses from the "C" core to the armature is not valid.
- In this PLECS model use is made of an electrical equivalent circuit to provide the winding MMF.

1.6.5 Flux-Linkage and Self-Inductance

The term flux-linkage is often required when dealing with the electrical equations which link to the magnetic circuit. The flux-linkage refers to the amount of flux linked to the coil. Each winding turn of the coil "sees" the circuit flux ϕ as can be observed from Fig. 1.11a and this means that the coil as a whole "sees" the product of the circuit flux and the number of turns. This quantity is referred to as the flux-linkage $\psi = n\phi$. Note that this is in fact a simplification and only holds for relatively simple examples as treated in this chapter. For example, if one observes Fig. 1.10a, it is hopefully clear that not all the turns are linked with the same circuit flux. Some flux lines stray in between the windings, forming the so-called stray or leakage flux. Leakage inductance and the effects of leakage flux will be considered in later chapters. However, the calculation of the flux-linkage and leakage inductance based on the geometry of magnetic circuits is beyond the scope of this book.

The relationship between flux-linkage and current is readily found by using Hopkinson's law, which states that the circuit flux is equal to the coil MMF divided by the total reluctance of the magnetic circuit. For the linear example treated above the flux-linkage can be written as

$$\psi = n \frac{\text{MMF}}{R_{\text{iron}} + R_{\text{airgap}}} \tag{1.11}$$

which can be further simplified using MMF $= n\, i$ to

$$\psi = \left(\frac{n^2}{R_{\text{iron}} + R_{\text{airgap}}} \right) i \tag{1.12}$$

where the term $\left(\frac{n^2}{R_{\text{iron}} + R_{\text{airgap}}} \right)$ is known as the coil inductance L (H). Hence the relationship between flux-linkage and current for a *linear* magnetic circuit is given by Eq. (1.13).

$$\psi = L\, i \tag{1.13}$$

Expression (1.13) is also represented in graphical form, see Fig. 1.14. Note that the inductance is determined by the geometry, material properties of the magnetic circuit, and the coil number of turns. Figure 1.14 shows a linear relationship between flux-linkage and current. Furthermore, the gradient of the slope is equal to the

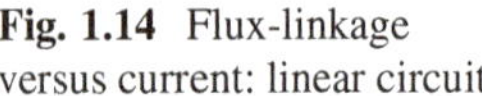

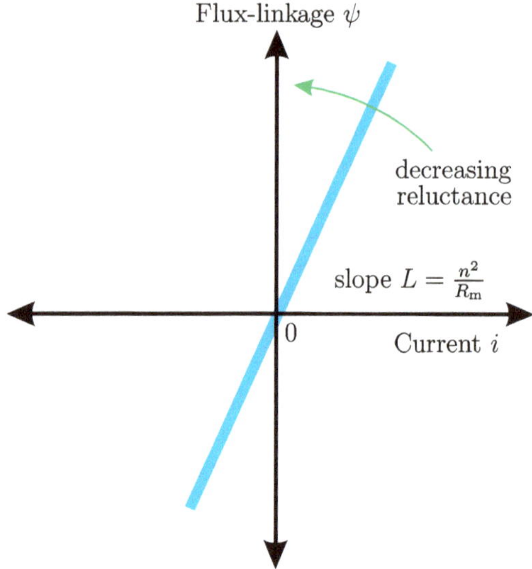

Fig. 1.14 Flux-linkage versus current: linear circuit

inductance. This means that the gradient of the function will increase in case the inductance increases, which in turn will take place when the total magnetic reluctance R_m of a magnetic circuit reduces. Zero magnetic reluctance (i.e., infinite inductance) corresponds to a flux-linkage current curve which is aligned with the vertical axis of this figure. This tells us that for a given flux there is *no* magnetizing current required.

1.6.6 Magnetic Saturation

The magnetic reluctance of steel (iron) is not constant when the flux density increases. When the flux density rises to levels typically approaching 2 T (Tesla or Vs/m^2), a marked increase in the magnetic reluctance of the steel occurs. This change in reluctance refers to a phenomenon called saturation, which in effect constrains the flux density in magnetic circuits using, for example, Si-steel to values below 2 T as may be observed from Fig. 1.15. The exact saturation level depends very much on the magnetic steel used. Cheaper steel or ferrites tend to have a lower saturation level. Note that the reluctance in air does *not* exhibit saturation.

The change in reluctance directly influences the flux-linkage current curve as an increasing R_m will reduce the slope of the $\psi(i)$ curve as the flux density B increases. Note that the latter is proportional to the circuit flux ϕ and flux-linkage ψ value. An example of a flux-linkage current curve for the linear and general case is given in Fig. 1.16.

Note that the notion of inductance is for the general case not really applicable as the gradient of the function is no longer constant. Hence, the term inductance

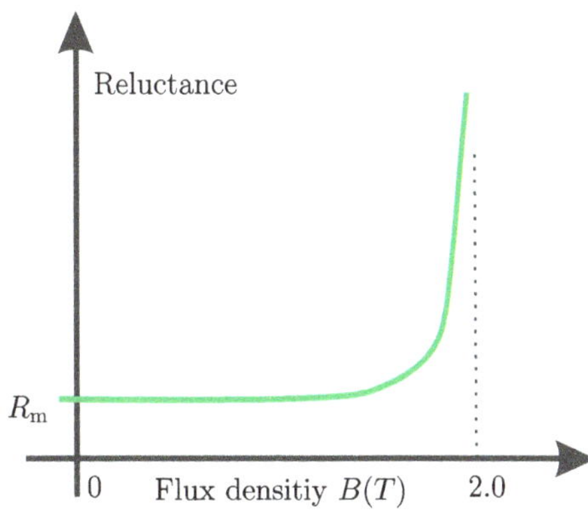

Fig. 1.15 Reluctance change due to saturation

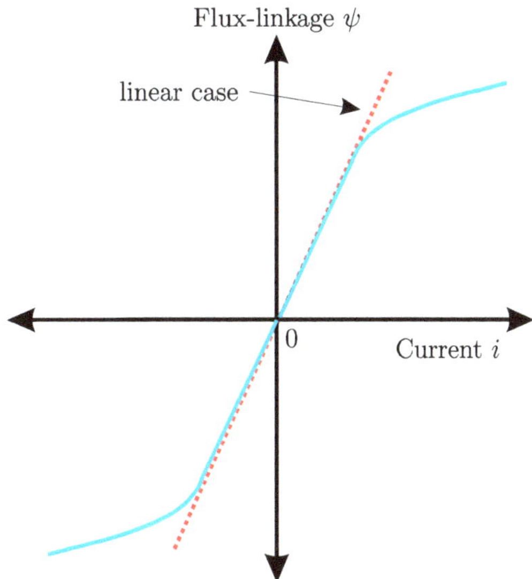

Fig. 1.16 Flux-linkage versus current: with saturation effects

is relevant when considering magnetically linear circuits. The so-called non-linear circuit analysis will require the use of the general flux-linkage current curve, which must be given or measured for the circuit to be analyzed. At a later stage an example of the use of this curve will be given.

1.7 Machine Sizing Principles

It is instructive at the end of this chapter to give the reader some insight into the concept of electrical machine sizing. This issue becomes important when faced with, for example, the task of choosing a certain machine size to accommodate a given load application.

In Fig. 1.8, we have introduced a single wire which was able to produce a force F_e when it was placed in a magnetic field and attached to a current source. This concept can be extended to electrical machines if we consider the latter in the form of a rotor and stator. The rotor, being the rotating component, is assumed to hold a set of n wires of thickness d at its circumference. For convenience of this calculation we will assume that the cross-section of these wires is square rather than round. A magnetic field with flux density B is assumed to be present in the airgap between the rotor and stator. Consequently, a resultant force F will be created on the surface of the rotor in case the rotor windings are made to carry a current i. If the rotor radius is set to r and its length to l, then the torque T_e produced by the machine will be equal to

$$T_e = rF \tag{1.14}$$

where $F = nBil$. It is instructive to introduce the concept of current density $j = {}^{i}\!/\!{}_{A}$ where A represents the cross-sectional area of a wire. If we consider the entire circumference of the rotor packed with n wires placed next to each other, then we can approximate the total coil area $A_w = nA \approx d2\pi r$. Use of this approximation together with the expression for F_e allows us to approximate expression (1.14) as

$$T_e \approx kBj \underbrace{\pi r^2 l}_{V_r} \tag{1.15}$$

where k is a machine constant ($k = 2d$ in this case). V_r represents the rotor volume. Equation (1.15) is significant in that it tells us that the torque is proportional to the product of flux density B, current density j, and rotor volume V_r.

In this chapter, we have already shown that magnetic saturation places a constraint on the flux density value we can practically use. Furthermore, current density values are typically constrained to values less than 10 A/mm^2 given thermal considerations. Hence, it follows that the rotor size and consequently the total size of a machine will need to be chosen to meet a certain torque requirement. The ratio between torque and rotor volume, known as *TRV* [8], is therefore an important figure for machine sizing. For industrial machines this value is typically in the order of 15–30 kNm/m^3. In linear-motors as well as rotating machines the same number can be interpreted as the maximum shear-stress of 15–30 kN/m^2 (thrust per unit area).

The overall size of the machine is determined by the stator volume, which is determined among other factors by the rotor volume V_r. A rough estimate of the stator volume V_s as given by Miller [8] is of the form

$$V_s \approx \frac{V_r}{srs^2} \tag{1.16}$$

where *srs* is a constant in the order of 0.6.

It is helpful to give a numerical example of such a sizing calculation. Consider a machine that must produce a torque of 70 Nm. If we assume that the rotor diameter is equal to its length, then the rotor diameter (and length) would be equal to 164 mm in case we assume a TRV of 20 kNm/m^3. The corresponding stator diameter would according to Eq. (1.16) be 259 mm which is a realistic expectation for such a machine. In reality, the machine length would be longer than the estimated value of 164 mm given the need to accommodate the stator winding at both ends of the machine as well as the rotor bearings and cooling fan-blades.

1.8 Tutorials

1.8.1 Tutorial 1: Magnetic Analysis of a Rotational Symmetric Structure

The model shown in Fig. 1.17 (cross-section shown) is rotational symmetric. A single $n = 1000$ turn coil is shown which carries a current of $i_{\mathrm{coil}} = 5$ A. The steel used has a permeability of $10^6\mu_0$, where μ_0 represents the permeability in vacuum (air). The key dimensions (in millimeters) are shown in Fig. 1.17.

The model in question was analyzed with a finite element package and gave the results as given in Table 1.2.

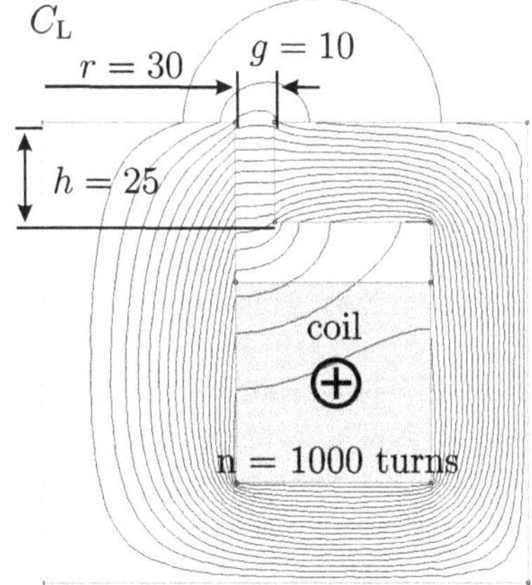

Fig. 1.17 Magnetic model

Table 1.2 Output finite element program

Output variable		Value
Flux density in airgap	B_a	0.62 T
Flux linked with coil	ψ	10.90 Wb
Self-inductance	L	2.19 H

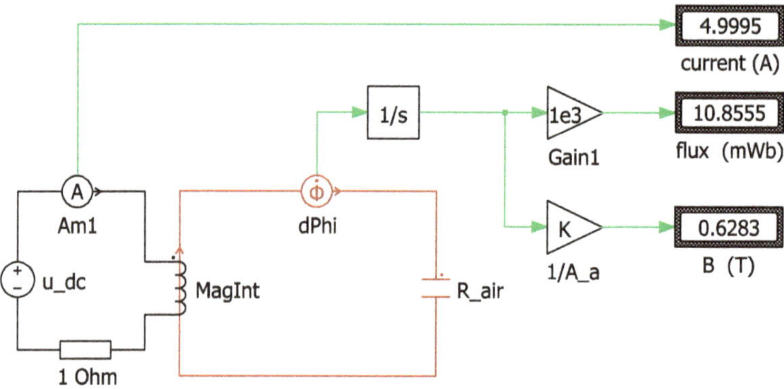

Fig. 1.18 PLECS magnetic model

Perform a “sanity check” on the results obtained from your magnetic (finite element) analysis by using an alternative method. An example solution to this type of problem is given below: Firstly, we know that the steel used has a high permeability which is very much larger than that of air. Consequently we can assume that the magnetic potential across the steel will be very much lower than that across the airgap. Hence, the first assumption to make is that the magnetic potential u_a across the airgap is approximately equal to the applied MMF, i.e., $u_a \simeq n i_{\text{coil}}$. The second critical issue here is to make a sensible judgment with respect to the cross-sectional area which the flux “sees” when crossing the airgap g. If there was no “fringing,” then the airgap cross-section would be equal to $A_c = 2 * \pi * (r + g/2)\,h$, where we have chosen a cylinder with height h and radius $r + g/2$. Observation of Fig. 1.17 shows that the flux crossing the airgap has in reality a larger cross-sectional area. The difficulty lies in finding a good estimate for this airgap area, which takes fringing into account. The so-called Carter factor is often used to allow for fringing effects. In our calculation, we will assume that this factor is equal to $C = 2$, this is based on the fact that we observe a significant number of flux lines at twice the “h” value. The magnetic reluctance is then found using Eq. (1.6b) with $l_{\text{airgap}} = g$ and $A_{\text{airgap}} = C A_c$. This in turn leads to the circuit flux $\phi = \text{MMF}_a / R_{\text{airgap}}$, flux-linkage $\psi = n\phi$, and self-inductance $L = \psi / i_{\text{coil}}$ (see Fig. 1.18).

The results from the PLECS model agree very well with those obtained from the finite element program. The reason for this is that we have chosen a good estimate for the effective airgap cross-sectional area. In reality, estimating the effect of flux fringing without using magnetic analysis software is difficult.

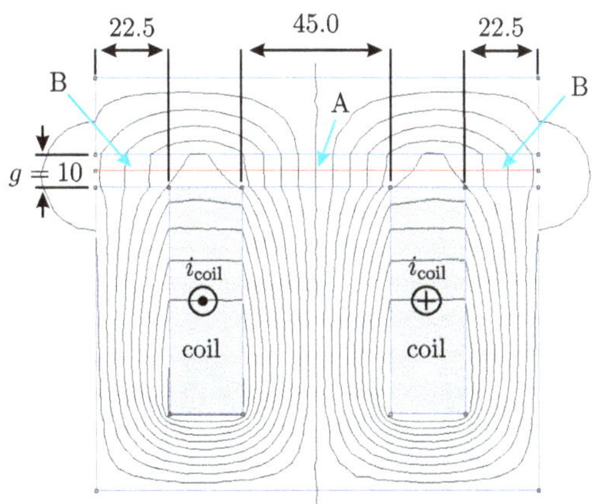

Fig. 1.19 Magnetic "E" core example

1.8.2 Tutorial 2: Magnetic Analysis of an "E" Core Type Structure

This tutorial considers an "E" core type structure as shown in Fig. 1.19. The distance between the "I" segment, which is also part of the total magnetic circuit and "E" core, is 10 mm. A 500 turn coil is wound around the center leg of the "E" core and carries a current of 20 A. The depth of both magnetic components is taken to be 20 mm. Furthermore, the magnetic material is taken to be magnetically ideal. Key dimensions (in mm) are shown in Fig. 1.19, which relate to the airgaps between the two magnetic components. The permeability of air is given in the previous tutorial.

The aim of the tutorial is to demonstrate the importance of finite element modeling and to emphasize that the results obtained with linear models should be used with care.

Perform a linear analysis of this problem and estimate with the aid of an equivalent magnetic circuit the flux density in the airgaps at locations "A" and "B," respectively. In addition, estimate the total flux ψ_{A} linked with the coil. A possible solution to this problem is as follows.

The equivalent magnetic circuit with the conditions specified, i.e., ideal magnetic material, is shown in Fig. 1.20.

The "MMF," which is equal to MMF $= n_{\mathrm{coil}}\, i_{\mathrm{coil}}$, must be equal to the sum of the magnetic potentials u_{A} and u_{B}. The reluctances R_{A} and R_{B} of the center and side legs of the "E" core require an estimation of the area of the flux which crosses between the two magnetic circuit components. A relatively large airgap is used in this example, which leads to considerable magnetic fringing. If we ignore fringing for the linear calculation, we are able to determine the reluctances with the aid of the dimensions given in Fig. 1.20. A possible PLECS implementation of this example is shown in Fig. 1.21. After running this model the following results appear: $B_{\mathrm{A}} = 0.628\,\mathrm{T}$, $B_{\mathrm{B}} = 0.628\,\mathrm{T}$, and $\psi_{\mathrm{A}} = 0.28\,\mathrm{Wb}$ (as shown on the display modules). The amplitudes are equal because the circuit flux in the side legs is half that of the

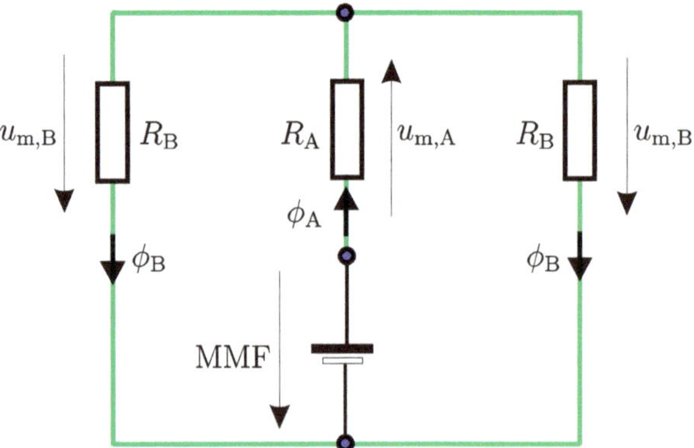

Fig. 1.20 Magnetic "E" core equivalent circuit

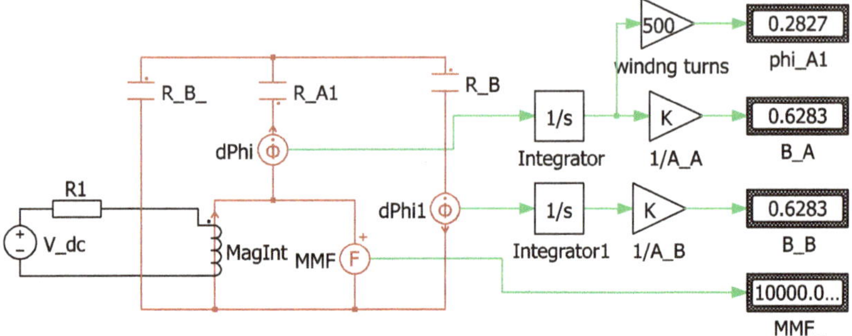

Fig. 1.21 PLECS model magnetic "E" core

center legs. Furthermore, the airgap areas of the side legs are half that of the center leg if flux fringing is ignored. In Fig. 1.21 use is made of integrators to calculate the circuit flux ϕ_A, ϕ_B from the incremental flux sensors present in the PLECS model. The flux-linkage of the winding is then found using $\psi_A = n\,\psi_A$ where $n = 500$ is the number of turns of the winding. Two gain modules are used to compute the flux density values B_A and B_B based on evaluation of expressions $B_A = {}^{\phi_A}/_{A_A}$ and $B_B = {}^{\phi_B}/_{A_B}$, where A_A, A_B are the respective cross-sectional areas of the airgaps.

An example of the absolute flux density plot as function of the position along an imaginary line (the length of which corresponds to the width of the "E" core), which passes through points "A" and "B" in the airgap, is shown in Fig. 1.22. The flux density plot according to Fig. 1.22 was obtained with the aid of a two-dimensional finite element package which was also used to calculate the flux distribution shown in Fig. 1.19. Observation of Fig. 1.22 shows that the flux density values at points "A" and "B" are markedly different when compared to the linear case. Furthermore, the coil flux-linkage was calculated with the same finite element program and the value was found to be 0.63 Wb, which is considerably higher than the value obtained via the linear analysis (as given above). The reasons for the differences between the two

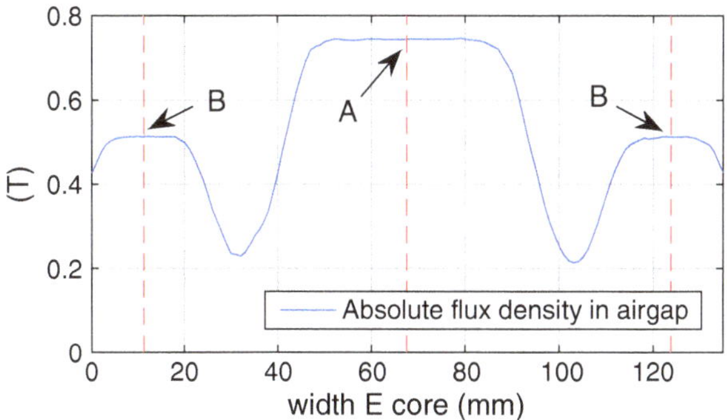

Fig. 1.22 Flux density plot along airgap in the "E" core

computation methods can, to a large extend, be attributed to the effects of fringing, i.e., the difficulty of determining an accurate analytical estimate of the flux area required to calculate the reluctances R_A and R_B. Furthermore, the assumption of an ideal magnetic material may also severely affect the result, in particular if saturation effects come into play. Finally, it is noted that even a two-dimensional finite element may not always be suitable, in which case a three-dimensional analysis may need to be undertaken.

Chapter 2
Simple Electro-Magnetic Circuits

2.1 Introduction

The simplest component which utilizes electro-magnetic interaction is the coil. A coil is an energy storage component, which stores energy in magnetic form. Air-cored coils are frequently used (for example, in loudspeaker filters), but coils with a core of (possibly gapped-) magnetic material are more common, because of their increased inductance (or reduced size), which may come at the cost of reduced maximum field strength and increased non-linearity. In this chapter we will develop a generic model of a coil with linear and non-linear self-inductance. Furthermore, the effect of coil resistance is considered. The use of phasors is introduced in this chapter as a means to verify simulation of such circuits when connected to a sinusoidal source.

2.2 Linear Inductance

The physical representation of the coil considered here is given in Fig. 2.1. The figure shows a coil with n turns which is wrapped around a toroidally shaped non-gapped magnetic core with cross-sectional area A_{m}. The permeability of the material is given as μ and the average flux path length is equal to l_{m}. Analog to Eq. (1.6), the magnetic reluctance of the circuit is: $R_{\mathrm{m}} = {}^{l_{\mathrm{m}}}/_{A_{\mathrm{m}}\mu}$ and the inductance is $L = {}^{n^2 \mu A_{\mathrm{m}}}/_{l_{\mathrm{m}}} = {}^{n^2}/_{R_{\mathrm{m}}}$.

Electronic supplementary material The online version of this chapter (doi: 10.1007/978-3-319-29409-4_2) contains supplementary material, which is available to authorized users.

© Springer International Publishing Switzerland 2016

A. Veltman et al., *Fundamentals of Electrical Drives*, Power Systems,
DOI 10.1007/978-3-319-29409-4_2

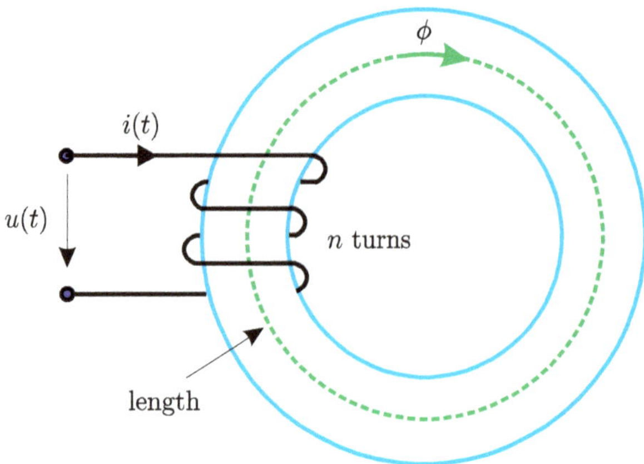

Fig. 2.1 Toroidal inductance

The relation between the magnetic flux and the current in the coil is described by the expression

$$\psi = Li \tag{2.1}$$

With Faraday's law

$$u = \frac{d\psi}{dt} \tag{2.2}$$

Equation (2.1) can be rewritten to the more familiar differential form of the coil's voltage terminal equation

$$u = L\frac{di}{dt} \tag{2.3}$$

Equation (2.3) can be integrated on both sides and rewritten as the general equation

$$i(t) = \frac{1}{L}\int_{-\infty}^{t} u(t)dt \tag{2.4}$$

The whole integrated history of the inductor voltage is reflected by the inductor current, so Eq. (2.4) can be expressed in a more practical form, starting at $t = 0$ with initial condition $i(0)$, according to

$$i(t) = \frac{1}{L}\int_{0}^{t} u(t)dt + i(0) \tag{2.5}$$

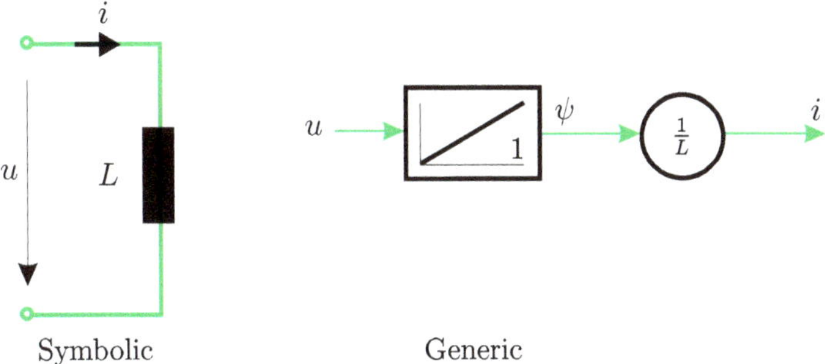

Fig. 2.2 Symbolic and generic model of a linear inductance

This integral form can be developed further

$$\Delta i = \frac{\Delta\psi}{L} \tag{2.6}$$

$$\underbrace{\psi(t) - \psi(0)}_{\Delta\psi} = \int_0^{t_0} u(t)\mathrm{d}t \tag{2.7}$$

introducing the concept of "incremental flux linkage" $\Delta\psi = \psi(t) - \psi(0)$. The equation basically states that a flux-linkage variation corresponds with a voltage-time integral (the so-called volt-second) when the resistance is zero.

A symbolic and generic model of the ideal coil is given in Fig. 2.2. With the model of Fig. 2.2, we will now simulate the time-response of a coil in reaction to a voltage pulse of magnitude $\hat{u}$ and duration T, starting at $t = t_0$, as displayed in Fig. 2.3. Integrating the supply voltage u over time gives the flux-linkage ψ in the coil, which linearly increases from 0 at $t = t_0$ to $\hat{u}T$ at $t = T$. The current is obtained by dividing the flux ψ by L.

2.3 Coil Resistance

In practical situations, the resistance of the coil wire can usually not be neglected. Wire resistance can simply be modeled as a resistor in series with the ideal coil. The modified symbolic model is shown in Fig. 2.4.

Figure 2.4 shows that the coil flux is no longer equal to the integrated supply voltage u. Instead, the variable u_{L} is introduced, which refers to the voltage across the "ideal" (zero resistance) inductance $u_{\mathrm{L}} = {}^{\mathrm{d}\psi}/_{\mathrm{d}t}$. The terminal equation for this circuit is now given by expression (2.8).

$$u = iR + \frac{\mathrm{d}\psi}{\mathrm{d}t} \tag{2.8}$$

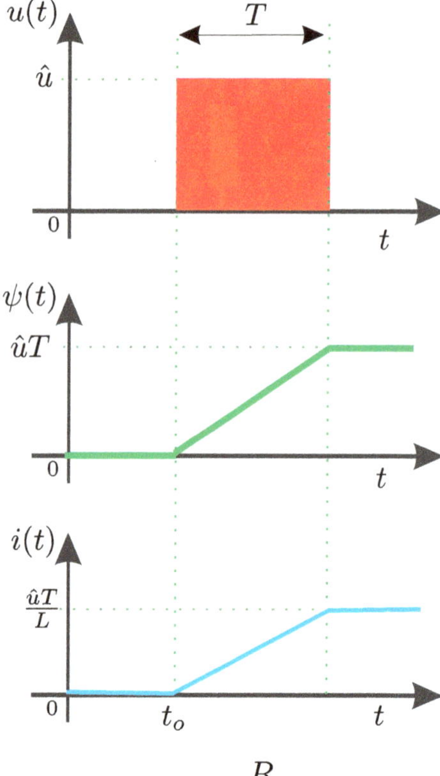

Fig. 2.3 Transient response of inductance

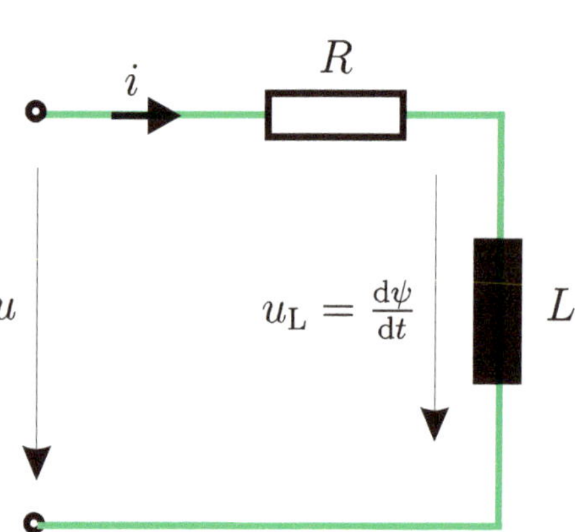

Fig. 2.4 Symbolic model of linear inductance with coil resistance

where R represents the coil resistance. The corresponding generic model of the lumped parameter "L, R" circuit is shown in Fig. 2.5. The generic model clearly shows how the inductor voltage u_{L} is decreased by the resistor voltage caused by the current through the coil.

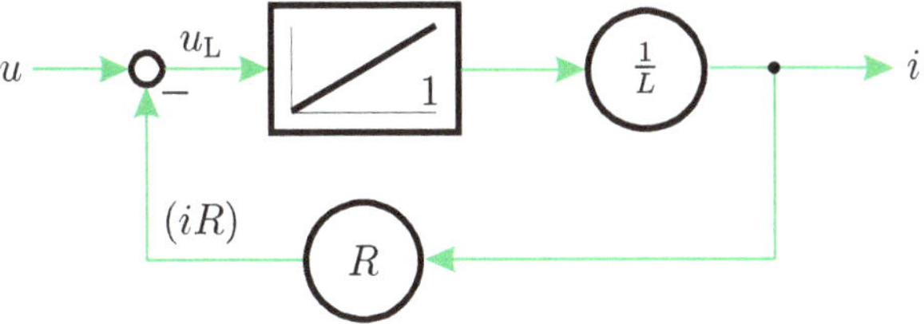

Fig. 2.5 Generic model of linear inductance with coil resistance

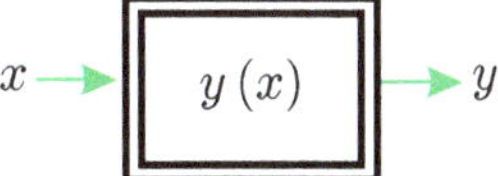

Fig. 2.6 Non-linear generic building block

2.4 Magnetic Saturation

As discussed in Chap. 1, the maximum magnetic flux density in magnetic materials is limited. Above the saturation flux density, the magnetic permeability μ drops and the material will increasingly behave like air, i.e., $\mu \rightarrow \mu_0$ when flux density is increased further. Since motors usually work at high flux density levels, with noticeable saturation, it is essential to incorporate saturation in our coil model.

The relationship between flux-linkage and current is in the magnetically linear case determined by the inductance, as shown in Fig. 1.14. In reality, the $\psi(i)$ relationship is only relatively linear over a limited region (in case the magnetic circuit contains "iron" (steel) core), as shown in Fig. 1.16. The generic model according to Fig. 2.5 needs to be revised in order to cope with the general case.

The generic building block for non-linear functions [7] is shown in Fig. 2.6. The double edged box indicates a non-linear module with input variable x and output variable y. The relationship between output and input is shown as $y(x)$ (y as a function of the input x). In some cases, a symbolic graph of the function that is implemented may also be shown on this building block.

The non-linear module has the coil flux ψ as input and the current i as output. Hence, the non-linear function of the module is described as $i(\psi)$, which expresses the current of the coil as a function of the coil flux. The terminal equation (2.8) remains unaffected by the introduction of saturation, only the gain module $^1/_L$ shown in Fig. 2.5 must be replaced by the non-linear module described above. The revised generic model of the coil is shown in Fig. 2.7.

2.5 Use of Phasors for Analyzing Linear Circuits

The implementation of generic circuits (such as those discussed in this chapter) in PLECS allows us to study models for a range of conditions. The use of a sinusoidal excitation waveform is of most interest given their use in electrical machines and

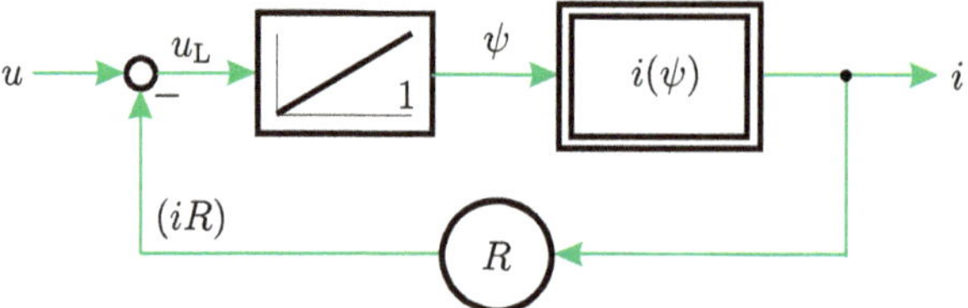

Fig. 2.7 Generic model of general inductance model with coil resistance

actuators. However, there must be a way to perform "sanity checks" on the results given by simulations. Analysis by way of phasors provides us with a tool to look at the *ac* steady-state results of linear circuits. The underlying principle of this approach lies with the fact that a sinusoidal excitation function, for example, the applied voltage, will cause a sinusoidal output function of the same frequency, be it that the amplitude and phase (with respect to the excitation function) will be different. For example, in the symbolic circuit shown in Fig. 2.4, the excitation function will be defined as $u(t) = \hat{u}\sin(\omega t)$, where $\hat{u}$ and ω represent the peak amplitude and angular frequency (rad/s), respectively. Note that the latter is equal to $\omega = 2\pi f$, where f represents the frequency in Hz. The output variables are the flux-linkage $\psi(t)$ and current $i(t)$ waveforms. Both of these will also be sinusoidal, be it that their amplitude and phase differ from the input signal $u(t)$. In general, a sinusoidal function can be described by

$$x(t) = \hat{x}\sin(\omega t + \rho) \tag{2.9}$$

This function can also be written in complex notation as

$$x(t) = \Im\left\{\hat{x}e^{j(\omega t+\rho)}\right\} \tag{2.10}$$

Equation (2.10) makes use of "Euler's rule" $e^{jy} = \cos y + j\sin y$. The imaginary part of this expression is defined as $\Im\left\{e^{jy}\right\} = \sin y$. $\Im\{\}$ is the imaginary operator, which takes the imaginary part from a complex number. Note that the analysis would be identical with $x(t)$ in the form of a cosine function. In the latter case it would be more convenient to use the real component of $\hat{x}e^{j(\omega t+\rho)}$, using the real operator $\Re\{\}$. Equation (2.10) can be rewritten to separate the time dependent component $e^{j\omega t}$ namely:

$$x(t) = \Im\left\{\underbrace{\hat{x}e^{j\rho}}_{\underline{x}}\, e^{j\omega t}\right\} \tag{2.11}$$

The time independent component in Eq. (2.11) is known as a "*phasor*" and is generally identified by the notation $\underline{x}$. In general the phasor will have a real and imaginary component and can therefore be represented in a complex plane.

In many cases it is also convenient to use the time differential of $x(t)$ namely $\mathrm{d}x/\mathrm{d}t$. The time differential of the function $x\left(t\right) = \Im\left\{\underline{x}\, e^{\mathrm{j}\omega t}\right\}$ is

$$\frac{\mathrm{d}x}{\mathrm{d}t} = \Im\left\{\mathrm{j}\omega\underline{x}\, e^{\mathrm{j}\omega t}\right\} \tag{2.12}$$

which implies that the differential of the phasor $\underline{x}$ is calculated by simply multiplying $\underline{x}$ with $\mathrm{j}\omega$.

2.5.1 *Application of Phasors to a Linear Inductance with Resistance Network*

As a first example of the use of phasors, we will analyze a coil with linear inductance and non-zero wire resistance, as shown in Fig. 2.4. We need to calculate the steady-state flux-linkage and current waveforms of the circuit. The differential equation set for this system is

$$u = iR + \frac{\mathrm{d}\psi}{\mathrm{d}t} \tag{2.13a}$$

$$\psi = Li \tag{2.13b}$$

The flux-linkage differential equation is found by substitution of Eq. (2.13b) into (2.13a) which gives

$$u = \frac{R}{L}\psi + \frac{\mathrm{d}\psi}{\mathrm{d}t} \tag{2.14}$$

The applied voltage will be $u = \hat{u}\sin\omega t$, hence the phasor representation of the input signal according to (2.11) is: $\underline{u} = \hat{u}$.

The flux-linkage will also be a sinusoidal function, albeit with different amplitude and phase: $\psi = \hat{\psi}\,\sin\left(\omega t + \rho_{\psi}\right)$. The parameters $\hat{\psi}$ *and* ρ_{ψ} are the unknowns at this stage. In phasor representation, the flux time function can be written as $\psi = \Im\left\{\underline{\psi}\, e^{\mathrm{j}\omega t}\right\}$ where $\underline{\psi} = \hat{\psi}\, e^{\mathrm{j}\rho_{\psi}}$.

Rewriting Eq. (2.14) using these phasors, we obtain

$$\underline{u} = \frac{R}{L}\underline{\psi} + \mathrm{j}\omega\underline{\psi} \tag{2.15}$$

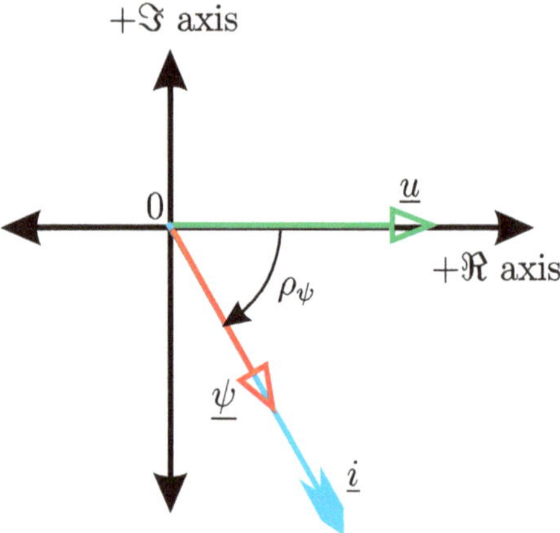

Fig. 2.8 Complex plane with phasors: $\underline{u}$, $\underline{\psi}$, $\underline{i}$

from which we can calculate the flux phasor by reordering, namely

$$\underline{\psi} = \frac{\underline{u}}{\left(\frac{R}{L} + \mathrm{j}\omega\right)} \tag{2.16}$$

The amplitude and phase angle of the flux phasor are now

$$\hat{\psi} = \frac{\hat{u}}{\sqrt{\left(\frac{R}{L}\right)^2 + \omega^2}} \tag{2.17a}$$

$$\rho_\psi = -\arctan\left(\frac{\omega L}{R}\right) \tag{2.17b}$$

and the corresponding current phasor is according to Eq. (2.13b): $\underline{i} = \underline{\psi}/L$.

The transformation of phasors back to corresponding time variable functions is carried out with the aid of Eq. (2.11). A graphical representation of the input and output phasors is given in the complex plane shown in Fig. 2.8.

2.6 Tutorials

2.6.1 Tutorial 1: Analysis of a Linear Inductance Model

In this chapter we analyzed a linear inductance and defined the symbolic and generic models as shown in Fig. 2.2. The aim of this tutorial is to build a PLECS model from this generic diagram. An example as to how this can be done is given in Fig. 2.9. Indicated in Fig. 2.9 is the inductance model in the form of an integrator and gain module. Also given are two "step" modules which, together with a "Sum" unit,

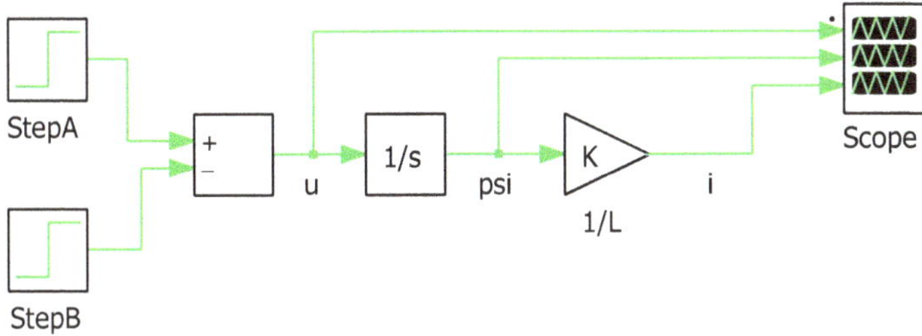

Fig. 2.9 PLECS model of linear inductance with excitation function

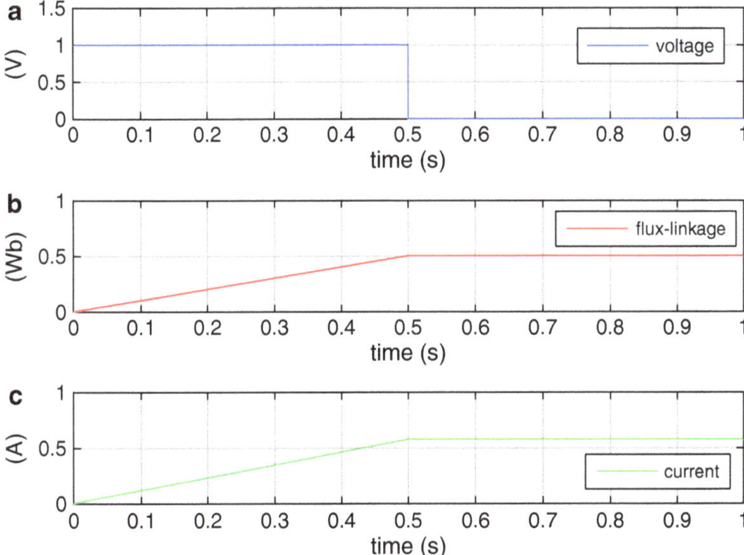

Fig. 2.10 PLECS results: ideal inductance simulation

generate a voltage pulse of magnitude 1 V. This pulse should start at $t = 0$ and end at $t = 0.5$ s. Build this circuit and also add a "Scope" module which allows you to display your data. In this exercise we look at the input voltage waveform, the flux-linkage, and current versus time functions. Once you have built the circuit you need to run this simulation. For this purpose you need to set the "stop time" (under Simulations/simulation parameters dialog window) to 1 s. The inductance value used in this case is $L = 0.87$ H, which should be set in the "Integrator" module dialog box. The results which should appear from your simulation after running this PLECS file are given in Fig. 2.10.

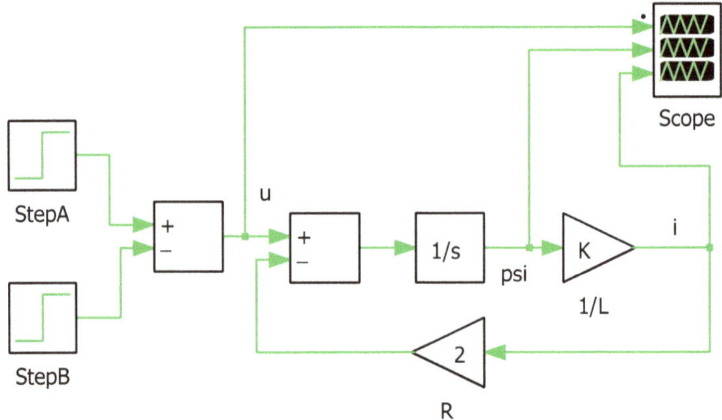

Fig. 2.11 PLECS model of linear inductance with resistance and excitation function

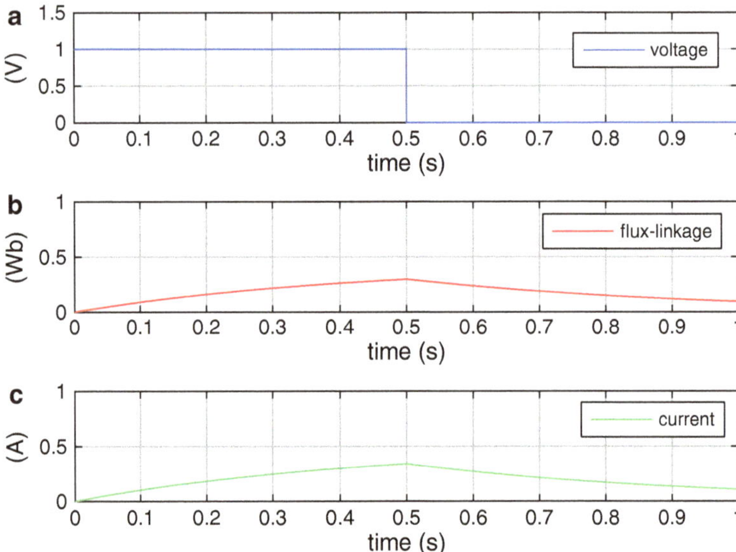

Fig. 2.12 PLECS results: inductance simulation, with coil resistance

The dynamic model as discussed above is to be extended to the generic model shown in Fig. 2.5. Add a coil resistance of $R = 2\,\Omega$ to the PLECS model given in Fig. 2.9. The new model should be of the form given in Fig. 2.11.

Run the simulation again, in which case the results should be of the form given in Fig. 2.12.

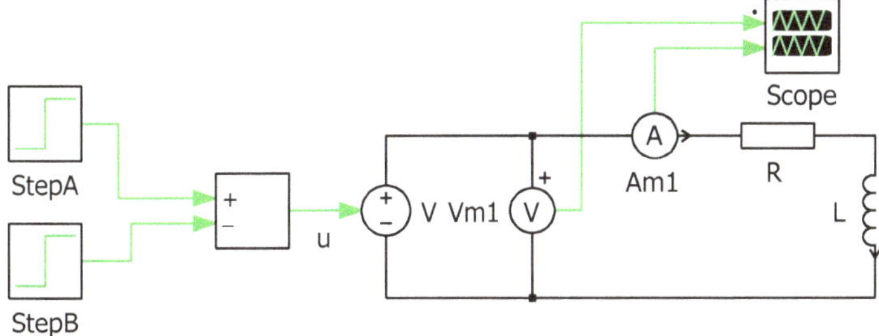

Fig. 2.13 PLECS "symbolic" model: linear inductance with coil resistance

2.6.2 Tutorial 2: Symbolic Model Analysis of a Linear Inductance Model

In this tutorial we will consider an alternative implementation of tutorial 1, based on the use of symbolic models (where possible) instead of "control" blocks as used in the previous case. Build a PLECS model of the symbolic model shown in Fig. 2.4 with the excitation and circuit parameters as discussed in tutorial 1. Note that "symbolic" modules in PLECS are known as "Electrical" blocks.

An example of a PLECS implementation is given in Fig. 2.13 on page 39. The "scope" module given in Fig. 2.13 displays the results of the simulation. The simulation results obtained with this simulation should match those given in Fig. 2.12 where it is noted that the flux plot is not shown in this case, given that it is not directly generated by a symbolic model. Furthermore, a "Voltmeter" (Vm1) and "Ammeter" (Am1) are used to measure the voltage and current, respectively.

2.6.3 Tutorial 3: Analysis of a Non-linear Inductance Model

In Sect. 2.4 we have discussed the implications of saturation effects on the flux-linkage/current characteristic. In this tutorial we aim to modify the simulation model discussed in the previous tutorial (see Fig. 2.11) by replacing the linear inductance component with a non-linear function module as shown in the generic model (see Fig. 2.7). In this case, the flux-linkage/current $\psi(i)$ relationship is taken to be of the form $\psi = \tanh(i)$ as shown in Fig. 2.14. Note that in this example the gradient of the flux-linkage/current curve becomes zero for currents in excess of ± 3 A. In reality, the gradient will be non-zero when saturation occurs.

The coil resistance of the coil is increased to $R = 100\,\Omega$. An example of a Simulink implementation is given in Fig. 2.15. The block diagram clearly shows the presence of the non-linear module used to implement the function $i(\psi)$. The non-linear module has the form of a "look-up" table which requires two vectors to be entered. Upon opening the dialog box for this module, provide the following

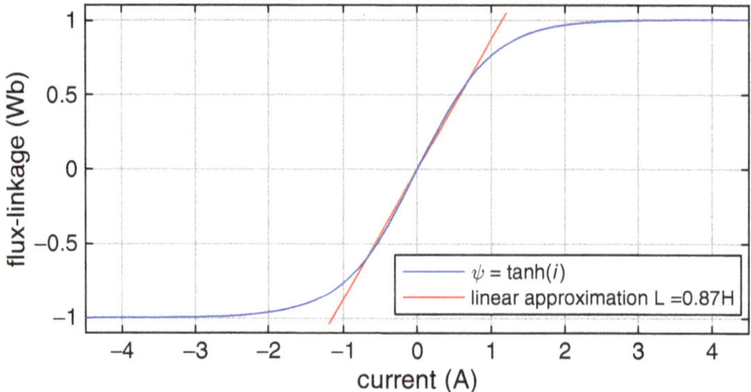

Fig. 2.14 Flux-linkage/current $\psi(i)$ relationship

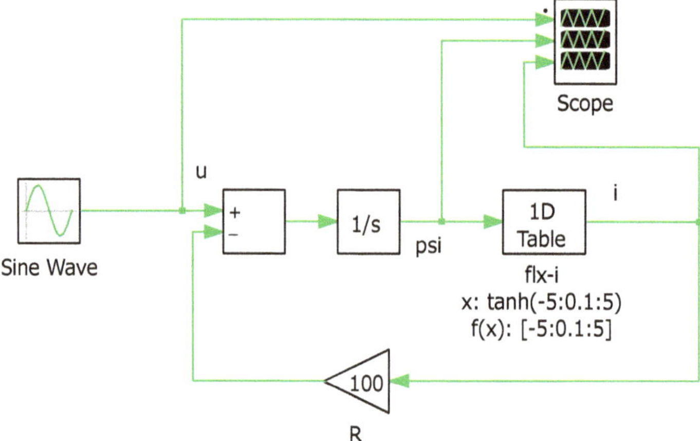

Fig. 2.15 PLECS model of non-linear inductance with sinusoidal excitation function

entries under: "vector of input values": set to `tanh([-5:0.1:5])` and "vector of output values": set to `[-5:0.1:5]`. Also given in Fig. 2.15 is a "sine wave" module, which in this case must generate the function $u = \hat{u}\cos\omega t$, where $\omega = 100\,\pi$ (rad/s) and $\hat{u}$ is initially set to $\hat{u} = 140\sqrt{2}\,\mathrm{V}$. Note that a cosine function is used. This means that in the "Sine Wave" dialog box (under "Phase") a phase angle entry is required, which must be set to $\pi/2$ (PLECS knows the meaning of "π" hence you can write this as "pi").

Once the new PLECS model has been completed, run this simulation for a time interval of 40 ms. For this purpose set the "stop time" (under Simulations/simulation parameters dialog window) to 40 ms. Save the results from the "Scope" module in the form of a "xxx.csv" file. An example of the results obtained with this simulation under the present conditions is given in Fig. 2.16. The results as given in Fig. 2.16 also include two "M-file" functions, which represent the results obtained via a phasor analysis to be discussed below.

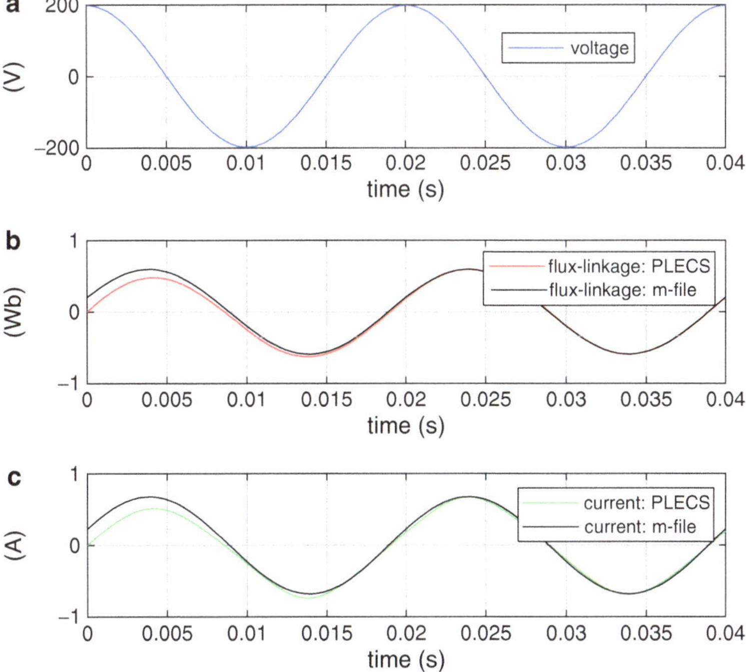

Fig. 2.16 PLECS/M-file results: inductance simulation, with coil resistance and non-linear $i(\psi)$ function

To obtain some idea as to whether or not the simulation results discussed in this tutorial are correct, we calculate the steady-state flux-linkage and current versus time functions by way of a phasor analysis. An observation of the current amplitude shows that, according to Fig. 2.14, operation is within the linear part of the current/flux-linkage curve. Assume a linear approximation of this function as shown in Fig. 2.14. This approximation corresponds to an inductance value of $L = 0.87\,\mathrm{H}$.

The input function $u = \hat{u}\cos\omega t$ may also be written as

$$u(t) = \Re\left\{\underbrace{\hat{u}}_{\underline{u}}\, e^{\mathrm{j}(\omega t)}\right\} \tag{2.18}$$

where in this case the phasor $\underline{u} = \hat{u} = 140\sqrt{2}\,\mathrm{V}$.

The actual phasor analysis must be done in MATLAB which also allows you to use complex numbers directly. For example, you can specify a phasor `xp=3+j*5` (in MATLAB form) and a reactance `X=100*pi*L`, where $L = 0.87\,\mathrm{H}$.

Write an M-file which will calculate the current and flux phasors. In addition calculate and plot the instantaneous current and flux versus time waveforms and add the results from the PLECS simulation (generated in the form of a "xxx.csv" file).

An example of such an M-file is given at the end of this tutorial, which also shows the code required to plot the results from the PLECS model.

The results obtained after running this M-file are shown in Fig. 2.16 (in "black"), together with the earlier PLECS results. A comparison between the results obtained via the PLECS model and phasor analysis (see Fig. 2.16) shows that the waveforms merge towards the end of the simulation time. In the first part of the simulation the transient effects dominate, hence the discrepancy between the simulation results and those calculated via a (steady-state *ac*) phasor analysis.

2.6.3.1 M-File Code

```
%Tutorial 3, chapter 2
close all
L=0.87;                                 %inductance value (H)
R=100;%resistance
dat = csvread('tut3ch2data.csv',1,0) % read in data from
    PLECS
close all
L=0.87;                                 %inductance value (H)
R=100;                                  %resistance
subplot(3,1,1)
plot(dat(:,1),dat(:,2));                % voltage input
xlabel(' (a) time (s)')
ylabel('voltage (V)')
grid
%%%%%%%%%%%%%%%%%%%%%%%%%%%%%%%%%%%%%%%%%%%%%%%%%%%%%
subplot(3,1,2)
plot(dat(:,1),dat(:,3),'r');            % flux-linkage
xlabel(' (b) time (s)')
ylabel('\psi (Wb)')
grid
%%%%%%%%%%%%%%%%%%%%%%%%%%%%%%%%%%%%%%%%%%%%%%%%%%%%%
subplot(3,1,3)
plot(dat(:,1),dat(:,4),'g');            % current
xlabel(' (c) time (s)')
ylabel(' current  (A)')
grid
%%%%%%%%%%%%%%%%%%%%%%%%%%%%%%%%%%%%%%%%%%%%%%%%%%%%%
%complex analysis
u_ph=140*sqrt(2);                       %voltage phasor
w=2*pi*50;                              %excitation frequency
                                           (rad/)
X=w*L;%reactance
i_ph=u_ph/(R+j*X);                      %current phasor
i_pk=abs(i_ph);                         %peak current value
i_rho=angle(i_ph);                      % angle current phasor
psi_ph=i_ph*L;%flux phasor
psi_pk=abs(psi_ph);                     %peak value flux
psi_rho=angle(psi_ph);                  %angle current phasor
%%%%%%%%%%plot results
```

```
time=[0:40e-3/100:40e-3];
i_t=i_pk*cos(w*time+i_rho);           %current/time function
psi_t=psi_pk*cos(w*time+psi_rho);     %flux/time function
subplot(3,1,3)
hold on
plot(time,i_t,'k');                   %add result to plot 3
legend('PLECS','m-file')
subplot(3,1,2)
hold on
plot(time,psi_t,'k');                  %add result to plot 2
legend('PLECS','m-file')
```

2.6.4 Tutorial 4: PLECS Based Analysis of a Non-linear Inductance Model with Revised Excitation Condition

It is instructive to repeat the analysis given in tutorial 3 by changing the peak supply voltage to $\hat{u} = 240\sqrt{2}\,\mathrm{V}$ in the PLECS model and M-file. An example of the results, which should appear after running your files, is given in Fig. 2.17.

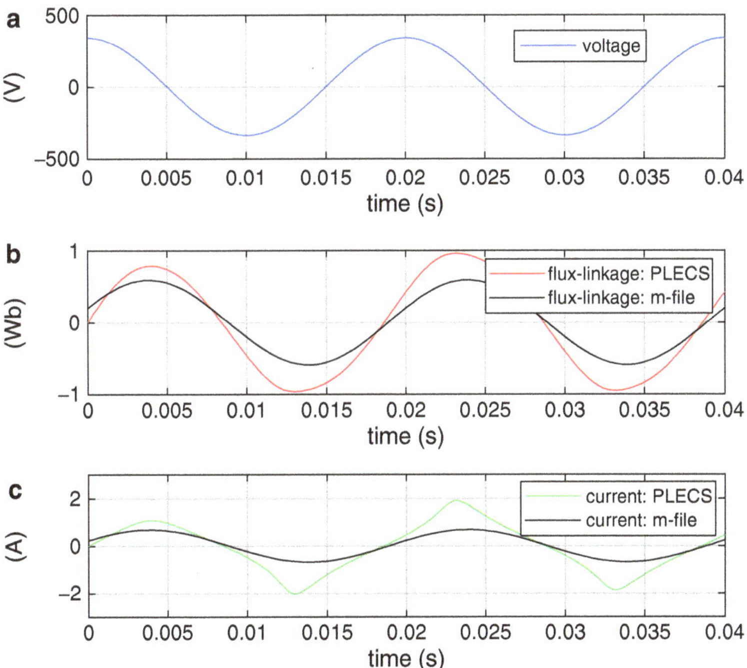

Fig. 2.17 PLECS/M-file results: induction simulation, with coil resistance, non-linear $i(\psi)$, and higher peak voltage

A comparison between the results obtained via the phasor analysis and PLECS simulation shows that the two are now decidedly different. The reason for the discrepancy is that the increased supply voltage level has increased the flux levels, which forces operation of the inductance into the non-linear regions of the flux-linkage/current curve. Note that the phasor analysis uses the same $L = 0.87\,\mathrm{H}$ inductance value. To prevent invalid conclusions, we must be aware that this *ac* phasor analysis tool is only usable for linear models.

2.6.5 Tutorial 5: PLECS Based Electro-magnetic Circuit Example

This tutorial makes use of the magnetic model introduced previously (see Sect. 1.8.1) which is to be connected to a 100 V, 50 Hz sinusoidal voltage source. The coil resistance R of the coil is assumed to be 500 Ω. Build a PLECS based model, which shown the magnetic structure and symbolic (electrical) circuit. Add a scope module to show: applied voltage, current, flux linked with the coil, and the coil MMF. Use the geometry parameters as defined in Sect. 1.8.1. The PLECS model according to Fig. 2.18 is an implementation of said problem. Readily observable are the "electrical" ("black" connections) and magnetic ("red" connections) components together with the meters used to measure voltage, current, and MMF. Furthermore, a meter `dPhi` is present, which measures the circuit flux differential $d\phi/dt$, hence a integrator must used to generate the circuit flux ϕ. The flux-linkage $\psi = n\phi$ is found by adding a gain module after the integrator with gain 1000, which is the number of turns of the coil. the simulation results by way of three "SCOPE" submodules. The results displays on the Scope module show the required variables for a time interval of 40 ms (Fig. 2.19).

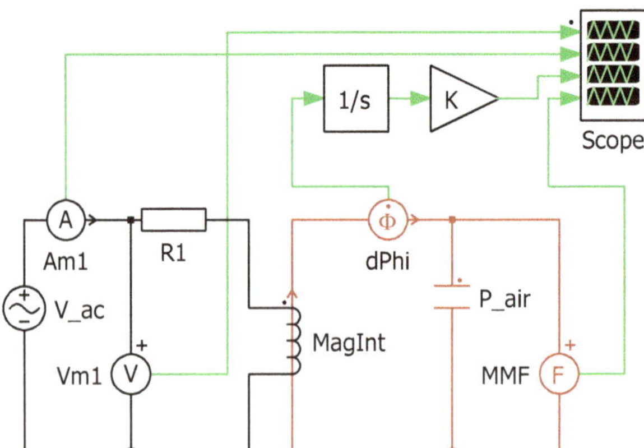

Fig. 2.18 PLECS simulation: electro-magnetic circuit example

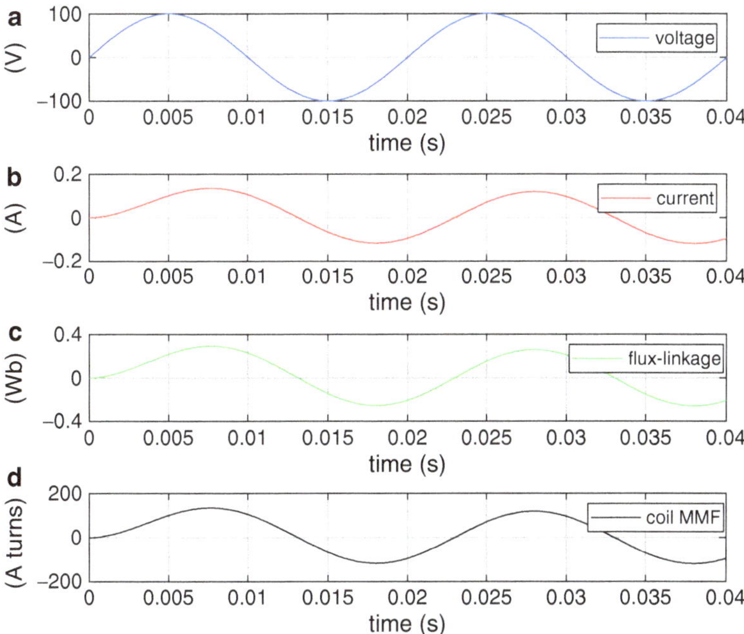

Fig. 2.19 Simulation results for electro-magnetic circuit example

It is instructive to briefly consider the results shown on the scope module:

- Current: this waveform lags the voltage waveform as expected because the coil has inductance and resistance.
- Flux linkage: this waveform is identical to the current waveform, but the magnitude is different. This is to be expected as the flux linkage is equal to $\psi = L i$, where L is the inductance which according to Sect. 1.8.1 was found to be 2.19 H.
- Coil MMF: this waveform is identical to the current waveform, but the magnitude is different. This is to be expected as the coil MMF is equal to MMF $= n\, i$, where n is the number of coil turns, set to 1000.

Chapter 3
The Transformer

3.1 Introduction

The aim of this chapter is to introduce the *ideal transformer* (ITF) concept. Initially, a single phase version is discussed, which forms the basis for a transformer model. This model will then be extended to accommodate the so-called magnetizing inductance and leakage inductance. Furthermore, coil resistances will be added to complete the model. Finally, a *universal* model will be shown which is fundamental to machine models. As in the previous chapters, symbolic and generic models will be used to support the learning process and to assist the readers with the development of PLECS models in the tutorial session at the end of this chapter.

Phasor analysis remains important as to be able to check the steady-state solution of the models when connected to a sinusoidal source.

3.2 Ideal Transformer Concept

The physical model of the transformer, shown in Fig. 3.1, replaces the toroidal shaped magnetic circuit used earlier. The transformer consists of an inner cylindrical rod and outer tube made of ideal magnetic material, i.e., infinite permeability. The inner bar and outer tube are each provided with windings n_2 and n_1 turns, respectively. The outer n_1 winding is referred to as the "primary" and carries a primary current i_1. The inner n_2 winding is known as the secondary winding and it carries a current i_2. The cross-sectional view shows the layout of the windings in the unity length inner rod and outer tube together with the assumed current

---

Electronic supplementary material The online version of this chapter (doi: 10.1007/978-3-319-29409-4_3) contains supplementary material, which is available to authorized users.

© Springer International Publishing Switzerland 2016

A. Veltman et al., *Fundamentals of Electrical Drives*, Power Systems,
DOI 10.1007/978-3-319-29409-4_3

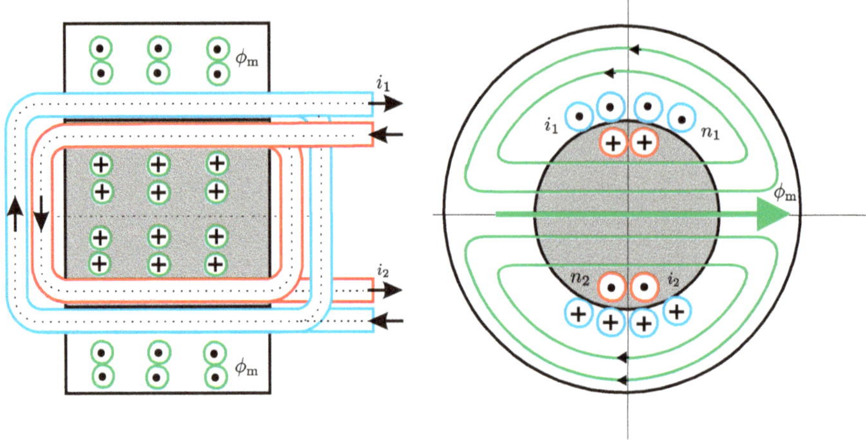

Fig. 3.1 Example real ITF model

polarity. Furthermore, a flux ϕ_m is shown in Fig. 3.1 which is linked with both coils. It is assumed at this stage that the total flux in the transformer is fully linked with both windings. In addition, the airgap between the inner rod and outer tube of the transformer is taken to be infinitely small at this stage.

The magnetic material of the transformer is, as mentioned above, assumed to have infinite permeability at this stage, which means that the reluctance R_m of the magnetic circuit is in fact zero. Consequently, the magnetic potential u_{core} across the iron core must be zero given that $u_{core} = \phi_m R_m$, where ϕ_m represents the circuit flux in the core, which is assumed to have a finite value. The fact that the total magnetic potential in the core must be zero shows us the basic mechanism of the transformer in terms of the interaction between primary and secondary current.

Let us assume a primary and secondary current as indicated in Fig. 3.1. Note carefully the direction of current flow in each coil. Positive current direction is "out of the page," negative "into the page." The MMF of the two coils can be written as

$$\mathrm{MMF}_{\mathrm{coil}\,1} = +n_1\, i_1 \tag{3.1a}$$

$$\mathrm{MMF}_{\mathrm{coil}\,2} = -n_2\, i_2 \tag{3.1b}$$

The MMFs of the two coils are purposely chosen to be in opposition, given that it is the "natural" current direction as will become apparent shortly. The resultant coil MMF "seen" by the magnetic circuit must be zero, because the magnetic potential $u_{iron} = 0$. This means that the following MMF condition holds:

$$n_1 i_1 - n_2 i_2 = 0 \tag{3.2}$$

Equation (3.2) is known as the basic ITF current relationship. This expression basically tells us that a secondary current i_2 must correspond with a primary current, according to $i_1 = {}^{n_2}/_{n_1} \cdot i_2$ [see Eq. (3.2)].

The second basic equation which exists for the ITF relates to the primary and secondary flux-linkage values. If we assume, for example, that a voltage source is connected to the primary, then a primary flux-linkage ψ_1 value will be present. This in turn means that the circuit flux ϕ_m will be equal to $\phi_\mathrm{m} = {}^{\psi_1}/_{n_1}$. The corresponding flux linked with the secondary is of the form $\psi_2 = n_2\,\phi_\mathrm{m}$. The relationship between primary and secondary flux-linkage values can therefore be written as

$$\psi_2 = \frac{n_2}{n_1}\psi_1 \tag{3.3}$$

The corresponding terminal voltage equations for the primary and secondary are of the form

$$u_1 = \frac{\mathrm{d}\psi_1}{\mathrm{d}t} \tag{3.4a}$$

$$u_2 = \frac{\mathrm{d}\psi_2}{\mathrm{d}t} \tag{3.4b}$$

These equations are similar to Eq. (2.2) which was developed for a single coil with zero resistance. A symbolic representation of the ITF is shown in Fig. 3.2.

The complete equation set of the ITF is given in Eq. (3.5).

$$u_1 = \frac{\mathrm{d}\psi_1}{\mathrm{d}t} \tag{3.5a}$$

$$u_2 = \frac{\mathrm{d}\psi_2}{\mathrm{d}t} \tag{3.5b}$$

$$\psi_2 = \frac{n_2}{n_1}\,\psi_1 \tag{3.5c}$$

$$i_1 = \frac{n_2}{n_1}\,i_2 \tag{3.5d}$$

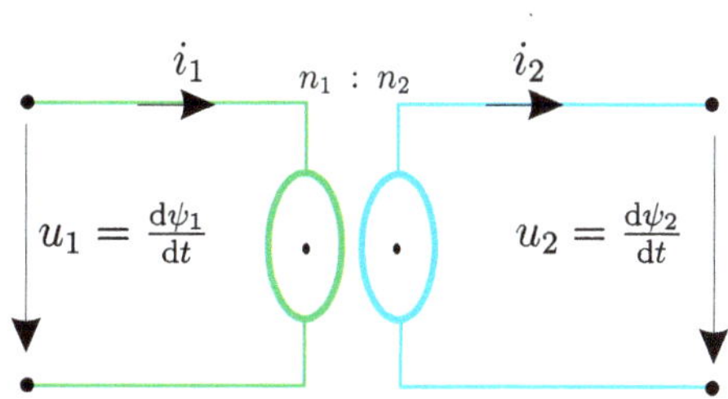

Fig. 3.2 Symbolic model of ITF

where n_2/n_1 represents the so-called winding or turns ratio of the ITF. Note that the current directions shown in Fig. 3.2 for primary and secondary are the same, i.e., pointing to the right. In some applications it is more convenient to reverse one or both current directions. The ITF model (Fig. 3.1) is directly linked to the symbolic model of Fig. 3.2 in terms of current polarities. If we choose the primary current "into" the ITF model then the secondary direction follows "naturally" (because of the reality that the total MMF must be zero), i.e., must come "out" of the secondary side of the model.

The generic diagram of the basic ITF module is linked to the flux-linkage and current relations given by Eqs. (3.5c) and (3.5d), respectively. The generic diagram that corresponds with Fig. 3.2 is given by Fig. 3.3a.

It is sometimes beneficial to replace the primary flux-linkage input ψ_1 with the primary current input i_1. This means that i_2 becomes an output. Under these circumstances we must also choose flux ψ_1 as an output and ψ_2 input (we cannot have both the flux and current on one side of an ITF as inputs or outputs). This version of the ITF module, named "ITF-Current" is given in Fig. 3.3b.

The instantaneous power is given as the product of voltage times current, i.e., $u_1 i_1$ and $u_2 i_2$. For the ITF model, power *into* the primary side corresponds to positive power ($p_{\text{in}} = u_1 i_1$). Positive output power for the ITF is defined as ($p_{\text{out}} = u_2 i_2$) *out* of the secondary as shown in Fig. 3.4.

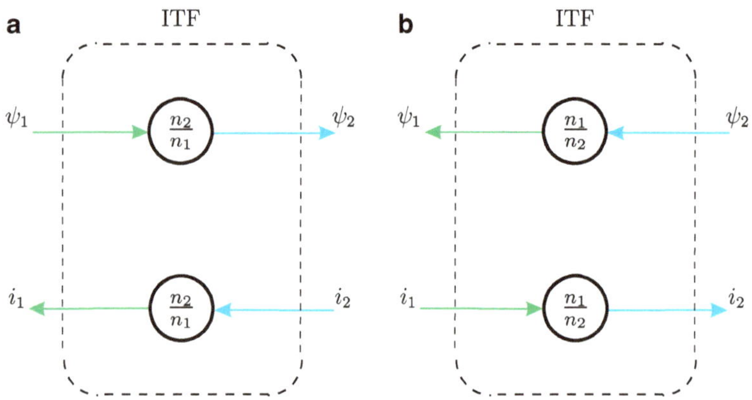

Fig. 3.3 Generic models of ITF. (**a**) ITF-Flux. (**b**) ITF-Current

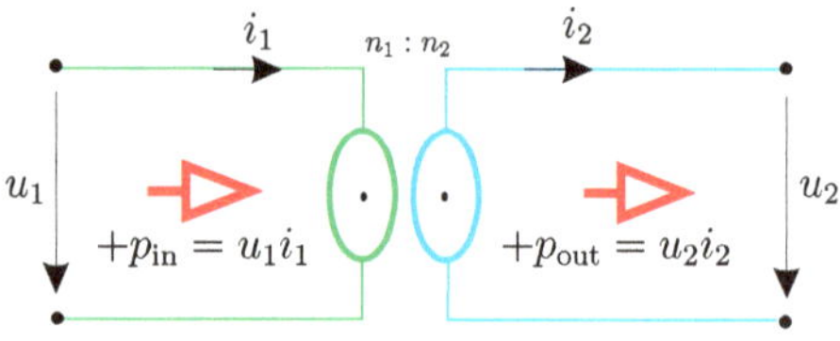

Fig. 3.4 Power convention ITF

3.3 Basic Transformer

The ITF module forms the corner stone in this book for transformer modeling and is also the stepping stone to the so-called IRTF module used for machine analysis. An example of the transformer connected to a resistive load is shown in Fig. 3.5.

In this example, an excitation voltage u_1 is assumed, which in turn corresponds to a secondary voltage u_2 across the load resistance. The ITF equation set is given in Eq. (3.6).

$$u_1 = \frac{\mathrm{d}\psi_1}{\mathrm{d}t} \tag{3.6a}$$

$$u_2 = \frac{\mathrm{d}\psi_2}{\mathrm{d}t} \tag{3.6b}$$

$$\psi_2 = \frac{n_2}{n_1}\,\psi_1 \tag{3.6c}$$

$$i_2' = \frac{n_2}{n_1}\,i_2 \tag{3.6d}$$

The ITF current on the primary side is renamed i_2' and is known as the *primary referred* secondary current. It is the current which is "seen" on the primary side, due to a current i_2 on the secondary side. In this case i_1 equals i_2', as may be observed in Fig. 3.5. The equation set of this transformer must be extended with the equation $u_2 = i_2 R_L$. A generic representation of the symbolic diagram according to Fig. 3.5 is given in Fig. 3.6.

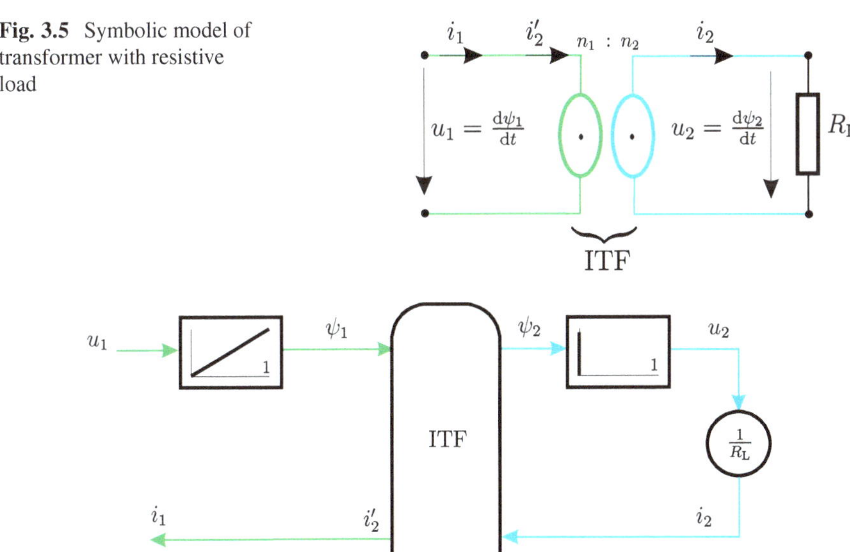

Fig. 3.5 Symbolic model of transformer with resistive load

Fig. 3.6 Generic model of transformer with load

The ITF module is shown as a "sub-module," which in fact represents the generic model according to Fig. 3.3a with the change that the current i_1 is renamed i'_2.

It is important to understand how the transformer functions. Basically, the integrated applied primary voltage gives a primary flux-linkage value, which in turn leads to a core flux ϕ_m. The core flux results in ψ_2, which is the flux linked with the secondary winding. The secondary flux-linkage time differential represents the secondary voltage which will cause a current i_2 in the load. The secondary current leads to an MMF equal to $n_2 i_2$ on the secondary side which must be countered by an MMF of $n_1 i_1$ on the primary side given that the magnetic reluctance of the transformer is considered to be zero in the ITF. Note that an open-circuited secondary winding ($R_L = \infty$) would correspond to a zero secondary and a zero primary current value. The fluxes are not affected as these are determined by the primary voltage, time, and winding ratio in this case (we have assumed that the primary coil is connected to a voltage source and the secondary to a load impedance).

Note that a differentiator module is used in the generic model shown in Fig. 3.6. Differentiators *should be avoided where possible* in actual simulations, given that simulations tend to operate poorly with such modules. In most cases, the use of a differentiator module in actual simulations is not required, because we can either implement the differentiator by alternative means or build models that avoid the use of such modules.

3.4 Transformer with Magnetizing Inductance

In electrical machines, airgaps are introduced in the magnetic circuit which, as was made apparent in Chap. 2, will significantly increase the total magnetic circuit reluctance R_m. Furthermore, in reality the magnetic material will have a finite permeability, which will further increase the overall magnetic circuit reluctance. The transformer according to Fig. 3.7 has an airgap between the primary and secondary windings. The core flux ϕ_m now needs to cross this airgap twice. Consequently, a given core flux will, according to Hopkinson's law, correspond to a non-zero magnetic circuit potential "u_M" in case $R_m > 0$. The required magnetic potential must be provided by the coil MMF which is connected to the voltage source. We have chosen the primary side to be excited with a voltage source, while the secondary side is connected to, for example, a resistive load. The implication of the above is that an MMF equal to $n_1 i_m$ must be provided via the primary winding. The current i_m is known as the *magnetizing current*, which is directly linked with the primary flux-linkage value ψ_1 and the so-called magnetizing inductance L_m. The relationship between these variables is of the form

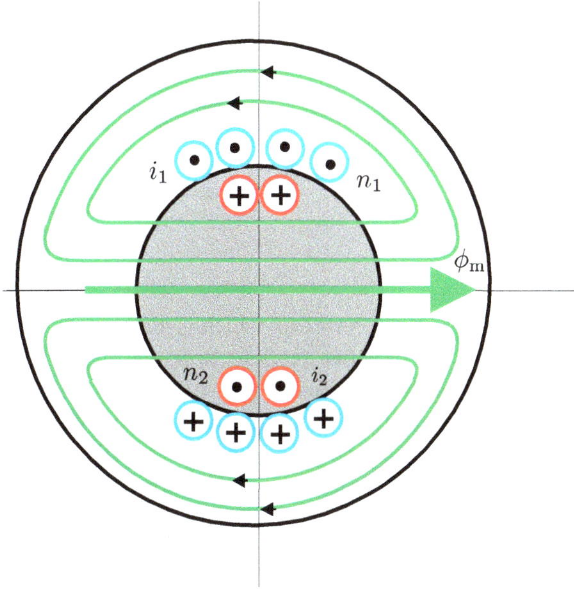

Fig. 3.7 Transformer model with finite airgap

$$i_{\mathrm{m}} = \frac{\psi_1}{L_{\mathrm{m}}} \tag{3.7}$$

Note that the magnetizing inductance is directly linked with the magnetic reluctance R_{m} namely $L_{\mathrm{m}} = {}^{n_1^2}/_{R_{\mathrm{m}}}$, as was discussed in Chap. 2. Zero magnetic reluctance corresponds to an infinite magnetizing inductance and, according to Eq. (3.7), zero magnetizing current i_{m}.

The presence of a core MMF requires us to modify Eq. (3.2) because the sum of the coil MMF's no longer equals zero. The revised MMF equation is now of the form

$$n_1 i_1 - n_2 i_2 = n_1 i_{\mathrm{m}} \tag{3.8}$$

which may also be rewritten as:

$$i_1 = i_{\mathrm{m}} + \underbrace{\frac{n_2}{n_1} i_2}_{i_2'} \tag{3.9}$$

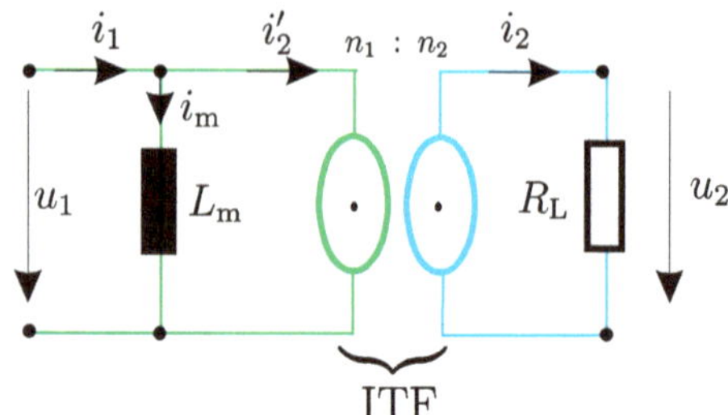

Fig. 3.8 Symbolic model of transformer with load and finite L_m

In expression (3.9) the variable i'_2 is shown, which is the primary referred secondary current, as introduced in the previous section. Note that in the magnetically ideal case (where $i_m = 0$) the primary current is given as $i_1 = i'_2$. The ITF equation set according to Eq. (3.6) remains directly applicable to the revised transformer model.

The symbolic transformer diagram according to Fig. 3.5 must also be revised to accommodate the presence of the magnetizing inductance on either the primary or secondary side of the ITF. The revised symbolic diagram is given in Fig. 3.8.

The complete equation set, which is tied to the symbolic transformer model according to Fig. 3.8, is given as

$$u_1 = \frac{d\psi_1}{dt} \tag{3.10a}$$

$$u_2 = \frac{d\psi_2}{dt} \tag{3.10b}$$

$$i_1 = i_m + i'_2 \tag{3.10c}$$

$$i_m = \frac{\psi_1}{L_m} \tag{3.10d}$$

$$u_2 = i_2 R_L \tag{3.10e}$$

For modeling a system of this type, it is important to be able to build a generic model which is directly based on Fig. 3.8 and the corresponding equation set (3.10). The generic module of the transformer as given in Fig. 3.9 is directly based on the earlier model given in Fig. 3.6. Shown in Fig. 3.9 is an ITF sub-module which is in fact of the form given in Fig. 3.3a, with the provision that the current output i_1 (of the ITF module) is now renamed i'_2, which is known as the primary referred secondary current.

3.5 Steady-State Analysis

The model representations discussed so far are dynamic, which means that they can be used to analyze a range of excitation conditions, which includes transient as well as steady-state. Of particular interest is to determine how such systems behave

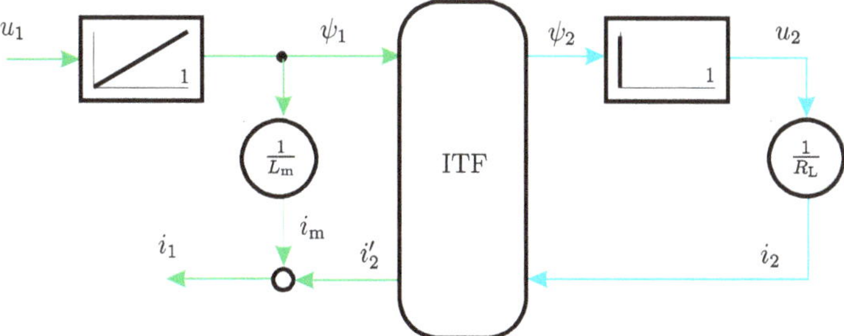

Fig. 3.9 Generic model of transformer with load and finite L_m

when connected to a sinusoidal voltage source. Systems, such as a transformer with resistive loads, will (after being connected to the excitation source) initially display some transient behavior but will then quickly settle down to their steady-state.

As discussed earlier, the steady-state analysis of linear systems connected to sinusoidal excitation sources is of great importance. Firstly, it allows us to gain a better understanding of such systems by making use of phasor analysis tools. Secondly, we can use the outcome of the phasor analysis as a way to check the functioning of our dynamic models once they have reached their steady-state.

3.5.1 Steady-State Analysis Under Load with Magnetizing Inductance

In steady-state, the primary excitation voltage is of the form $u_1 = \hat{u}_1 \cos \omega t$, which corresponds to a voltage phasor $\underline{u}_1 = \hat{u}_1$. The supply frequency is equal to $\omega = 2\pi f$ where f represents the frequency in Hz.

The aim is to use complex number theory together with Eqs. (3.10) and (3.6) to analytically calculate the phasors: $\underline{\psi}_1$, $\underline{\psi}_2$, $\underline{i}_2$, $\underline{i}'_2$, $\underline{i}_1$, and $\underline{u}_2$. The flux-linkage phasor is directly found using Eq. (3.10a), which in phasor form is given by $\underline{u}_1 = \mathrm{j}\omega\underline{\psi}_1$. The corresponding flux-linkage phasor on the secondary side of the ITF module is found using (3.6d), which leads to $\underline{\psi}_2 = {}^{n_2}/_{n_1} \cdot \underline{\psi}_1$. The secondary voltage equation (3.10b) gives us the secondary voltage (in phasor form) $\underline{u}_2 = \mathrm{j}\omega\underline{\psi}_2$, which in turn allows us to calculate the secondary current phasor according to $\underline{i}_2 = {}^1/_{R_L} \cdot \underline{u}_2$. This phasor may also be written in terms of the primary voltage phasor $\underline{u}_1 = \hat{u}_1$ as

$$\underline{i}_2 = \left(\frac{n_2}{n_1}\right)\frac{\underline{u}_1}{R_L} \tag{3.11}$$

The corresponding primary referred secondary current phasor is found using Eqs. (3.6d), (3.11) which gives $\underline{i}'_2 = ({}^{n_2}/_{n_1})^2\, {}^{\underline{u}_1}/_{R_L}$. The primary current phasor is found using (3.10c), where the magnetizing current phasor $\underline{i}_m$ is found using (3.10d) namely $\underline{i}_m = {}^1/_{L_m} \cdot \underline{\psi}_1$. The resultant primary current phasor may also be written as

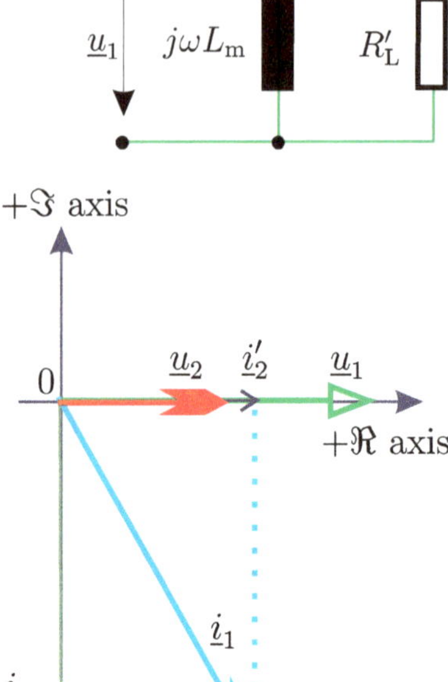

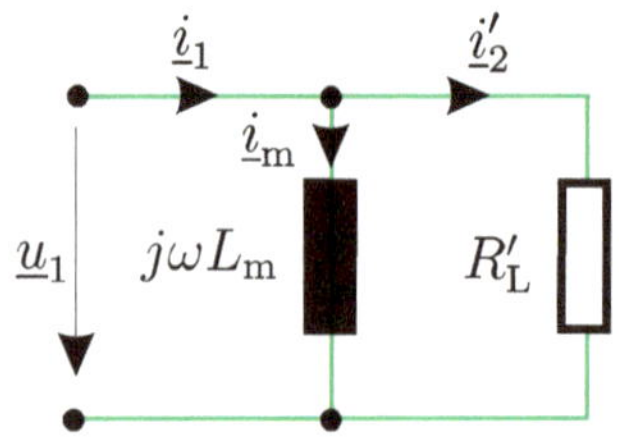

Fig. 3.10 Primary referred phasor model of transformer with load and finite L_m

Fig. 3.11 Phasor diagram of transformer with load and finite L_m

$$\underline{i}_1 = \frac{\underline{u}_1}{\mathrm{j}\omega L_\mathrm{m}} + \left(\frac{n_2}{n_1}\right)^2 \frac{\underline{u}_1}{R_\mathrm{L}} \tag{3.12}$$

It is instructive to consider Eq. (3.12) in terms of an equivalent circuit model as shown in Fig. 3.10.

The diagram shows the load resistance in its so-called transformer primary side "referred" form $R'_\mathrm{L} = ({n_1}/{n_2})^2 R_\mathrm{L}$. Hence, we are able to determine the currents (in phasor form) directly from this diagram. A phasor diagram of the transformer with a resistive load R_L and magnetizing inductance L_m, which corresponds with the given phasor analysis and equivalent circuit (Fig. 3.10), is shown in Fig. 3.11.

Some interesting observations can be made from this diagram. Firstly, the primary and secondary voltages are in phase. Secondly, the magnetizing current phasor lags the primary voltage phasor by $\pi/2$. The primary referred secondary current phasor $\underline{i}'_2$ is in phase with $\underline{u}_2$ because we have a resistive load. Furthermore, the primary current is found by adding (in vector form) the phasors $\underline{i}'_2$ and $\underline{i}_\mathrm{m}$. Note that the primary current equals the magnetizing current when the load resistance is removed, i.e., $R_\mathrm{L} = \infty$.

The corresponding steady-state time function of, for example, the current i_1 can be found by using

$$i_1(t) = \Re\left\{\underline{i}_1\, e^{j\omega t}\right\} \tag{3.13}$$

where $\underline{i}_1$ is found using (3.12).

3.6 Three Inductance Model

The model according to Fig. 3.7 assumes that all the flux is linked with both coils. In reality this is not the case as may be observed from Fig. 3.12. This diagram shows two flux contributions: $\phi_{\sigma 1}$ and $\phi_{\sigma 2}$, which are known as the primary and secondary leakage flux components, respectively. The leakage fluxes physically arise from the fact that *not all* flux of each coil is "seen" by both.

Consequently, these components are *not* linked with both coils and they are represented by primary and secondary leakage inductances

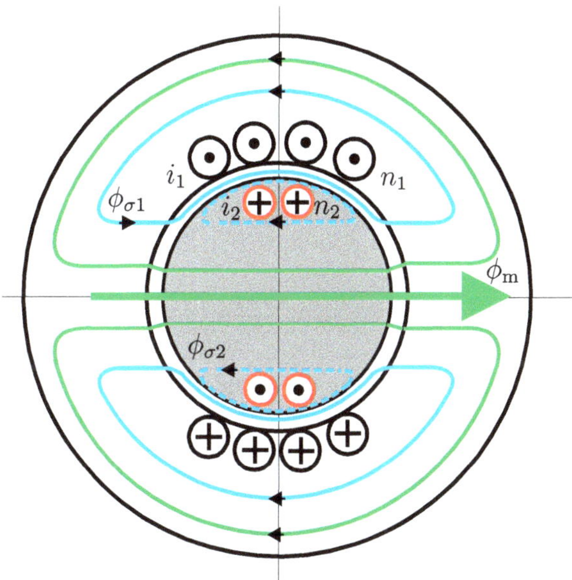

Fig. 3.12 Transformer model with finite airgap and leakage

$$L_{\sigma 1} = \frac{\psi_{\sigma 1}}{i_1} \tag{3.14a}$$

$$L_{\sigma 2} = \frac{\psi_{\sigma 2}}{i_2} \tag{3.14b}$$

where the primary and secondary leakage flux-linkage values are given by $\psi_{\sigma 1} = n_1 \phi_{\sigma 1}$ and $\psi_{\sigma 2} = n_2 \phi_{\sigma 2}$, respectively. The total flux-linkage seen by the primary and secondary is thus equal to

$$\psi_1 = \psi_{\rm m} + \psi_{\sigma 1} \tag{3.15a}$$

$$\psi_2 = \psi'_{\rm m} - \psi_{\sigma 2} \tag{3.15b}$$

The core flux $\phi_{\rm m}$ which is linked with the primary coil is now renamed $\psi_{\rm m} = n_1 \phi_{\rm m}$. Similarly, the core flux $\phi_{\rm m}$ which is linked with the secondary coil gives us the flux-linkage $\psi'_{\rm m} = n_2 \phi_{\rm m}$. The terminal equations for the transformer in its current form are given as

$$u_1 = \frac{{\rm d}\psi_{\rm m}}{{\rm d}t} + L_{\sigma 1} \frac{{\rm d}i_1}{{\rm d}t} \tag{3.16a}$$

$$u_2 = \frac{{\rm d}\psi'_{\rm m}}{{\rm d}t} - L_{\sigma 2} \frac{{\rm d}i_2}{{\rm d}t} \tag{3.16b}$$

where

$$u_1 = \frac{{\rm d}\psi_1}{{\rm d}t} \tag{3.17a}$$

$$u_2 = \frac{{\rm d}\psi_2}{{\rm d}t} \tag{3.17b}$$

The symbolic representation of the transformer according to Fig. 3.8 must be extended to include the leakage inductance. The revised symbolic model as given in Fig. 3.13 clearly shows the leakage inductances. A generic model of the transformer

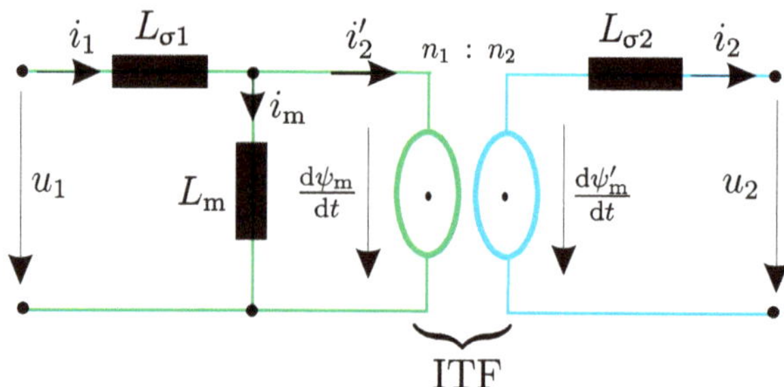

Fig. 3.13 Symbolic representation, transformer with magnetizing and leakage inductance

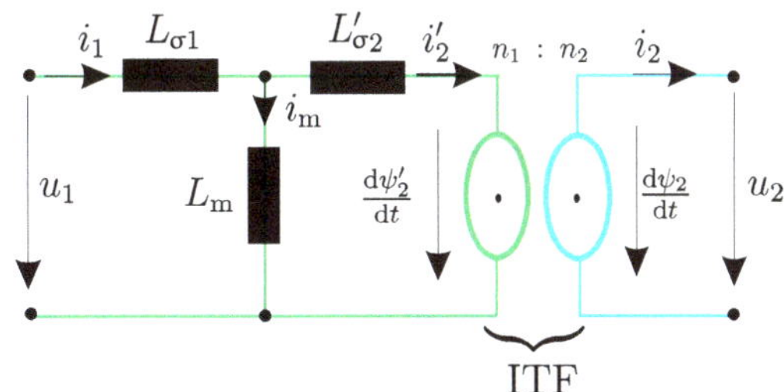

Fig. 3.14 Symbolic representation, transformer with magnetizing inductance, primary leakage inductance, and referred secondary leakage inductance

with leakage inductances can be developed by firstly moving the secondary leakage inductance $L_{\sigma 2}$ to the primary side of the ITF. In this case the referred value $L'_{\sigma 2}$ equals $\left({}^{n_1}/_{n_2}\right)^2 L_{\sigma 2}$, as may be observed from Fig. 3.14. The relationship between the flux-linkages ψ_1, ψ'_2 and currents i_1, i'_2 can be written in matrix format as

$$\begin{bmatrix} \psi_1 \\ \psi'_2 \end{bmatrix} = \begin{bmatrix} L_1 & -L_m \\ L_m & -L'_2 \end{bmatrix} \begin{bmatrix} i_1 \\ i'_2 \end{bmatrix} \tag{3.18}$$

where $L'_2 = L_m + L'_{\sigma 2}$. For the development of an ITF based model it is helpful to invert Eq. (3.18) which gives

$$\begin{bmatrix} i_1 \\ i'_2 \end{bmatrix} = \underbrace{\frac{1}{L_1 L'_2 - L_m^2} \begin{bmatrix} L'_2 & -L_m \\ L_m & -L_1 \end{bmatrix}}_{[L^{-1}]} \begin{bmatrix} \psi_1 \\ \psi'_2 \end{bmatrix} \tag{3.19}$$

which, together with Eq. (3.17), leads to the generic model given in Fig. 3.15 that corresponds to the symbolic model given in Fig. 3.14. Clearly observable in Fig. 3.15 is a module L^{-1}, which represents the inverse matrix equation (3.19) required to calculate the currents i_1, i'_2 from the flux-linkage variables ψ_1 and ψ'_2. It is further noted that the model is *not suited* to simulate the no-load situation (with a passive load), i.e., open-circuited secondary winding, since current i_2 is an output and thus cannot be forced to zero by any load.

3.7 Universal ITF Based Transformer Model

The problem with the three-inductance model as discussed in Sect. 3.6 lies within the fact that it is extremely difficult to determine individual values for the two leakage inductances in case access to the secondary side of the model is not possible. For transformers this is not an issue (when the winding ratio is known) but, for

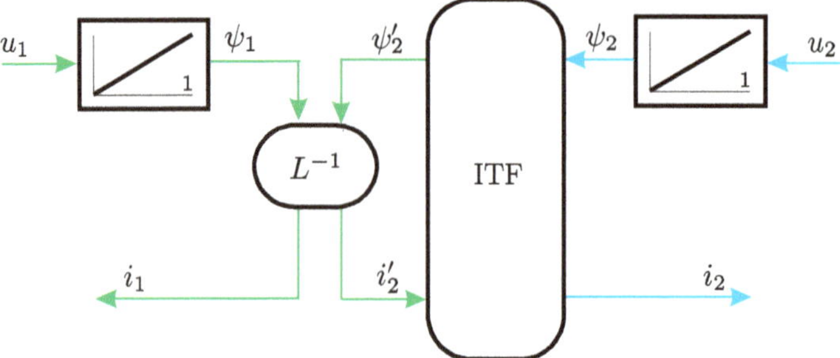

Fig. 3.15 Generic representation, transformer with magnetizing and leakage inductances

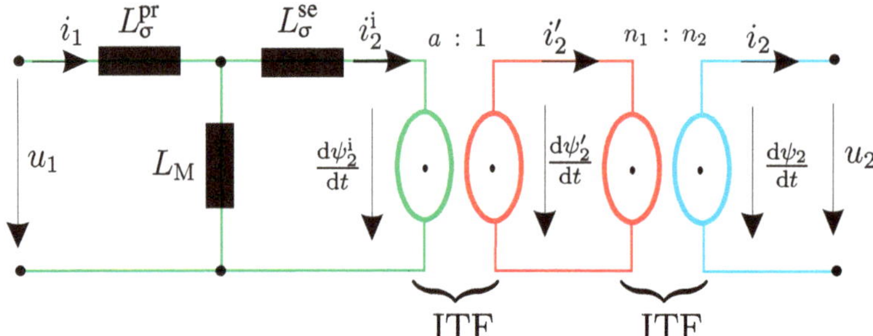

Fig. 3.16 Symbolic representation, universal transformer model

example, for squirrel cage asynchronous machines, to be discussed later, this is certainly the case. A triple inductance model of the transformer is in fact not needed given that its behavior can be perfectly modeled by a two-inductance based model as will become apparent shortly. For the purpose of the analysis it is helpful to introduce a new set of inductance parameters L_M, L_σ^{pr}, and L_σ^{se} and a second ITF module with winding ratio a:1, as shown in Fig. 3.16 to replace the three element inductance network of the model (shown previously in Fig. 3.14). The new set of inductance parameters is a function of the so-called transformation factor a, which can (as will be shown by the ensuing analysis) be judicially chosen to zero either one of the leakage inductances L_σ^{pr} or L_σ^{se}, thus creating a two-inductance model. Furthermore, the impedance as viewed from either side of the revised inductance network must correspond to the values found in the original inductance network and should not be affected by changes in the transformation factor "a". The transformation process is initiated by considering the equation matrix (3.18) which is linked to the model given in Fig. 3.14. In particular, it is helpful to consider the relationship between the primary flux-linkage ψ_1 and currents i_1, i_2', which can also be written as

$$\psi_1 = L_1 i_1 - L_m i_2' - aL_m i_1 + aL_m i_1 \tag{3.20}$$

This expression may also be written as

$$\psi_1 = \underbrace{(L_1 - aL_\mathrm{m})}_{L_\sigma^\mathrm{pr}} i_1 + \underbrace{aL_\mathrm{m}}_{L_\mathrm{M}} \left(i_1 - i_2^i\right) \tag{3.21}$$

where the new parameters L_σ^pr and L_M represent a generalized leakage inductance and magnetizing inductance, respectively. Furthermore, a scaled secondary current i_2^i is introduced in Eq. (3.21) according to

$$i_2^i = \frac{i_2'}{a} \tag{3.22a}$$

$$\psi_2^i = a\,\psi_2' \tag{3.22b}$$

Also shown in Eq. (3.22) is a secondary flux-linkage variable ψ_2^i, which represents the scaled (by the transformation factor a) flux-linkage variable ψ_2'. The choice of scaling these two variables ψ_2^i, i_2^i is such that the product of the current and flux variables, as well as the impedances remain unaffected by the scaling. In the "universal" model (see Fig. 3.16), Eq. (3.22) is represented by the ITF module with winding ratio a:1. Equation (3.22b) and the relationship between the secondary flux-linkage ψ_2' and currents i_1, i_2', as defined in Eq. (3.18) form the basis for the second part of the transformed model. Using these two equations to compute the scaled secondary flux-linkage variable ψ_2^i gives

$$\psi_2^i = aL_\mathrm{m} i_1 - a^2 L_2' i_2^i - aL_\mathrm{m} i_2^i + aL_\mathrm{m} i_2^i \tag{3.23}$$

This expression may also be rewritten as

$$\psi_2^i = \underbrace{aL_\mathrm{m}}_{L_\mathrm{M}} \left(i_1 - i_2^i\right) - \underbrace{\left(a^2 L_2' - aL_\mathrm{m}\right)}_{L_\sigma^\mathrm{se}} i_2^i \tag{3.24}$$

where a second leakage inductance parameter L_σ^se is introduced. The resultant flux-linkage based equation set as given by Eqs. (3.21), (3.24) can also be written as

$$\psi_1 = L_\sigma^\mathrm{pr} i_1 + L_\mathrm{M} i_\mathrm{M} \tag{3.25a}$$

$$\psi_2^i = L_\mathrm{M} i_\mathrm{M} - L_\sigma^\mathrm{se} i_2^i \tag{3.25b}$$

where $i_\mathrm{M} = i_1 - i_2^i$ represents the scaled magnetizing current. The flux equation set contains a set of leakage inductances and magnetizing inductance as shown in Fig. 3.16, which are function of the transformation variable "a" as mentioned earlier.

Some simplification of the model shown in Fig. 3.16 may be achieved by combining the two ITF modules to a single unit with winding ratio

$$k = a\frac{n_1}{n_2} \tag{3.26}$$

The new set on inductances shown in Fig. 3.16 is conveniently summarized in Eq. (3.27)

$$L_\sigma^{pr} = L_m\left(\frac{L_1}{L_m} - a\right) \tag{3.27a}$$

$$L_\sigma^{se} = aL_2'\left(a - \frac{L_m}{L_2'}\right) \tag{3.27b}$$

$$L_M = aL_m \tag{3.27c}$$

Observation of Eq. (3.27) shows that the transformation variable "a" is bound by the condition

$$\frac{L_m}{L_2'} \le a \le \frac{L_1}{L_m} \tag{3.28}$$

on the grounds that the leakage inductances L_σ^{pr}, L_σ^{se} must be greater or equal to zero.

The model according to Fig. 3.16 is reduced to a two-inductance model in case the value of the transformation factor is set to either $a = {}^{L_m}/_{L_2'}$ (which gives $L_\sigma^{se} = 0$) or $a = {}^{L_1}/_{L_m}$ (which gives $L_\sigma^{pr} = 0$). Both "two inductance" model configurations will be discussed in the next two subsections. Note that theoretically a model with a variable transformation factor a could be used, in which case a "universal" generic model similar to Fig. 3.15 can be found. Such a universal model would require a transformation factor a dependent inverse matrix module and an ITF module with winding ratio k. Note that a value of $a = 1$ corresponds to the case where the universal model is exactly equal to the model discussed in Sect. 3.6.

3.7.1 Primary Leakage Inductance Based Model

A *two-inductance model* with a leakage inductance L_σ^{pr} located on the primary side of the transformer can be obtained by considering the universal model shown in Fig. 3.16, with a transformation factor $a = {}^{L_m}/_{L_2'}$. Under these conditions, the

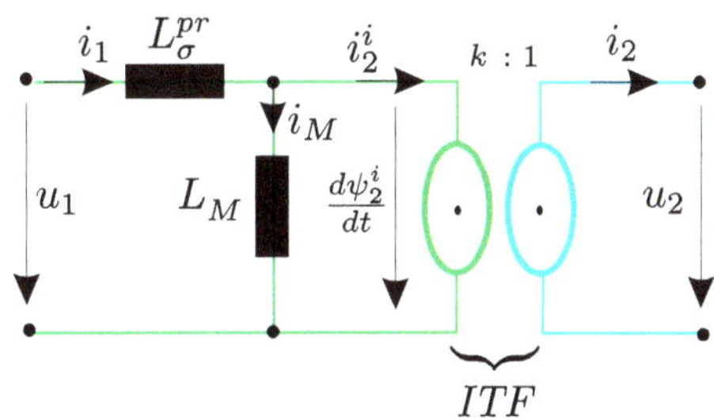

Fig. 3.17 Primary leakage inductance based symbolic transformer model

leakage inductance L_σ^{se} will be zero as may be observed from Eq. (3.27b). The symbolic model given in Fig. 3.17 corresponds to the symbolic model given in Fig. 3.16 with $a = {}^{L_\text{m}}/{}_{L_2'}$. A single ITF transformer with winding ratio k according to Eq. (3.26) is introduced instead of the two ITF modules shown in Fig. 3.16. The inductance values and winding ratio for the so-called *primary leakage inductance based model* or *secondary flux-linkage based model* may be found with the aid of Eqs. (3.27), (3.26) and substitution of $a = {}^{L_\text{m}}/{}_{L_2'}$ which gives

$$k = \frac{L_\text{m}}{L_2'}\frac{n_1}{n_2} \tag{3.29a}$$

$$L_\text{M} = \kappa^2 L_1 \tag{3.29b}$$

$$L_\sigma^\text{pr} = L_1\left(1 - \kappa^2\right) \tag{3.29c}$$

in which the so-called *coupling factor* κ is introduced, according to

$$\kappa = \sqrt{\frac{L_\text{m}^2}{L_1 L_2'}} \tag{3.30}$$

The equation set which corresponds to the two-inductance symbolic model of Fig. 3.17 is given in Eq. (3.31).

$$u_1 = \frac{\text{d}\psi_1}{\text{d}t} \tag{3.31a}$$

$$u_2 = \frac{\text{d}\psi_2}{\text{d}t} \tag{3.31b}$$

$$\psi_1 = i_1 L_\sigma^\text{pr} + \psi_2^i \tag{3.31c}$$

$$\psi_2^i = i_M L_M \tag{3.31d}$$

$$i_M = i_1 - i_2^i \tag{3.31e}$$

$$\psi_2^i = k\,\psi_2 \tag{3.31f}$$

$$i_2 = k\,i_2^i \tag{3.31g}$$

Equations (3.31f) and (3.31g) represent those implemented by the ITF module.

3.7.2 Secondary Leakage Inductance Based Model

The two-inductance model described in the previous subsection is particularly useful when the excitation is provided (by means of a current source, as is often the case for electrical drives) from the primary side of the ITF. Similarly the model described in the previous section is useful when a voltage source is connected to the secondary side, because this simplifies the task of computing primary currents.

In some cases, where, for example, a voltage source is connected to the primary side it is convenient to consider an alternative two-inductance model, which is also useful in case excitation (by means of a current source) is provided by the secondary side. A two-inductance model with a leakage inductance L_σ^{se} may be realized by considering the universal model shown in Fig. 3.16 with a transformation factor of $a = {}^{L_1}/_{L_\mathrm{m}}$. Under these conditions, the leakage inductance L_σ^{pr} will be zero as may be observed from Eq. (3.27b). The symbolic model given in Fig. 3.18 corresponds to the symbolic model given in Fig. 3.16 with $a = {}^{L_1}/_{L_\mathrm{m}}$. A single ITF transformer with winding ratio k as defined by Eq. (3.26) is introduced instead of the two ITF modules shown in Fig. 3.16. The inductance values and winding ratio for the so-called *secondary leakage inductance based model* or *primary flux based model* may be found with the aid of Eqs. (3.27), (3.26) and substitution of $a = {}^{L_1}/_{L_\mathrm{m}}$, which gives

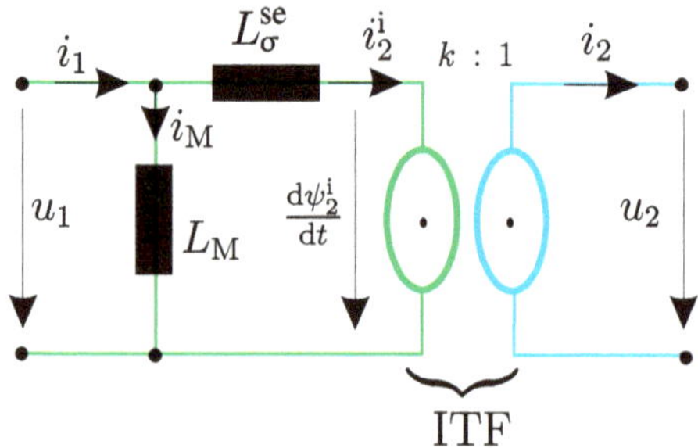

Fig. 3.18 Secondary leakage inductance based symbolic transformer model

$$k = \frac{L_1}{L_m}\frac{n_1}{n_2} \tag{3.32a}$$

$$L_M = L_1 \tag{3.32b}$$

$$L_\sigma^{se} = L_1\left(\frac{1}{\kappa^2} - 1\right) \tag{3.32c}$$

where κ is defined in Eq. (3.30). It is emphasized that this "alternative" two-parameter model is referred to as a *primary flux based model* because the primary flux linkage equals the (transformed) magnetizing flux in this model. In this model the leakage inductance is oriented towards the secondary terminals of the transformer. In this book the primary flux based ITF model is not used. Henceforth, whenever reference is made to a "two-inductance model," we will assume by default the configuration given by Fig. 3.17.

3.8 Mutual and Self-inductance Based Model

A model often used in the field of communication systems is of the form given in Fig. 3.19. The positive current directions for this type of model are inwards as is customary for this configuration. The aim of this section is to establish the link, in terms of the parameters which exist between this model and the model in Fig. 3.17.

Mutual and self-inductance based transformer models are defined in terms of the parameters L_1 and L_2, respectively. In addition, the magnetic coupling between the primary and secondary coils is defined in terms of the so-called *mutual inductance* parameter M, as indicated in Fig. 3.19. The mutual inductance can be defined from either the primary or secondary side.

From the primary side the mutual inductance is defined as the ratio of the flux linked with the secondary winding and current in the primary side, i.e., $M = \psi_2/i_1$, with the condition $i_2 = 0$. Vice versa, the mutual coupling can also be defined as the ratio between the primary flux linkage and secondary current, i.e., $M = \psi_1/i_2$,

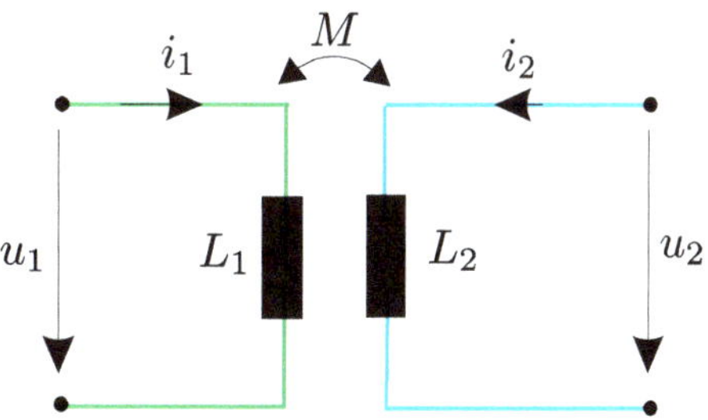

Fig. 3.19 Mutual coupling type model of transformer

with the condition $i_1 = 0$. The actual mutual inductance value remains unchanged whether viewed from the primary or secondary side as will become apparent shortly. As such, the mutual inductance is also (like L_1, L_2) an "independent" parameter.

The basic equation set which applies to Fig. 3.19 can be written as

$$u_1 = L_1 \frac{di_1}{dt} + M \frac{di_2}{dt} \tag{3.33a}$$

$$u_2 = M \frac{di_1}{dt} + L_2 \frac{di_2}{dt} \tag{3.33b}$$

The revised model states that the flux-linkage time differential, on for example, the primary side consists of a term due to primary self-inductance to which we must now add a *mutual inductance* term.

By determining the equivalent (to Eq. (3.33)) *primary based ITF transformer* model (see Fig. 3.17) equation set, we will be able to show how the parameters of the mutual coupling based model are linked to the ITF based model. A suitable starting point for this analysis is expression (3.31a), which remains unaffected by the model in use. From Eq. (3.31c) we can deduce that the term $d\psi_1/dt$ may also be written as

$$\frac{d\psi_1}{dt} = L_\sigma^{pr} \frac{di_1}{dt} + \frac{d\psi_2^i}{dt} \tag{3.34}$$

The primary referred flux-linkage ψ_2^i can according to Eq. (3.31d) also be written as $\psi_2^i = L_M i_M$, in which the magnetizing current can also be expressed as $i_M = i_1 - i_2^i$. Given the above, we can with the aid of Eqs. (3.31f) and (3.31g) rewrite Eq. (3.34) in the following form:

$$\frac{d\psi_1}{dt} = \underbrace{\left(L_\sigma^{pr} + L_M\right)}_{L_1} \frac{di_1}{dt} - \underbrace{\left(\frac{L_M}{k}\right)}_{M} \frac{di_2}{dt} \tag{3.35}$$

A comparison between Eqs. (3.35) and (3.33a) (right-hand side) shows that they differ only in terms of the minus sign between the two terms. The reason for this is that the secondary current sign convention between the two models is in opposition. If we consider the parameters present in both equations (in front of the current differential terms), then it is hopefully apparent that the primary self-inductance and mutual inductance terms may be expressed, using k from Eq. (3.29) as

$$L_1 = L_\sigma + L_M \tag{3.36a}$$

$$M = \frac{L_M}{k} \tag{3.36b}$$

Note that the mutual inductance may also be expressed in terms of the coupling factor [(Eq. (3.30)] which gives

$$M = \kappa \sqrt{L_1 L_2} \tag{3.37}$$

A similar type of analysis, as shown above, may also be carried out with respect to expressing the flux differential term $\mathrm{d}\psi_2/\mathrm{d}t$ in terms of ITF parameters and current differential terms. An analysis of this type, an exercise left to the reader, shows that this flux differential can be written as

$$\frac{\mathrm{d}\psi_2}{\mathrm{d}t} = \underbrace{\left(\frac{L_\mathrm{M}}{k}\right)}_{M} \frac{\mathrm{d}i_1}{\mathrm{d}t} - \underbrace{\left(\frac{L_\mathrm{M}}{k^2}\right)}_{L_2} \frac{\mathrm{d}i_2}{\mathrm{d}t} \tag{3.38}$$

A comparison between Eqs. (3.38) and (3.33b) (right-hand side) and taking into account the secondary current direction in both models show that the secondary self-inductance may be expressed as

$$L_2 = \frac{L_\mathrm{M}}{k^2} \tag{3.39}$$

Note from Eq. (3.38) that a mutual inductance term is also present and indeed of the form given by expression (3.36b).

3.9 Two-Inductance Model with Coil Resistance

The remaining extension which has to be made with respect to the model shown in Sect.,3.7.1 is concerned with the introduction of the primary and secondary coil resistances R_1 and R_2, respectively. Use of these parameters requires a change to Eqs. (3.31a) and (3.31b) which are now of the form

$$u_1 - R_1 i_1 = \frac{\mathrm{d}\psi_1}{\mathrm{d}t} \tag{3.40a}$$

$$u_2 + R_2 i_2 = \frac{\mathrm{d}\psi_2}{\mathrm{d}t} \tag{3.40b}$$

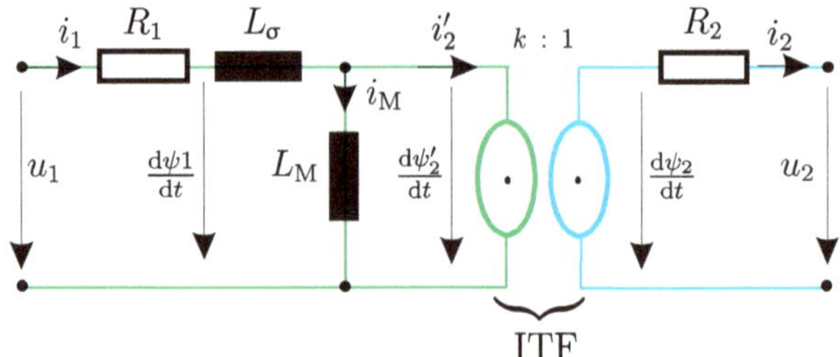

Fig. 3.20 Symbolic representation of a four parameter transformer model

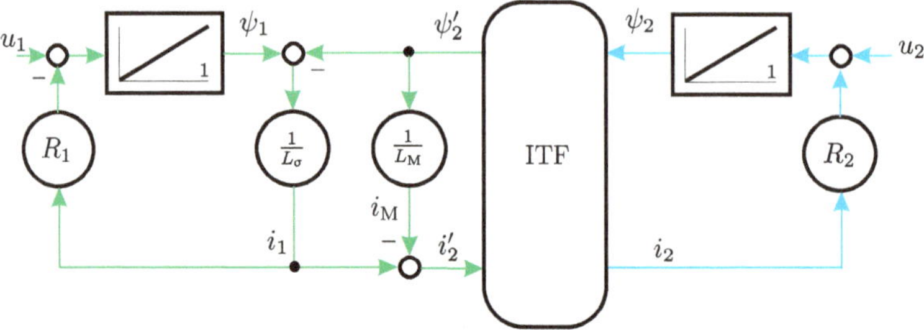

Fig. 3.21 Generic representation of a four parameter transformer model

Note that the variables ψ_2^i *and* i_2^i used in Fig. 3.17 will in future be shown as ψ'_2 and i'_2, respectively. The notation of a variable x' implies, in general, a primary or secondary "referred" value. Furthermore, for the sake of readability, the inductance L_σ^{pr} will, in the following be simply referred to as L_σ, when referring to this "two-inductance" model.

The symbolic representation of the transformer with resistances is given in Fig. 3.20. A generic form of the four parameter single phase transformer, as given in Fig. 3.21, can be directly converted to a Simulink or PLECS type model *without* the need for a differentiator module. Note that the ITF module according to Fig. 3.3b is now used in Fig. 3.21. The flux and current relations for the ITF are now of the form given in Eqs. (3.31f) and (3.31g), where the modified winding ratio k is defined in Eq. (3.29a). Hence, the ITF outputs are the primary referred secondary flux-linkage value ψ'_2 and the secondary current i_2. It is by virtue of the fact that we have reversed the signal flow in the ITF (by using the ITF *current* model, see Fig. 3.3b) that we are able to avoid the use of a differentiator on the secondary side.

3.9.1 Phasor Analysis of Revised Model, with Resistive Load

The steady-state analysis of the model according to Fig. 3.20 is carried out along the lines of the previous model given in Sect. 3.5.1. As with the previous case, a primary supply voltage $u_1 = \hat{u}_1 \cos \omega t$ is assumed, which corresponds with a phasor $\underline{u}_1 = \hat{u}_1$. Prior to undertaking this analysis it is helpful to summarize in phasor form the complete equation set (including the ITF) for this system.

$$\underline{u}_1 = \underline{i}_1 R_1 + \mathrm{j}\omega \underline{\psi}_1 \tag{3.41a}$$

$$\underline{i}_1 = \frac{1}{L_\sigma}\left(\underline{\psi}_1 - \underline{\psi}'_2\right) \tag{3.41b}$$

$$\underline{u}_2 = \mathrm{j}\omega \underline{\psi}_2 - \underline{i}_2 R_2 \tag{3.41c}$$

$$\underline{u}_2 = \underline{i}_2 R_\mathrm{L} \tag{3.41d}$$

$$\underline{i}'_2 = \underline{i}_1 - \underline{i}_\mathrm{M} \tag{3.41e}$$

$$\underline{i}_\mathrm{M} = \frac{\underline{\psi}'_2}{L_\mathrm{M}} \tag{3.41f}$$

$$\underline{\psi}'_2 = k\, \underline{\psi}_2 \tag{3.41g}$$

$$\underline{i}_2 = k\, \underline{i}'_2 \tag{3.41h}$$

The analysis of this type of circuit is aimed at finding the primary current phasor $\underline{i}_1$ as a function of the circuit parameters and the input (known) voltage phasor $\underline{u}_1$. The required expression can be obtained by use of Eq. (3.41). An alternative approach is possible in this case by realizing that the ITF module in fact acts as an impedance converter. For example, if we consider the impedance $Z_2 = \underline{e}_2/\underline{i}_2$ (where we have ignored the sign convention), then the equivalent impedance on the primary side, known as the primary referred secondary impedance, will be equal to $Z'_2 = \underline{e}'_2/\underline{i}'_2$ where $\underline{e}'_2 = \mathrm{j}\omega \underline{\psi}'_2$.

The relationship between the two impedances is found by using Eqs. (3.41g) and (3.41h) which gives $Z'_2 = k^2 Z_2$. In this case, we can simply move the secondary elements (coil resistance and load resistance) to the primary side, provided we multiply the value by a factor k^2. This process is known as building a primary referred model of the transformer, which greatly simplifies the steady-state analysis. The result of moving the secondary circuit elements to the primary side of the ITF is shown in Fig. 3.22. Note that the inductances are represented in a phasor circuit as $\mathrm{j}\omega L$.

The primary current phasor is then found by determining the equivalent impedance at the primary terminals, which according to Fig. 3.22 is of the form

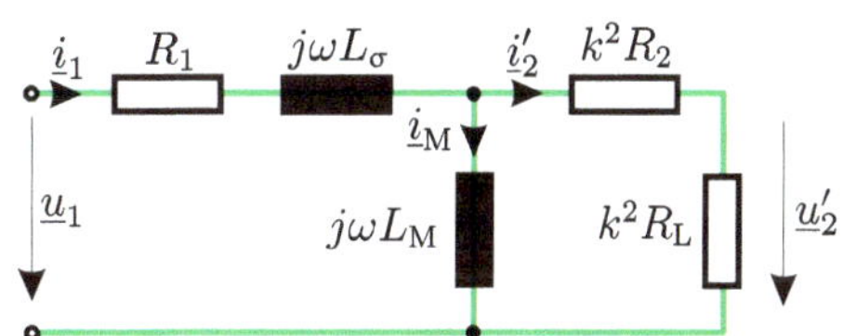

Fig. 3.22 Equivalent primary referred model of transformer

$$Z_{\text{prim}} = R_1 + \text{j}\omega L_\sigma + \frac{\text{j}\omega L_\text{M}\, k^2\,(R_2 + R_\text{L})}{\text{j}\omega L_\text{M} + k^2\,(R_2 + R_\text{L})} \tag{3.42}$$

The current phasor is then calculated using $\underline{i}_1 = \hat{u}_1/Z_{\text{prim}}$. Once this phasor is defined we can determine, with the aid of Eq. (3.41), the remaining phasors of this circuit. For example, the secondary current phasor is directly found using Eq. (3.41h). The corresponding voltage phasor $\underline{u}_2$ can with the aid of Eq. (3.41g) be written as $\underline{u}_2 = \underline{u}_2'/k$. Note in this context that the product $\underline{u}_2\,\underline{i}_2$ is equal to $\underline{u}_2'\,\underline{i}_2'$, i.e., the transformation is power invariant.

3.10 Tutorials

3.10.1 *Tutorial 1: PLECS Model of a Single Phase Zero Leakage Current Transformer*

The single phase transformer shown in Fig. 3.23 is to be used for current measurement. The secondary winding is connected to a resistance R_L, hence the voltage across u_2 will be a function of the primary current i_1, which is given or has to be measured. The primary and secondary coil resistance, as well as the leakage inductance of the transformer, are ignored. The transformer has a magnetizing inductance L_m and a winding ratio of $n_1/n_2 = 1$.

Goal of this tutorial is to build a PLECS model diagram of the ITF based transformer circuit with the primary current i_1 as input variable. Differentiator modules may *not* be used in this example. Use a "Scope" module to show the variables $i_1 and u_2$ as function of time. The current function is of the form $i_1 = 10\sin(\omega t)$, with $\omega = 100\pi$ rad/s. The magnetizing inductance L_m and load resistance R_L are equal to 100 mH and 5 Ω, respectively. Set your simulation "run time" to 60 ms.

An example of an implementation of this problem is given in Fig. 3.24. The results of the simulation in the form of the input current and output voltage are given in Fig. 3.25. The M-file given below is used to process the results obtained from the simulation scope module (via an export of data to a file `tut1ch3data.csv`).

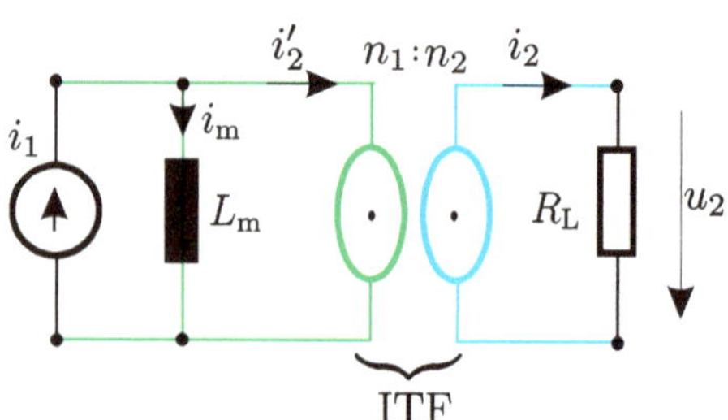

Fig. 3.23 Current transformer with resistance R_L

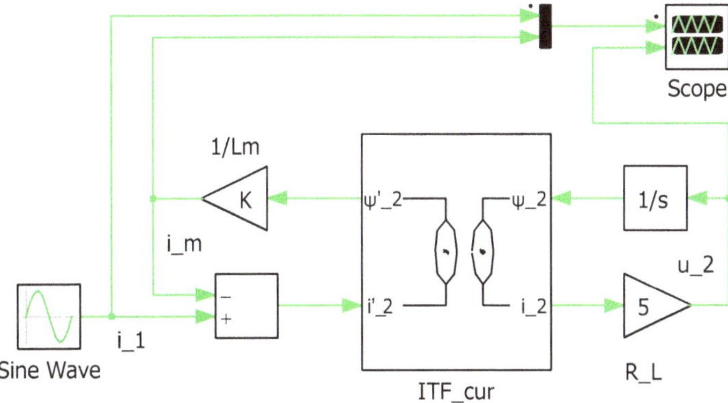

Fig. 3.24 Simulation of current transformer

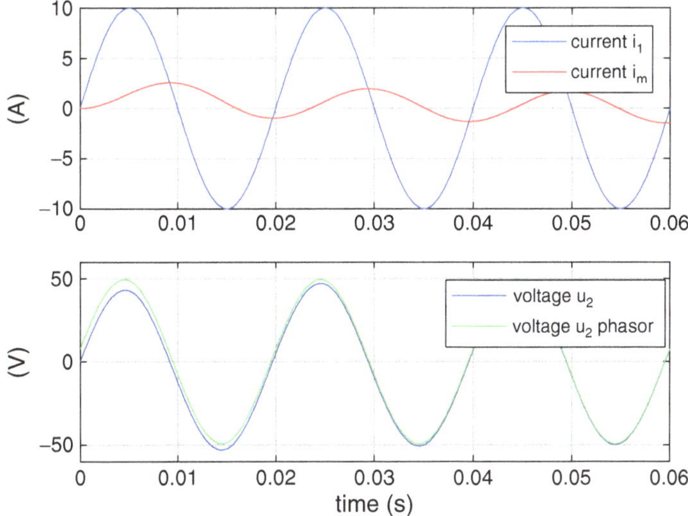

Fig. 3.25 PLECS results: current transformer tutorial

It is instructive to check the results from the simulation by a steady-state *phasor analysis*. The same M-file also shows the phasor analysis for this problem and the corresponding output voltage waveform is also added to the results shown in Fig. 3.25.

M-file Code

```
%Tutorial 1, chapter 3
close all
datout = csvread('tut1ch3data.csv',1,0) % read in data from
```

```
  PLECS
subplot(2,1,1)
plot(datout(:,1),datout(:,2));% i1
hold on
plot(datout(:,1),datout(:,3),'r') %im
grid
legend('current i_1','current i_m')
ylabel('(A)')
subplot(2,1,2)
plot(datout(:,1),datout(:,4))  % u_2
grid
hold on
%%%%%calculation phasors
i1_ph=10;                          %current phasor i1
RL=5; Lm=100e-3; w=100*pi;
%%%%%%%%%%%%%%
XL=j*w*Lm;
u2_ph=i1_ph*RL/(1+RL/XL);
u2_hat=abs(u2_ph);                 % amplitude
u2_rho=angle(u2_ph);               %angle between u2_ph and
                                       i2_ph
%%%%%%plot time wave form u2(t)
t=[0:0.1e-3:60e-3];
u2_t=u2_hat*sin(w*t+u2_rho);
plot(t,u2_t,'g');                  %plot u2(t)
legend('voltage u_2','voltage u_2 phasor')
xlabel('time (s)')
ylabel('(V)')
ylim([-60 60])
```

An observation of Fig. 3.25 shows that the output voltage waveform from the simulation is aligned with the output obtained from the phasor analysis after a time interval of approximately 30 ms. Therefore, a "transient" effect is present which cannot be "seen" with the phasor analysis. A further observation of Fig. 3.25 shows that, under steady-state conditions, a phase angle difference exists between input and output waveforms, which is caused by the presence of the magnetizing inductance L_{m} and corresponding magnetizing current i_{m}. The latter is also shown in Fig. 3.25 (top subplot), where it is noted that the output voltage is determined by the expression $u_2 = (i_1 - i_{\mathrm{m}})\, {}^{n_1}\!/_{n_2}\, R_{\mathrm{L}}$. Consequently, the presence of a non-zero magnetizing current will cause an output error given that the ideal voltage output should be $u_2^{\mathrm{ideal}} = i_1 {}^{n_1}\!/_{n_2}\, R_{\mathrm{L}}$. It can be shown that the output phasor $\underline{u}_2$ may be expressed in terms of the input current phasor $\underline{i}_1$ and parameters R_{L} and L_{m}.

$$\underline{u}_2 = \frac{R_{\mathrm{L}}\, \underline{i}_1}{\left(1 + \frac{R_{\mathrm{L}}}{\mathrm{j}\omega L_{\mathrm{m}}}\right)} \tag{3.43}$$

The denominator of Eq. (3.43) shows that the phase angle is equal to: $\arctan\left({}^{R_{\mathrm{L}}}\!/_{\omega L_{\mathrm{m}}}\right)$. Hence, during the design/manufacture of transformers for this purpose it is prudent to maximize the L_{m} value and limit the size of R_{L}.

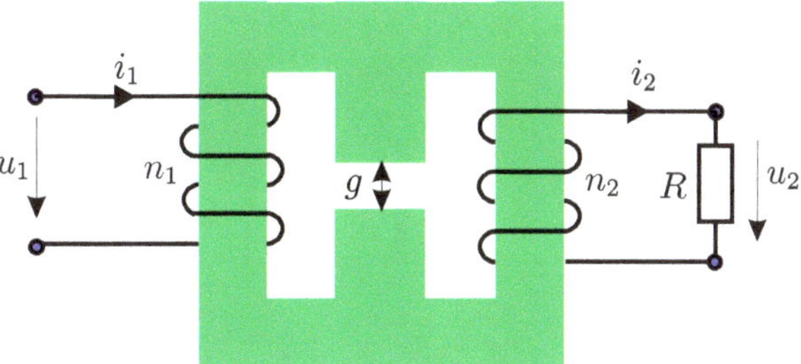

Fig. 3.26 High leakage transformer

3.10.2 Tutorial 2: PLECS Model of a Single Phase High Leakage Transformer

An interesting and educative transformer configuration, as shown in Fig. 3.26, is considered here. Unique to this transformer concept are two magnetic "fingers" separated by an airgap, which adds an inductance component in series with the leakage inductance. The transformer has a primary and secondary winding with $n_1 = 10$ and $n_2 = 20$ turns, respectively. Coil resistance of both coils may be ignored. The primary winding is connected to an AC voltage source defined as $u_1 = 10\cos(2\pi ft)$, with $f = 10\,\mathrm{kHz}$, while the secondary coil is attached to a load resistance of $R = 50\,\Omega$. The transformer magnetic material is assumed to be ideal (infinite permeability) and an airgap of $g = 0.05\,\mathrm{mm}$. The core cross-sectional area on either side of the airgap g is assumed to be $A_g = 50\,\mathrm{mm}^2$. Build a PLECS model of the transformer using "electrical/magnetic" blocks and a sinusoidal excitation source as defined above. Add a scope module which shows: primary/secondary voltages, primary/secondary currents, circuit flux in both coils, MMF across the airgap, and corresponding airgap flux.

The PLECS model given in Fig. 3.27 shows the two electrical circuits linked with the primary and secondary coil. In addition, a set of magnetic blocks is shown which are used to represent the airgap `P_air` and the electrical/magnetic interface coil modules. A set of meters is used to measure the flux in the two coils and airgap as well as the MMF due to the coils. The results obtained with the PLECS model shown in Fig. 3.28 are briefly discussed below. The excitation voltage ("blue" coil voltage plot) causes the coil circuit flux ("blue" plot) which lags the latter by $\pi/2$ rad/s. Due to the presence of the airgap part of the flux due to primary coil crosses the airgap ("red" circuit flux plot). The difference between the primary coil flux and airgap coil flux is the secondary coil circuit flux ("green" circuit flux plot). This in turn causes the coil voltage u_2 ("red" coil voltage plot) which leads the latter by $\pi/2$ rad/s. A load resistance R is connected to the secondary coil hence a current i_2 is generated ("red" current plot) that corresponds to an airgap MMF (airgap MMF plot). Both MMF and secondary current waveforms are in phase with the secondary voltage plot. Because an ideal magnetic material is used the MMF generated by the secondary coil must

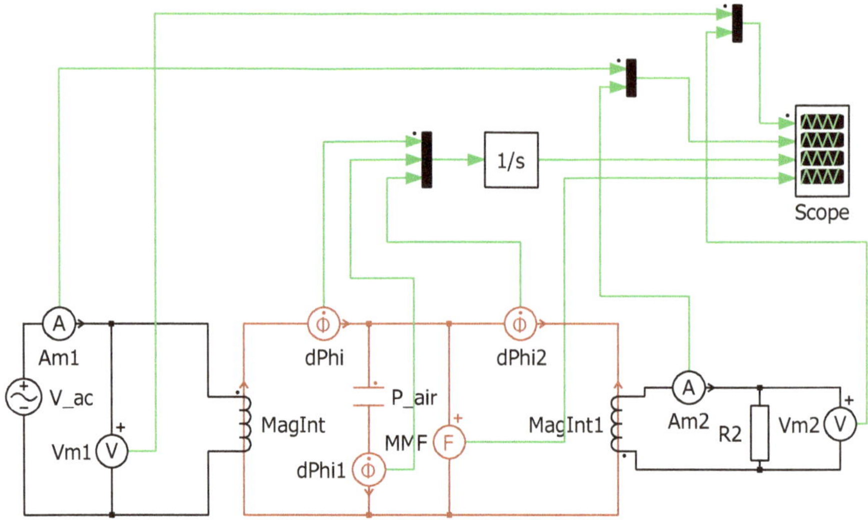

Fig. 3.27 PLECS model: high leakage transformer

be compensated by an equal but opposite MMF in the primary coil which causes the primary current i_1 ("blue" current plot). The amplitude of both currents (which must be in in phase) differ because the turns ratio between primary and secondary differs by a factor 2 in this case.

In the sequel to this tutorial a phasor analysis is to be performed which generates the steady-state secondary voltage for the model according to Fig. 3.26. Plot this result together with the PLECS generated secondary coil voltage waveform shown in Fig. 3.28.

The results of this analysis given in Fig. 3.29 shown that the calculated voltage ("green" plot) merges with the PLECS generated waveform u_2 after approximately 30 μs. At the conclusion of this tutorial the M-file is given which shows the phasor analysis and mathematical handling required to arrive at Fig. 3.29. Central to the analysis is a series network which consists of an inductance due to the airgap and the primary referred resistance R. Attached to this network is the phase representation of the excitation voltage $\underline{u}_1 = 10$, while the voltage across the resistor is the primary referred secondary coil phasor. The inductance is calculated using n_1^2/R_m, where R_m is the magnetic reluctance due to the airgap. After computation of the output phasor u2_phr a transformation to the time domain is undertaken. Furthermore, scaling by a factor n_2/n_1 is required to generate the coil voltage u_2 function from the referred value.

```
%Tutorial 2, chapter 3
    close all
    datout = csvread('tut2ch3data.csv',1,0) % read in
      data from PLECS
     plot(datout(:,1),datout(:,3),'r')  % u_2
    grid
```

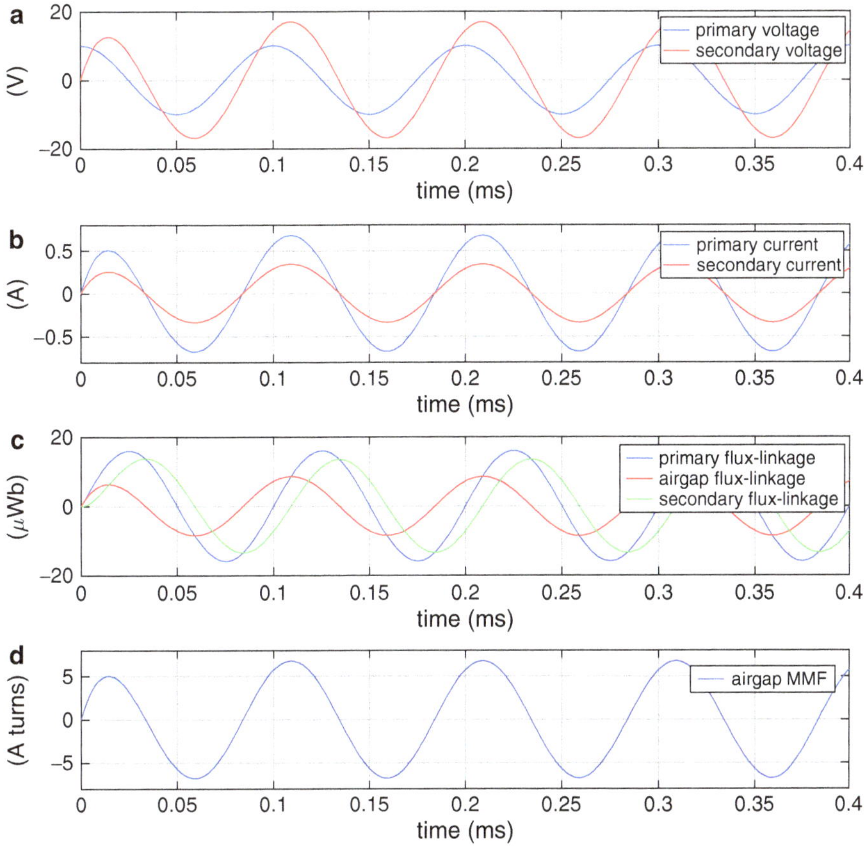

Fig. 3.28 PLECS model: high leakage transformer

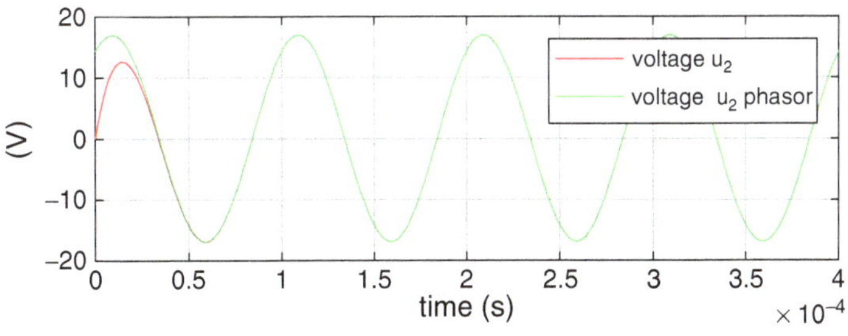

Fig. 3.29 Calculated and PLECS derived secondary coil voltage $u_2(t)$

```
hold on
%%%%%%calculation phasors
u1_ph=10   %voltage phasor u_1
f=10e3;    % excitation frequency
```

```
R=50;; w=2*pi*f;
%%%%%%%%%% magnetic
uo=4*pi*1e-7;  %permeability in air
n1=10;         %number of turns primary
n2=20;         %number of turns secondary
g=0.05e-3;     % airgap
Ag=50e-6;      %  airgap crossection, same as core
Rm=g/(uo*Ag)   % reluctance airgap
Lsig=n1^2/Rm  %  leakage inductance
%%%%%%%%%%%%%%%%%%%%%%%%%%
%%%  build primary referred model
Xsig=j*w*Lsig;      %primary reactance
Rr=(n1/n2)^2*R;     %referred load resistance
%%%%%%%%%%%%%% phase analysis
u2_phr=u1_ph/(1+Xsig/Rr);
u2_hatr=abs(u2_phr);                        % amplitude
u2_rho=angle(u2_phr); %angle between u2_phr and u1_ph
u2_hat=(n2/n1)*u2_hatr; % actual secondary voltage
%%%%%%%plot time wave form u2(t)
t=[0:1e-6:400e-6];
u2_t=u2_hat*cos(w*t+u2_rho);
plot(t,u2_t,'g');                          %plot u2(t)
legend('voltage u_2','voltage u_2 phasor')
xlabel('time (s)')
ylabel('(V)')
```

3.10.3 Tutorial 3: PLECS Model of a Single Phase Transformer with Leakage Inductance

A 50 Hz, 660 V/240 V supply transformer is considered in this tutorial. The object is to determine the parameters of the transformer in question using data obtained from a no-load and short-circuit test. In the second part of this tutorial, a PLECS based dynamic model is to be built. This model will then be used to examine the behavior of the transformer under load conditions. A phasor analysis is also performed so that the steady-state results obtained with the PLECS model can be verified. The symbolic model of the transformer, as given in Fig. 3.30, is based on the symbolic model discussed in this chapter (see Fig. 3.20). The model used in this tutorial is extended by the addition of a resistance R_{M} placed across the terminals of the primary. The power dissipated in this resistance represents the so-called iron losses(due to eddy currents and hysteresis) in the transformer. In three-inductance models, shown in Fig. 3.13, this resistance is usually connected in parallel with the inductance L_{m}. Positioning this resistance across the primary terminals (or the terminals which corresponds to the supply side) simplifies the generic model at the price of a marginal reduction in accuracy. Under steady-state

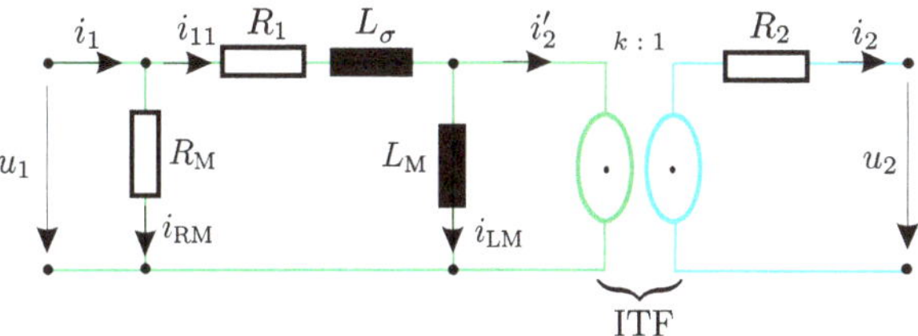

Fig. 3.30 Transformer model with iron losses

conditions the power (in Watts) dissipated in a resistance is given by $P_R = I^2R = {}^{U^2}/_{R}$, where $UandI$ represent the RMS voltage and current seen by the resistance. The transformer turns ratio was identified prior to this tutorial by the voltage ratio 660/240, which represents the rated primary and secondary RMS voltage values of this unit. To determine the parameters of this transformer, a "no-load" test can be carried out where the primary (terminal 1 side) was connected to a 660 V, 50 Hz sinusoidal supply source and the secondary side open circuited. The measured current and power under these conditions were found to be 0.2 A (RMS) and 20 W, respectively. Furthermore, the voltage across the secondary winding was found to be 240 V (RMS). A "short-circuit" test was also carried out, where the secondary winding was connected to a 8 V (RMS), 50 Hz voltage source, which gives a rated secondary current of 20 A (RMS) with the primary winding short-circuited. Note that in this example the short-circuit test is carried out from the secondary side (voltage source connected to the secondary winding), which is often done in case the primary voltage is relatively high, as is the case here. The secondary power was also measured under these circumstances and found to be 20 W. On the basis of these experimental tests the parameters of the model according to Fig. 3.30 can be identified with reasonable accuracy.

The solution to this problem requires a phasor analysis, given that the experimental data was obtained under "steady-state AC conditions." Under "no-load" conditions we can simplify the model according to Fig. 3.30 by assuming that the voltage drop across the primary resistance and leakage inductance is small in relation to the applied primary voltage. The primary current in phasor representation (under no-load only) is of the form $\underline{i}_1 = \underline{i}_{RM} + \underline{i}_{LM}$, where $\left(\underline{i}_{RM}, \underline{i}_{LM}\right)$ represent the current through components R_M and L_M, respectively. The RMS current through the resistance R_M is found using $I_{RM} = {}^{P_1}/_{U_1} = {}^{20}/_{660}$, where P_1 and U_1 represent the measured no-load power and RMS voltage, respectively. The winding ratio k is found using $k \simeq {}^{U_1}/_{U_2}$. The current through L_M is found using $I_{LM} = \sqrt{(I_1)^2 - (I_{RM})^2}$, where I_1 represents the measured (RMS) no-load primary current. Note that I_{RM} has already been calculated. On the basis of these calculations and no-load data, the following parameters are obtained:

$$k \simeq \frac{U_1}{U_2} \tag{3.44a}$$

$$R_{\mathrm{M}} \simeq \frac{(U_1)^2}{P_1} \tag{3.44b}$$

$$L_{\mathrm{M}} \simeq \frac{U_1}{\omega\sqrt{(I_1)^2 - \left(\frac{P_1}{U_1}\right)^2}} \tag{3.44c}$$

where U_1, U_2, I_1, and P_1, shown in Eq. (3.44), represent the no-load experimental data.

Under short-circuit test conditions the model according to Fig. 3.30 can be simplified by ignoring the magnetizing current and iron losses. The reason for this is that the secondary voltage is very low compared to normal operation. Hence, the current which flows in R_{M} and L_{M} is negligible in comparison to the rated primary current. Consequently, the currents which flow in these components can be ignored when the transformer is exposed to a short-circuit test. For the calculation linked to this test, we will refer to the applied secondary voltage and measured secondary current to the primary side. This means that we can consider (for calculation purposes) the short-circuit problem from the primary side. The total resistance R_{p} as seen from the primary side consists of the primary resistance R_1 to which we must add the primary referred secondary resistance $R_2' = k^2 R_2$. The leakage reactance ωL_σ completes this series network, which is excited by a primary referred secondary voltage $U_2' = kU_2$. The primary referred secondary current is equal to $I_2' = I_2/k$. The short-circuit impedance Z_{p}, as seen from the primary side, is equal to $Z_{\mathrm{p}} = U_2'/I_2' = \sqrt{R_{\mathrm{p}}^2 + (\omega L_\sigma)^2}$. The impedance Z_{p} can therefore be found on the basis of the applied secondary voltage U_2, calculated winding ratio k and measured current I_2. In addition, the short-circuit power P_2 was measured which may be written as $P_2 = \left(I_2'\right)^2 R_{\mathrm{p}}$. From this equation the total resistance as "seen" from the primary side can be obtained. The individual resistance values cannot be found (unless they are measured directly with the aid of an Ohm meter) from these measurements. Typically, the assumption made in this case is that $R_2' = R_1$. The parameters, which are obtained from the short-circuit measurements, are calculated as follows:

$$R_{\mathrm{p}} \simeq k^2 \frac{P_2}{I_2^2} \tag{3.45a}$$

$$R_1 \simeq \frac{R_{\mathrm{p}}}{2} \tag{3.45b}$$

$$R_2 \simeq \frac{R_{\mathrm{p}}}{2\,k^2} \tag{3.45c}$$

$$L_\sigma \simeq \frac{1}{\omega}\sqrt{\left(k^2 \frac{U_2}{I_2}\right)^2 - R_{\mathrm{p}}^2} \tag{3.45d}$$

where U_2, I_2, and P_2 shown in Eq. (3.45) represent the short-circuit experimental data. The winding factor k was calculated using Eq. (3.44a). The first part of the M-file shown below calculates the parameters for this transformer based on the no-load and short-circuit test data.

M-file Code

```
%Tutorial 3, part 1, chapter 3
%no-load data
U1_n=660;                                 % RMS primary voltage
I1_n=0.2;                                 % RMS primary current
U2_n=240;                                 % RMS secondary voltage
P1_n=20;                                  % noload measured power
%%%%%%%%%%%%%%
w=2*pi*50;%frequency rad/s
%%%%parameters from noload data
k=U1_n/U2_n;                              %winding ratio
RM=U1_n^2/P1_n;                           % resistance
LM=1/w*U1_n/sqrt(I1_n^2-(P1_n/U1_n)^2);   %inductance LM
%%%%%%%%%%%%%%%%%%%%
%%short circuit data
U2_s=8;                                   % secondary RMS
                                           % short circuit voltage
I2_s=20;                                  % secondary RMS rated current
P2_s=20;                                  % secondary power
%%%%%parameters from this data
Rp=k^2*P2_s/I2_s^2;                       %total primary resistance
R1=Rp/2;                                  % primary resistance
R2=Rp/(2*k^2);                            %secondary resistance
Lsigma=1/w*sqrt((k^2*U2_s/I2_s)^2-Rp^2);  %leakage inductance
```

The following parameters were obtained after running this M-file (Table 3.1).

The second part of this tutorial is concerned with the development of a dynamic model of the transformer in question. A load resistance R_L is connected to the secondary winding and its value will be set to: $R_L = 2000\,\Omega$, $0\,\Omega$, and $10\,\Omega$, respectively. In the first case (*a*), a high load resistance is chosen to approximate the (secondary) open-circuit case. The second case (*b*), $R_L = 0$ corresponds to the case where the secondary winding is short-circuited. The primary RMS voltage for this example is set to $8\,k = 22$ V, which represents the voltage which would need to be applied to the primary side in case the short-circuit test was carried

Table 3.1 Parameters for single phase transformer

Parameters		Value
Winding ratio	k	2.75
Loss resistance	R_M	21.78 kΩ
Magnetizing inductance	L_M	10.62 H
Primary resistance	R_1	0.189 Ω
Secondary resistance	R_2	0.025 Ω
Leakage inductance	L_σ	9.6 mH

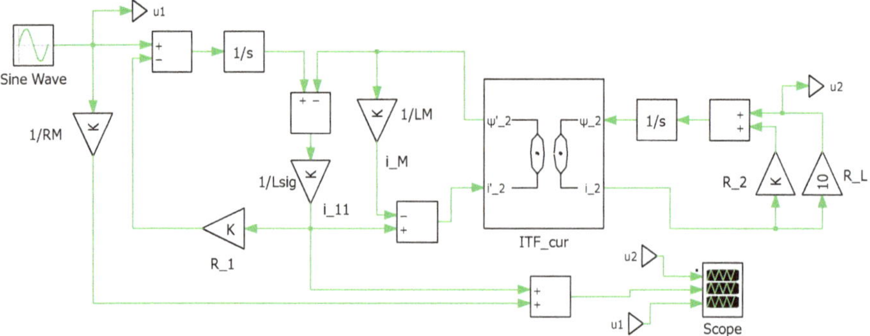

Fig. 3.31 PLECS model: transformer with iron losses

Table 3.2 Simulation data

Simulation setup	U_1 (RMS) V	R_L (Ω)	File name
No-load	660	2000	D1
Short-circuit	22	0	D2
Load	660	10	D3

out from that winding rather than from the secondary side. By considering these two cases we are able to check the simulation model in steady state AC conditions against the experimental data from the no-load and short-circuit test. The third load resistance value $R_L = 10\,\Omega$ has been chosen arbitrarily to demonstrate the operation of the transformer under load conditions.

Implementation of the PLECS model, as shown in Fig. 3.31, is directly based on the generic model of Fig. 3.21. However, a resistance $1/R_M$ is present on the primary side of the ITF module. On the secondary side, a two-resistance network is present which consists of the two series connected resistances: R_2 and R_L. The input to the secondary integrator is the flux differential $d\psi_2/dt$ which must, according to Fig. 3.21, be equal to $d\psi_2/dt = u_2 + i_2 R_2$, with $u_2 = i_2 R_L$, given that a load resistance is connected to the secondary winding of the transformer. The simulation run time is set to 100 ms and needs to be executed three times with different R_L,U_1 values as indicated in Table 3.2. This implies that, prior to each simulation run, the appropriate R_L and U_1 values must be set in the PLECS model. After running the simulation rename the output file in the MATLAB workspace to the file name given in Table 3.2. For example, "D1=datout" (for the no-load simulation). The results from the three simulations, as represented by the files D1, D2, and D3, need to processed to show the results in the form of the input voltage $u_1 = U_1\sqrt{2}\cos(\omega t)$, current i_1, and secondary voltage u_2. An example of an M-file, which can process this data, is as follows:

M-file Code

```
%Tutorial 3, part 2, chapter 3
    close all
    D1 = csvread('tut3D1.csv',1,0) % read in data from PLECS
    D2 = csvread('tut3D2.csv',1,0) % read in data from PLECS
    D3 = csvread('tut3D3.csv',1,0) % read in data from PLECS
    subplot(3,1,1)
    %%%% primary voltage
    plot(D1(:,1),D1(:,4));                % u1 open loop U1=660
    grid; hold on
    plot(D2(:,1),D2(:,4),'r');            % u1 shortcircuit U1=22
    plot(D3(:,1),D3(:,4),'g');            % u_1 load RL=10, U1=660
    legend('voltage u_1 OC','voltage u_1 SC','voltage u_1 Load',...
        'Location', 'NorthEastOutside')
    xlabel(' (a) time (s)')
    ylabel('(V)')
    subplot(3,1,2)
    %%%% primary current
    plot(D1(:,1),D1(:,3));                % open loop U1=660
    grid; hold on
    plot(D2(:,1),D2(:,3),'r');            % shortcircuit U1=22
    plot(D3(:,1),D3(:,3),'g');            % load RL=10, U1=660
    legend('current i_1 OC','current i_1 SC','current i_1 Load',...
        'Location', 'NorthEastOutside')
    xlabel(' (b) time (s)')
    ylabel('(A)')
    %%%%%%%%%% load  RL 10 Ohm
    subplot(3,1,3)
    plot(D1(:,1),D1(:,2));               % open loop U1=660
    grid; hold on
    plot(D2(:,1),D2(:,2),'r');           % shortcircuit U1=22
    plot(D3(:,1),D3(:,2),'g');           % load RL=10, U1=660
    legend('voltage u_2 OC','voltage u_2 SC','voltage u_2 Load',...
        'Location', 'NorthEastOutside')
    xlabel(' (c) time (s)')
    ylabel('(V)')
```

The results obtained from these simulations are given in Fig. 3.32. It is left to the reader to run these PLECS/MATLAB files and analyze the results in detail. Some indication with respect to the correct functioning of the PLECS model can be made by observation of Fig. 3.32 and a comparison of the results according to Table 3.3. The results show that there is good agreement between the data for the no-load and short-circuit tests. However, a phasor analysis could be carried out to verify the steady-state results. This exercise is left to the reader.

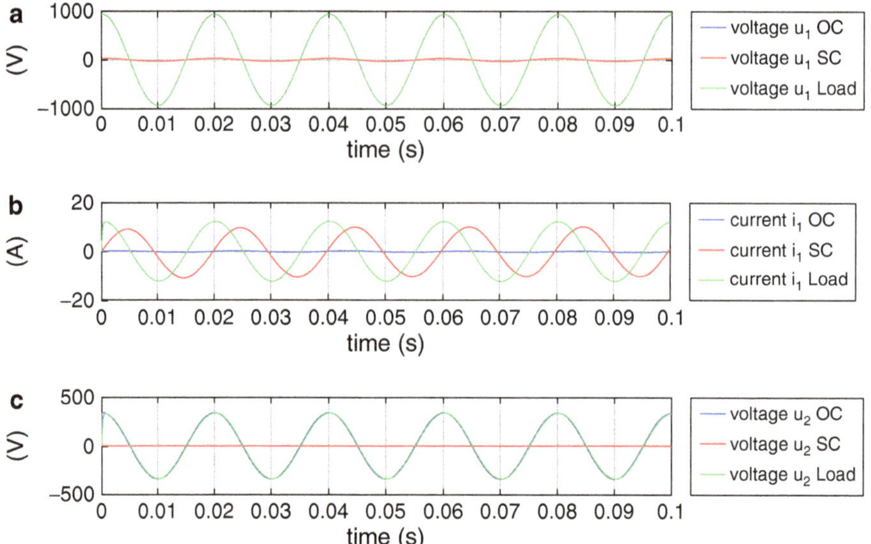

Fig. 3.32 PLECS results: transformer with iron losses

Table 3.3 Comparison simulation and experimental results

–	I_1(RMS) A	U_2(RMS) V
Simulation (no-load, $U_1 = 660\,\text{V}$)	0.210	239.8
Experimental (no-load, $U_1 = 660\,\text{V}$)	0.2	240
Simulation (short-circuit, $U_1 = 22\,\text{V}$)	7.24	0
Experimental (short-circuit, $U_2 = 8\,\text{V}$)	7.27	0
Simulation ($R_L = 10\,\Omega$, $U_1 = 660\,\text{V}$)	8.70	238.4
Experimental ($R_L = 10\,\Omega$, $U_1 = 660\,\text{V}$)	–	–

Chapter 4
Three-Phase Circuits

4.1 Introduction

The majority of electrical drive systems in use are powered by a so-called three-phase (three-wire) supply. The main reason for this is that a more efficient (moving from a two- to a three-wire system increases the transmitted power by 73 %) energy transfer from supply to the load, such as a three-phase AC machine, is possible in comparison with a single (two-wire) AC circuit. The load, being the machine acting as a motor, is formed by three phases. Each phase winding has two terminals, yielding a total of six terminal-bolts, usually configured as sketched in Fig. 4.1. *The phase impedances are assumed to be equal.* The terminal layout as shown in Fig. 4.1 has been purposely chosen to allow the user to readily connect the machine's phase windings in two distinct configurations. The star and delta configurations are depicted in Figs. 4.3 and 4.9, respectively. Voltages and currents in the different configurations are identified by the subscripts S1, S2, and S2 when the machine is in star, also called Wye or Y-configuration. Subscripts D1, D2, and D3 apply to the delta or Δ configuration.

The voltages/currents, identified by the subscripts R, S, and T, are linked to the supply source, which is usually a power electronic converter or the three-phase grid. Figure 4.2 shows an example of a three-phase voltage supply which generates three voltages (of arbitrary shape) u_R, u_S, and u_T that are defined with respect to the 0 V (neutral) of this system. In this chapter, we will look into modeling three-phase circuits, and in this context introduce a new set of building blocks as required to move (in both directions) from machine phase variables to supply variables for either star or delta connected machines.

Electronic supplementary material The online version of this chapter (doi: 10.1007/978-3-319-29409-4_4) contains supplementary material, which is available to authorized users.

© Springer International Publishing Switzerland 2016

A. Veltman et al., *Fundamentals of Electrical Drives*, Power Systems,
DOI 10.1007/978-3-319-29409-4_4

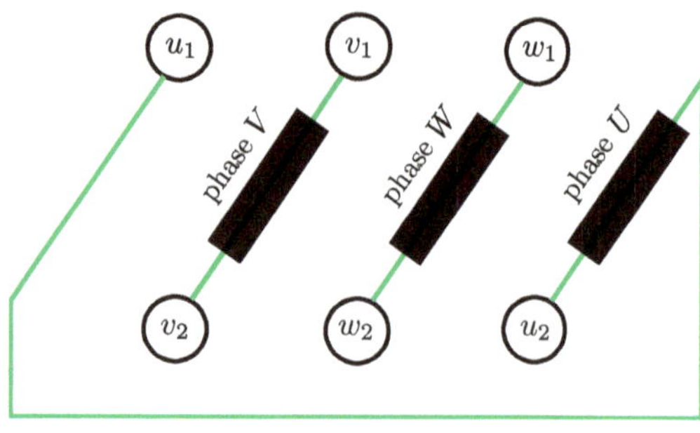

Fig. 4.1 Connector on a three-phase machine, with a total of six terminals

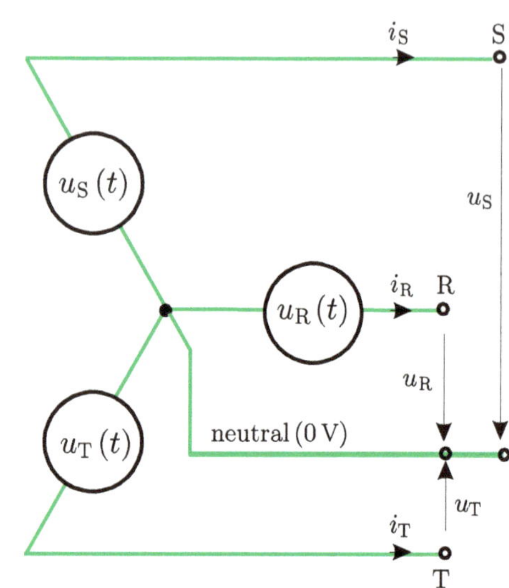

Fig. 4.2 Supply convention (voltage sources shown)

The so-called space vectors are introduced as an important tool to simplify the dynamic analysis of three-phase circuits. In the sequel to this chapter, the link between phasors and space vectors is made in order to examine three-phase circuits under steady-state conditions in case the supply is deemed to be sinusoidal in nature. Finally, a set of tutorials will be provided which serve to reinforce the concepts outlined in this chapter.

4.2 Star/Wye Connected Circuit

The term "star" or "wye" connected circuit refers to the configuration shown in Fig. 4.4, where the machine phases are connected in such a manner that a common "star" or "neutral" point is established. In inverter fed three-phase systems, this star point is usually *not* connected to the neutral or 0 V ground reference point of the supply. For the "star" connected configuration the lower three terminals v_2, w_2, and u_2 are interconnected as shown by the red lines in Fig. 4.3. This figure also shows how the R, S, and T supply is connected to the machine terminals.

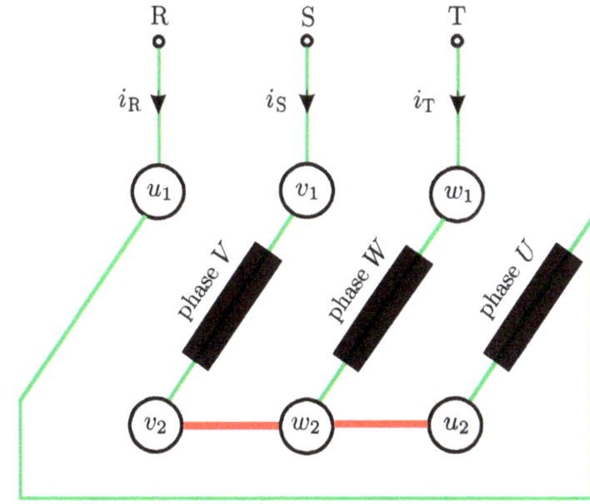

Fig. 4.3 Three-phase machine, star (Y) connected

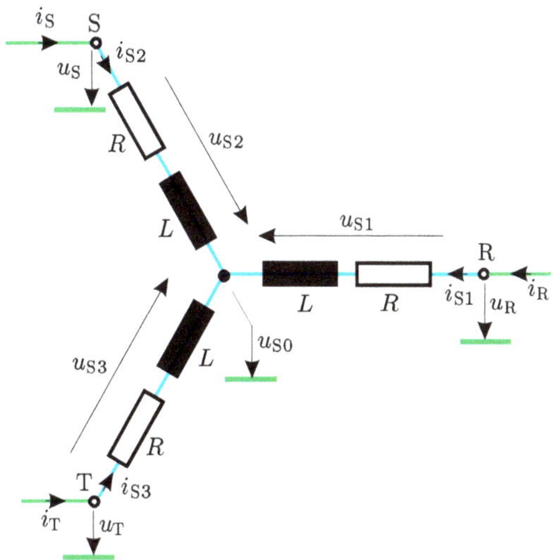

Fig. 4.4 Star/Wye connected according to Fig. 4.3

The supply voltages u_R, u_S, and u_T and u_{S0} are defined with respect to the 0 V of the supply source (see Fig. 4.2). Note (again) that the supply voltages are instantaneous functions of time and need not be sinusoidal. Furthermore, the sum of the three voltages does not and indeed will not usually be zero when a power electronic converter is used as a supply source. On the basis of Kirchhoff's voltage and current laws and observation of Fig. 4.4, we will determine the relationships that exist between supply and phase variables.

With respect to the phase variables, the following expressions are valid

$$i_{S1} + i_{S2} + i_{S3} = 0 \tag{4.1a}$$

$$u_{S1} + u_{S2} + u_{S3} = 0 \tag{4.1b}$$

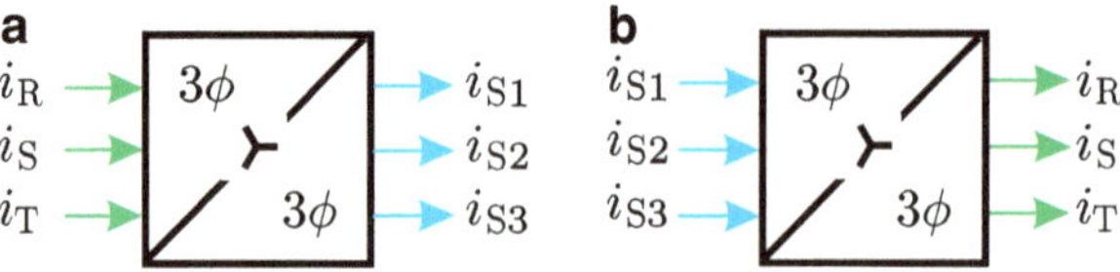

Fig. 4.5 Current conversions: star connected. (**a**) Supply to phase. (**b**) Phase to supply

Note that Eq. (4.1b) states that the sum of the phase voltages is zero. This is indeed the case here because the *phase impedances are deemed to be equal (symmetric load)*. The supply currents i_R, i_S, and i_T are in this case equal to the phase currents i_{S1}, i_{S2}, and i_{S3}, respectively. Hence, the building block as shown in Fig. 4.5a has a transfer matrix as given by Eq. (4.2).

$$\begin{bmatrix} i_{S1} \\ i_{S2} \\ i_{S3} \end{bmatrix} = \begin{bmatrix} 1 & 0 & 0 \\ 0 & 1 & 0 \\ 0 & 0 & 1 \end{bmatrix} \begin{bmatrix} i_R \\ i_S \\ i_T \end{bmatrix} \tag{4.2}$$

In the following analysis we will also discuss the inverse, i.e., the transfer function and building block(s) needed to return from phase to supply variables. This approach is instructive because cascading the two modules must give the original supply waveforms. In this case, the inverse is the unity matrix as represented by Eq. (4.3) and building block as represented by Fig. 4.5b.

$$\begin{bmatrix} i_R \\ i_S \\ i_T \end{bmatrix} = \begin{bmatrix} 1 & 0 & 0 \\ 0 & 1 & 0 \\ 0 & 0 & 1 \end{bmatrix} \begin{bmatrix} i_{S1} \\ i_{S2} \\ i_{S3} \end{bmatrix} \tag{4.3}$$

The conversion of supply to phase voltages is according to Fig. 4.4 of the form given by Eq. (4.4).

$$\begin{bmatrix} u_{S1} \\ u_{S2} \\ u_{S3} \end{bmatrix} = \begin{bmatrix} 1 & 0 & 0 \\ 0 & 1 & 0 \\ 0 & 0 & 1 \end{bmatrix} \begin{bmatrix} u_R \\ u_S \\ u_T \end{bmatrix} - u_{S0} \begin{bmatrix} 1 \\ 1 \\ 1 \end{bmatrix} \tag{4.4}$$

in which the voltage u_{S0} given in Eq. (4.4) is the potential of the star point with respect to the 0 V reference of the supply. The voltage u_{S0} is the so-called zero sequence component and can be found with the aid of Eqs. (4.1b) and (4.4) which leads to

$$u_{S0} = \frac{u_R + u_S + u_T}{3} \tag{4.5}$$

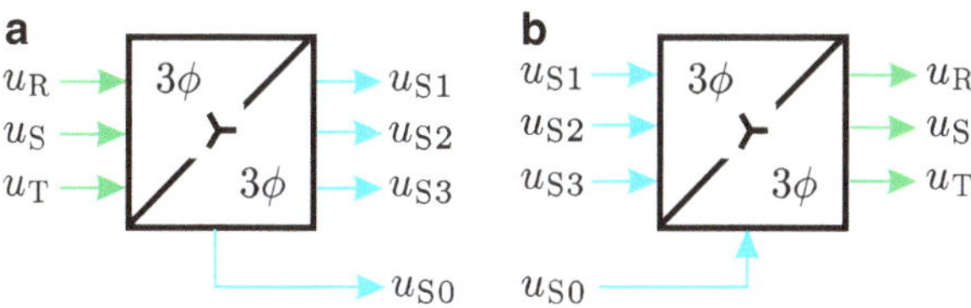

Fig. 4.6 Voltage conversions: star connected. (**a**) Supply to phase. (**b**) Phase to supply

The conversion module which represents Eq. (4.4) is given by Fig. 4.6a. An important observation from Fig. 4.6a is that this module has a fourth output, the voltage u_{S0}, which is obtained from u_R, u_S, and u_T and the assumption of *symmetrical* machine impedances. The inversion follows directly from Fig. 4.4 and is of the form

$$\begin{bmatrix} u_R \\ u_S \\ u_T \end{bmatrix} = \begin{bmatrix} 1 & 0 & 0 \\ 0 & 1 & 0 \\ 0 & 0 & 1 \end{bmatrix} \begin{bmatrix} u_{S1} \\ u_{S2} \\ u_{S3} \end{bmatrix} + u_{S0} \begin{bmatrix} 1 \\ 1 \\ 1 \end{bmatrix} \tag{4.6}$$

In Eq. (4.6), the value of u_{S0} can be chosen freely, hence the supply voltages u_R, u_S, and u_T are not unique for a given set of phase voltages u_{S1}, u_{S2}, and u_{S3}. The conversion module is given in Fig. 4.6b.

4.2.1 Modeling Star Connected Circuit

The single phase R-L circuit model has been discussed earlier and the generic implementation given in Fig. 2.5 on page 33 needs to be duplicated three times, as shown in Fig. 4.7. Note that the three-phase R-L model shown in Fig. 4.7 is a simplified representation of an AC machine. In reality, mutual coupling terms exist between the phases which severely complicates the three-phase circuit model. At a later stage in this chapter, an alternative approach to modeling three-phase circuits will be given, which is able to handle more complex circuits than the R-L concept considered here. The combined conversion process with all the building blocks needed to arrive at the supply currents, on the basis of a given set of supply voltages, is given in Fig. 4.8.

4.3 Delta Connected Circuit

The term "delta" connected circuit refers to the configuration shown in Fig. 4.10. In the terminal box on the machine, the terminals pairs (u_1, v_2), (v_1, w_2), and (w_1, u_2) are interconnected, as shown by three red lines in Fig. 4.9. The delta connection is

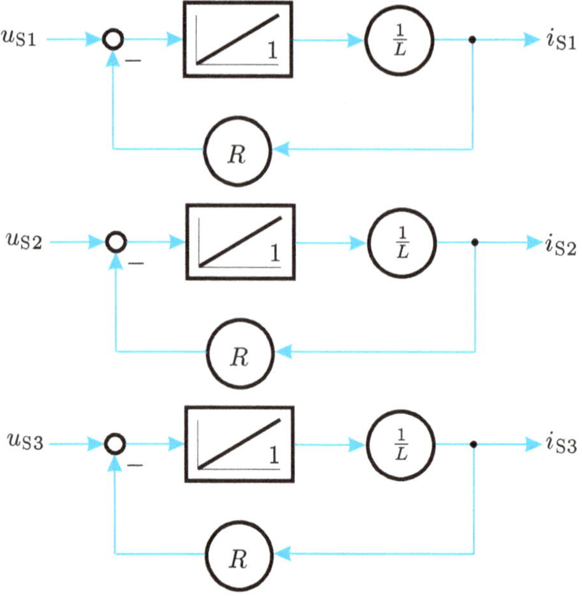

Fig. 4.7 Generic three-phase R-L model

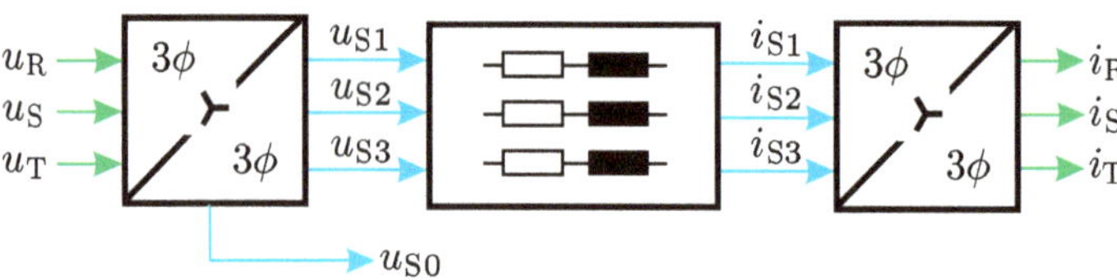

Fig. 4.8 Star connected circuit model

often used in applications with relatively low supply voltages. Furthermore, delta connected machines are commonly used in high power applications (typically from about 0.5 MW upwards). In both cases the use of a delta connected machine is beneficial in terms of maximizing the power output for a given voltage level.

The supply voltages u_R, u_S, and u_T are defined with respect to the 0 V of the supply source in the same manner as discussed in Sect. 4.2. It is re-emphasized that the supply voltages are instantaneous functions of time and need not be sinusoidal. Furthermore, the sum of the three supply voltages does not need to be zero. On the basis of Kirchhoff's voltage and current laws and observation of Fig. 4.10 we can again determine the relationships that exist between supply and phase variables.

With respect to the phase variables the following expressions are introduced.

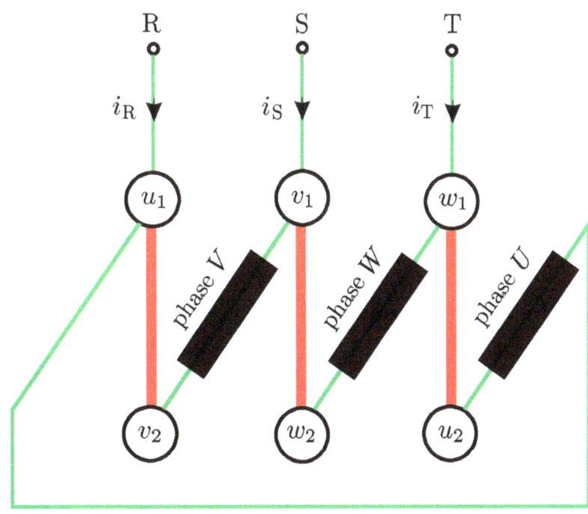

Fig. 4.9 Three-phase machine, delta (Δ) connected

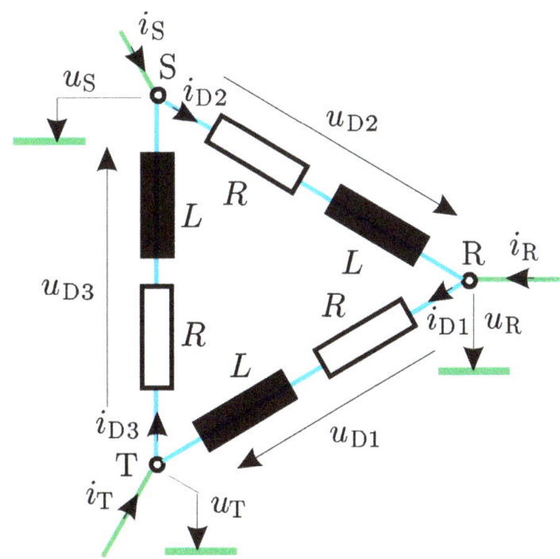

Fig. 4.10 Delta connected according to Fig. 4.9

$$u_{D1} + u_{D2} + u_{D3} = 0 \tag{4.7a}$$

$$i_{D1} + i_{D2} + i_{D3} = 3\, i_{D0} \tag{4.7b}$$

where i_{D0} represents a so-called zero sequence current. The phase voltage expression follows directly from Kirchhoff's voltage law, whereas the phase current expression (4.7b) is based on the presence of a non-zero loop current i_{D0}. In the circuit model given in Fig. 4.10, no such current will exist. However, if, for example, a voltage source is introduced in each phase leg, which has a third harmonic component, then a non-zero loop current i_{D0} will be generated, hence $i_{D1} = i_{D0}$, $i_{D2} = i_{D0}$, and $i_{D3} = i_{D0}$. Under these conditions the sum of these phase currents is equal to $3i_{D0}$ as shown by Eq. (4.7b). Measurements from a practical system with substantial loop current are shown on page 107, where the current i_{D0} is found using Eq. (4.7b), which may also be written as

$$i_{D0} = \frac{i_{D1} + i_{D2} + i_{D3}}{3} \tag{4.8}$$

The relationship between supply currents i_R, i_S, and i_T and phase currents i_{D1}, i_{D2}, and i_{D3} is in this case found using Kirchhoff's current law and observation of Fig. 4.10. For example, the current i_R may be expressed as $i_R = i_{D1} - i_{D2}$. If we extend this analysis to all three phases, the transfer matrix according to Eq. (4.9) appears.

$$\begin{bmatrix} i_R \\ i_S \\ i_T \end{bmatrix} = \begin{bmatrix} 1 & -1 & 0 \\ 0 & 1 & -1 \\ -1 & 0 & 1 \end{bmatrix} \begin{bmatrix} i_{D1} \\ i_{D2} \\ i_{D3} \end{bmatrix} \tag{4.9}$$

The conversion module which represents Eqs. (4.8) and (4.9) is given by Fig. 4.11a. Observation of Fig. 4.11a shows that this module has a fourth output, namely the current i_{D0} [see Eq. (4.8)], which is required to facilitate the conversion from supply currents to phase currents. This conversion follows from Fig. 4.10 and it is instructive to initially consider the process by which an expression for the branch current i_{D1} is formed. From Fig. 4.10 the following expressions can be found

$$i_{D1} = i_R + i_{D2} \tag{4.10a}$$

$$i_{D1} = -i_T + i_{D3} \tag{4.10b}$$

Adding Eqs. (4.10a) and (4.10b) gives

$$2i_{D1} = i_R - i_T + \underbrace{(i_{D2} + i_{D3})}_{-i_{D1}+3i_{D0}} \tag{4.11}$$

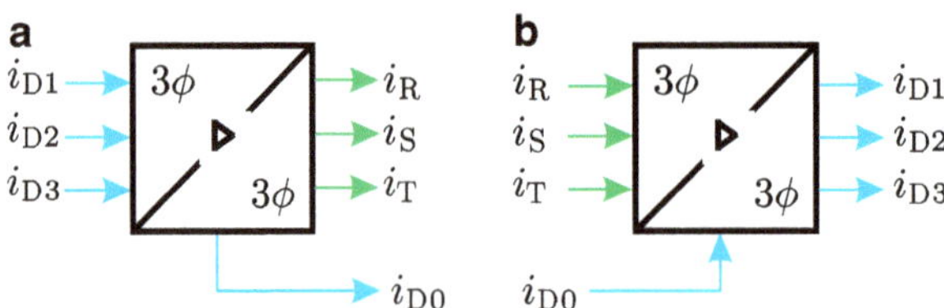

Fig. 4.11 Current conversions: delta connected. (**a**) Phase to supply. (**b**) Supply to phase

where the term $(i_{D2} + i_{D3})$ can, according to Eq. (4.7b), also be written as $(-i_{D1} + 3i_{D0})$, which leads to $i_{D1} = {}^{1}\!/_{3}\,(i_R - i_T) + i_{D0}$. It is noted that this expression is in fact not an explicit function for i_{D1} given that i_{D0} is also a function of the currents i_{D1}, i_{D2}, and i_{D3}. This means that the conversion from supply to phase current can only be made if the current i_{D0} is known, i.e., obtained from the "delta" phase to supply current conversion module mentioned earlier (see Fig. 4.11a). The exception to this rule is the case where the sum of the phase currents will be zero, as is the case when the latter are sinusoidal, of equal amplitude, and displaced by an angle of ${}^{2\pi}\!/_{3}$ with respect to each other. If we extend this single phase analysis for i_{D1} to all three phases, the conversion matrix, as given by expression (4.12) and module (Fig. 4.11b), appears.

$$\begin{bmatrix} i_{D1} \\ i_{D2} \\ i_{D3} \end{bmatrix} = \begin{bmatrix} \frac{1}{3} & 0 & -\frac{1}{3} \\ -\frac{1}{3} & \frac{1}{3} & 0 \\ 0 & -\frac{1}{3} & \frac{1}{3} \end{bmatrix} \begin{bmatrix} i_R \\ i_S \\ i_T \end{bmatrix} + i_{D0} \begin{bmatrix} 1 \\ 1 \\ 1 \end{bmatrix} \tag{4.12}$$

The conversion of supply voltages to phase voltage is according to Fig. 4.10 of the form given by Eq. (4.13).

$$\begin{bmatrix} u_{D1} \\ u_{D2} \\ u_{D3} \end{bmatrix} = \begin{bmatrix} 1 & 0 & -1 \\ -1 & 1 & 0 \\ 0 & -1 & 1 \end{bmatrix} \begin{bmatrix} u_R \\ u_S \\ u_T \end{bmatrix} \tag{4.13}$$

The conversion module which represents Eq. (4.13) is given by Fig. 4.12a. Figure 4.12a has a fourth output, the voltage u_{S0}, as found using Eq. (4.5), which is again required to facilitate the conversion from phase voltage to supply voltages. The inversion follows directly from Fig. 4.10 and can be made more translucent by initially considering a single phase conversion first. An observation of Fig. 4.10 shows that the following two expressions may be found which contain the voltage u_R.

$$u_R = u_{D1} + u_T \tag{4.14a}$$

$$u_R = -u_{D2} + u_S \tag{4.14b}$$

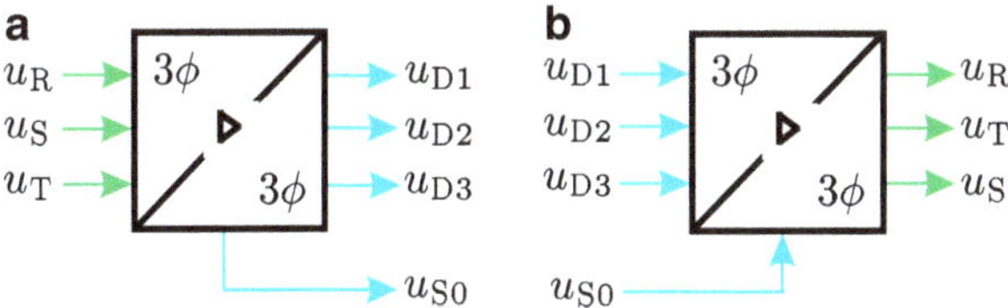

Fig. 4.12 Voltage conversions: delta connected. (**a**) Supply to phase. (**b**) Phase to supply

Adding Eqs. (4.14a) and (4.14b) gives

$$2u_{\mathrm{R}} = u_{\mathrm{D1}} - u_{\mathrm{D2}} + \underbrace{(u_{\mathrm{T}} + u_{\mathrm{S}})}_{-u_{\mathrm{R}}+3u_{\mathrm{S0}}} \tag{4.15}$$

where the term $(u_{\mathrm{T}} + u_{\mathrm{S}})$ can according to Eq. (4.5) also be written as $(-u_{\mathrm{R}} + 3u_{\mathrm{S0}})$, which leads to $u_{\mathrm{R}} = {}^{1}\!/_{3}\,(u_{\mathrm{D1}} - u_{\mathrm{D2}}) + u_{\mathrm{S0}}$. It is noted (as was the case for the "current" conversion) that this expression is in fact not an explicit expression for u_{R} given that u_{S0} is also a function of the voltages u_{R}, u_{S}, and u_{T}. This means that the conversion from phase to supply voltages can only be made if the voltage u_{S0} is known. In the case where the sum of the supply voltages is zero, as is the case when the latter are sinusoidal, of equal magnitude, and displaced by an angle of ${}^{2\pi}\!/_{3}$ with respect to each other the voltage u_{S0} will be zero. If we extend our single phase analysis shown above for u_{R} to the remaining two phases, the conversion matrix as given by expression (4.16) and building module (Fig. 4.12b) appears.

$$\begin{bmatrix} u_{\mathrm{R}} \\ u_{\mathrm{S}} \\ u_{\mathrm{T}} \end{bmatrix} = \begin{bmatrix} \frac{1}{3} & -\frac{1}{3} & 0 \\ 0 & \frac{1}{3} & -\frac{1}{3} \\ -\frac{1}{3} & 0 & \frac{1}{3} \end{bmatrix} \begin{bmatrix} u_{\mathrm{D1}} \\ u_{\mathrm{D2}} \\ u_{\mathrm{D3}} \end{bmatrix} + u_{\mathrm{S0}} \begin{bmatrix} 1 \\ 1 \\ 1 \end{bmatrix} \tag{4.16}$$

4.3.1 Modeling Delta Connected Circuit

The three-phase R-L generic circuit model shown in Fig. 4.7 for the star connected phase configuration is directly applicable here with the important difference that the current/voltage phase variables u_{S1}, u_{S2}, u_{S3}, i_{S1}, i_{S2}, and i_{S3} must be replaced by the variables u_{D1}, u_{D2}, u_{D3}, i_{D1}, i_{D2}, and i_{D3} given that we are dealing with a delta connected load. The inputs to this module will be the phase voltages from the delta connected circuit and the outputs are the three-phase currents. The conversion process needed to arrive at the supply currents given a set of supply voltages is shown in Fig. 4.13.

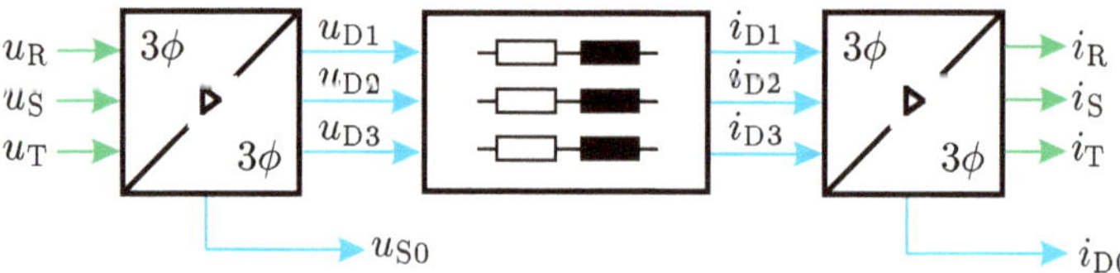

Fig. 4.13 Delta connected circuit model

4.4 Space Vectors

The question as to why we need "space vectors" comes down to the difficulty of handling complex three-phase systems as was mentioned earlier. It will be shown that the introduction of a space vector type representation for a three-phase system leads to considerable simplification.

The space vector formulation is in its general form given by Eq. (4.17).

$$\vec{x} = C\left\{x_{\mathrm{R}} + x_{\mathrm{S}}\,\mathrm{e}^{\mathrm{j}\gamma} + x_{\mathrm{T}}\,\mathrm{e}^{\mathrm{j}2\gamma}\right\} \tag{4.17}$$

with $\gamma = {}^{2\pi}/_{3}$. The variables x_{R}, x_{S}, and x_{T} represent three instantaneous time dependent supply variables. These may, for example, be the three supply voltages u_{R}, u_{S}, and u_{T}, or in fact any other variable. Furthermore, these variables are *real* functions of time and do *not* need to be sinusoidal. The constant C is a scalar and its value will be defined later.

The space vector $\vec{x}$ itself is both complex and time dependent. The space vector is represented in a complex plane which at present is assumed to be stationary. The space vector can, according to Eq. (4.18), also be written in terms of a real x_{α} and imaginary x_{β} component with $\mathrm{j} = \sqrt{-1}$.

$$\vec{x} = x_{\alpha} + \mathrm{j}x_{\beta} \tag{4.18}$$

Figure 4.14 shows the space vector in the complex plane. Note that x_{α} is equal to $\Re\left\{\vec{x}\right\}$, while x_{β} may be written as $\Im\left\{\vec{x}\right\}$. An observation of Eqs. (4.17) and (4.18) shows that the space vector deals with a transformation process, in which a linear combination of the three supply variables $x_{\mathrm{R}}, x_{\mathrm{S}}$, and x_{T} is converted to a two-phase x_{α}, and x_{β} form. It is noted that this three-dimensional to two-dimensional representation is possible because the former is in fact over dimensioned, when

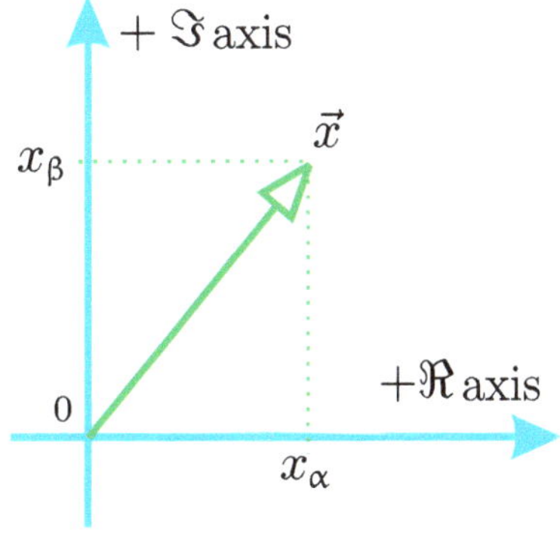

Fig. 4.14 Space vector representation in a complex plane

dealing with symmetrical systems, as is the case in this book. In other words, one variable is superfluous, namely the zero sequence component.

It is important to realize that the space vector amplitude $\left(|\vec{x}|\right)$ and argument $(\arctan {}^{x_\beta}/_{x_\alpha})$ can be a function of time. We may therefore see non-continuous changes of both argument and amplitude in many cases such as three-phase PWM.

It is instructive at this stage to give an example based on Eq. (4.17) with $C = 1$. In this case we will plot the space vector for three cases. Each case corresponds to one of the three supply variables of Eq. (4.17) being non-zero.

Case 1: $x_R > 0$, $x_S = 0$, and $x_T = 0$, the space vector is then of the form $\vec{x} = x_R$. This corresponds to $x_\alpha = x_R$ and $x_\beta = 0$.

Case 2: $x_S > 0$, $x_R = 0$, and $x_T = 0$, the space vector is then of the form $\vec{x} = x_S\,e^{j\gamma}$. This corresponds to $x_\alpha = x_S \cos\gamma$ and $x_\beta = x_S \sin\gamma$ which may also be written as $x_\alpha = -{}^1/_2\, x_S$ and $x_\beta = {}^{\sqrt{3}}/_2\, x_S$.

Case 3: $x_T > 0$, $x_R = 0$, and $x_S = 0$, the space vector is then of the form $\vec{x} = x_T\,e^{j2\gamma}$. This corresponds to $x_\alpha = x_T \cos 2\gamma$ and $x_\beta = x_T \sin 2\gamma$, which may be written as $x_\alpha = -{}^1/_2\, x_T$ and $x_\beta = -{}^{\sqrt{3}}/_2\, x_T$.

Figure 4.15 shows the three cases considered above where it is assumed that $x_T > x_S > x_R$. It is left as an exercise to the reader to consider the case where one of the supply variables (in the example above) is, for example, a sinusoidal function of time.

4.5 Amplitude Versus Power Invariant Space Vectors

In this section we will consider the prudent choices we can make with respect to the value of the constant C [see Eq. (4.17)]. In support of this discussion we will consider three supply voltages, which are assumed sinusoidal and of the form given

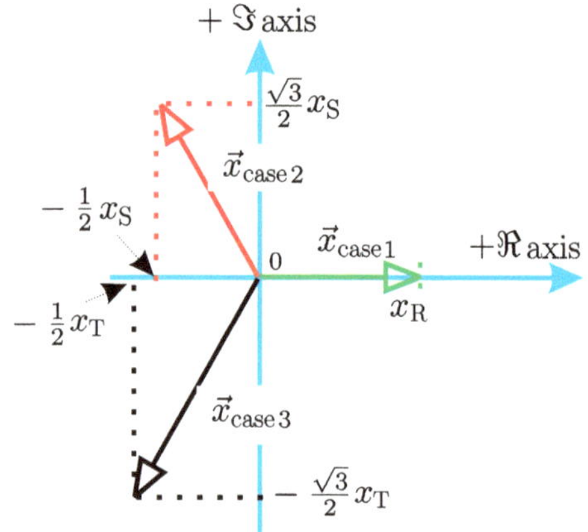

Fig. 4.15 Space vector example, for three case studies

by Eq. (4.19). Note that this assumption does not undermine the generality of using space vectors for waveforms which are not sinusoidal nor for that matter does the sum of the three waveforms need to be zero as will become apparent shortly.

$$u_R = \hat{u}\cos(\omega t) \tag{4.19a}$$

$$u_S = \hat{u}\cos(\omega t - \gamma) \tag{4.19b}$$

$$u_T = \hat{u}\cos(\omega t - 2\gamma) \tag{4.19c}$$

The phase shift of the waveforms in Eq. (4.19) is represented by the variable $\gamma = {}^{2\pi}\!/_{3}$. The process of finding a space vector form for the three voltages u_R, u_S, and u_T, as defined by Eq. (4.19), is readily realized by substituting said equation into (4.17) which gives

$$\vec{u} = C\hat{u}\left(\cos(\omega t) + \cos(\omega t - \gamma)\,e^{j\gamma} + \cos(\omega t - 2\gamma)\,e^{j2\gamma}\right) \tag{4.20}$$

Expression (4.20) may be developed further by making use of the expression $\cos y = \frac{e^{jy}+e^{-jy}}{2}$ which, after some manipulation (which the reader should look at carefully), gives

$$\vec{u} = C\hat{u}\left(\frac{3}{2}\,e^{j\omega t} + \frac{1}{2}\underbrace{\left\{e^{-j\omega t} + e^{-j(\omega t - 2\gamma)} + e^{-j(\omega t - 4\gamma)}\right\}}_{\text{vector sum is zero}}\right) \tag{4.21}$$

The second term in the right-hand side of Eq. (4.21) is zero, given that this term is formed by three vectors of the same amplitude which are phase shifted with respect to each other by an angle γ. This means that the vector sum of these three vectors equals zero. Consequently, the voltage space vector is reduced to

$$\vec{u} = \frac{3}{2}\,C\hat{u}\,e^{j\omega t} \tag{4.22}$$

The voltage space vector is thus a function of time (argument ωt) and its amplitude is equal to $\frac{3}{2}\,C\hat{u}$. Presented in a complex plane, the voltage vector end point will move on a circle, as shown in Fig. 4.16.

The analysis, to determine the voltage space vector from the three-phase voltages, can also be given for the currents. It is left to the reader to undertake this exercise in detail. In summary, you must consider the three current variables according to Eq. (4.23)

$$i_R = \hat{i}\cos(\omega t + \rho) \tag{4.23a}$$

$$i_S = \hat{i}\cos(\omega t + \rho - \gamma) \tag{4.23b}$$

$$i_T = \hat{i}\cos(\omega t + \rho - 2\gamma) \tag{4.23c}$$

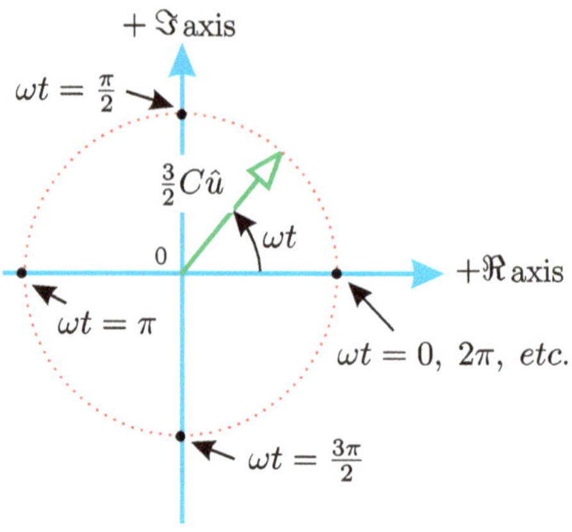

Fig. 4.16 Voltage space vector as function of time

and use Eq. (4.17), which then leads to Eq. (4.24).

$$\vec{i} = \frac{3}{2}\, C\, \hat{i}\, e^{j(\omega t + \rho)} \tag{4.24}$$

The issue of choosing a suitable value for the constant C, as defined by Eq. (4.17), is considered here. In the example discussed in Sect. 4.4 this constant was conveniently set to unity but this is not a suitable value as will become apparent shortly.

There are in fact two prudent values which may be assigned to the constant C. The first option we have, and the one used in this book, is to set this constant to a value $C = 2/3$. Space vectors which abide with this C value are given in the so-called *amplitude invariant* form. The reason for this can be made clear by considering again Eqs. (4.22) and (4.24), with $C = 2/3$. With this value of C the equations are given as

$$\vec{u} = \hat{u}\, e^{j\omega t} \tag{4.25a}$$

$$\vec{i} = \hat{i}\, e^{j(\omega t + \rho)} \tag{4.25b}$$

An observation of Eq. (4.25) shows that the amplitude of the space vectors is now *equal* to the peak variable value, which is why this notation form is referred to as amplitude invariant.

The second notation form refers to a so-called *power invariant* notation form. For this notational form, the constant is chosen to be $C = \sqrt{\frac{2}{3}}$. The voltage and current space vectors given by Eqs. (4.22) and (4.24) will then be of the form

$$\vec{u} = \sqrt{\frac{3}{2}}\, \hat{u}\, e^{j\omega t} \tag{4.26a}$$

$$\vec{i} = \sqrt{\frac{3}{2}}\, \hat{i}\, e^{j(\omega t + \rho)} \tag{4.26b}$$

This means that the space vector amplitude of, for example, the voltage is of the form $|\vec{u}| = \sqrt{3/2}\,\hat{u}$. Hence the *space vector amplitude is a factor* $\sqrt{3/2}$ *larger* than the amplitude of the corresponding phase variables. The power invariant notation form can be made more plausible by considering the total instantaneous power in the two-phase system represented by the variables u_α, u_β and i_α, i_β which is of the form

$$p_{2\phi} = u_\alpha i_\alpha + u_\beta i_\beta \tag{4.27}$$

Equation (4.27) may be written in its space vector notation form, namely,

$$p_{2\phi} = \Re\left\{\vec{u}\left(\vec{i}\right)^*\right\} \tag{4.28}$$

where $\vec{u}$ and $\vec{i}$ represent the voltage and current space vectors. It is shown in Sect. 4.6 that the variables u_α, u_β and i_α, i_β given in Eq. (4.27) may also be expressed in terms of the three phase variables u_R, u_S, and u_T and i_R, i_S, and i_T which allows Eq. (4.27) to be written as

$$p_{2\phi} = \frac{3}{2}\,C^2 p_{3\phi} \tag{4.29}$$

where $p_{3\phi} = u_R i_R + u_S i_S + u_T i_T$ represents the total instantaneous power of a three-phase system. Note that it can be shown that Eq. (4.29) is valid if the sum of the supply currents i_R, i_S, and i_T and/or the sum of the supply voltages u_R, u_S, and u_T is equal to zero. In this book, the sum of the supply currents is always taken to be zero, which implies that only three wires are present between the supply and the machine (load).

Equation (4.29) is significant as it conveys the meaning of the term "power invariant." Namely, that the instantaneous power of a two-phase system is equal to that of a three-phase system in case the constant C is chosen to be equal to $C = \sqrt{2/3}$. Note that the use of an amplitude invariant space vector notation where the constant C is chosen as

$$C = \frac{2}{3} \tag{4.30}$$

causes the instantaneous power of a three-phase system to be scaled by a factor $3/2$. When calculating the output power of a three-phase electrical machine using amplitude invariant space vectors, this constant must be factored in to "correct" the power calculation. A more detailed discussion on the concept of "power" in single and three-phase systems is given in Chap. 5.

4.6 Application of Space Vectors for Three-Phase Circuit Analysis

In this section, the aim is to introduce the space vector concept in the star and delta connected three-phase circuits, as discussed in Sects. 4.2 and 4.3.

Two common transfer modules, namely, from three phase to space vector and vice versa need to be discussed. The first case concerns the conversion of phase x_{S1}, x_{S2}, and x_{S3} (star connected), x_{D1}, x_{D2}, and x_{D3} (delta connected) or supply x_R, x_S, and x_T, variables, given in a general form as x_a, x_b, and x_c, to a space vector form $\vec{x}_{abc} = x_\alpha + jx_\beta$. The sum of the three scalar variables is of the form $x_a + x_b + x_c = 3x_0$ where x_0 represents a zero sequence component which may have a non-zero value.

According to Eq. (4.17) the relationship between vector and scalar variables may be written as

$$\vec{x}_{abc} = C\left\{x_a + x_b\, e^{j\gamma} + x_c\, e^{j2\gamma}\right\} \tag{4.31}$$

When we equate the real and imaginary components of Eq. (4.31), the conversion matrix according to Eq. (4.32) and Fig. 4.17a is found.

$$\begin{bmatrix} x_\alpha \\ x_\beta \end{bmatrix} = \begin{bmatrix} C & -\frac{C}{2} & -\frac{C}{2} \\ 0 & \frac{C\sqrt{3}}{2} & -\frac{C\sqrt{3}}{2} \end{bmatrix} \begin{bmatrix} x_a \\ x_b \\ x_c \end{bmatrix} \tag{4.32}$$

In this expression the value of $C = 2/3$ should be used for amplitude invariant space vector representations. The building block according to Fig. 4.17a has a fourth output which is the zero sequence variable x_0 introduced earlier as

$$x_0 = \frac{x_a + x_b + x_c}{3} \tag{4.33}$$

The process of finding the inverse conversion process, which allows us to move from space vector variables to scalar variables, is considered next. A suitable starting point for this conversion is Eq. (4.32), which gives an expression for x_α, namely,

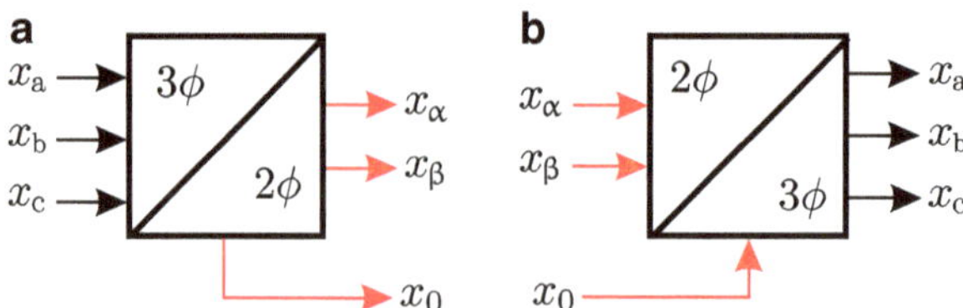

Fig. 4.17 Conversion from and to space vector format: general case. (**a**) Scalar to space vector. (**b**) Space vector to scalar

$$x_{\alpha} = Cx_{a} - \frac{C}{2}(x_{b} + x_{c}) \tag{4.34}$$

The sum of the three-phase variables is according to Eq. (4.33) equal to $3x_0$ and this expression can also be written as $x_b + x_c = -x_a + 3x_0$. Substitution of this expression into (4.34) leads to

$$x_a = \frac{2}{3C} x_{\alpha} + x_0 \tag{4.35}$$

The variable x_b is also directly obtained from Eq. (4.32) in which we consider the second row, namely,

$$x_{\beta} = C\frac{\sqrt{3}}{2}(x_b - x_c) \tag{4.36}$$

Substitution of $x_c = -(x_b + x_a) + 3x_0$ and use of (4.35) give, after some manipulation,

$$x_b = -\frac{1}{3C} x_{\alpha} + \frac{1}{C\sqrt{3}} x_{\beta} + x_0 \tag{4.37}$$

The remaining variable x_c is found by use of $x_c = -(x_a + x_b) + 3x_0$ together with Eqs. (4.35) and (4.36). The resultant complete conversion in matrix form as given by Eq. (4.38) corresponds to the building block shown in Fig. 4.17b.

$$\begin{bmatrix} x_a \\ x_b \\ x_c \end{bmatrix} = \begin{bmatrix} \frac{2}{3C} & 0 \\ -\frac{1}{3C} & \frac{1}{C\sqrt{3}} \\ -\frac{1}{3C} & -\frac{1}{C\sqrt{3}} \end{bmatrix} \begin{bmatrix} x_{\alpha} \\ x_{\beta} \end{bmatrix} + x_0 \begin{bmatrix} 1 \\ 1 \\ 1 \end{bmatrix} \tag{4.38}$$

4.6.1 *Use of Space Vectors in Star Connected Circuits*

The conversion process from phase voltages and currents to a space vector form (in stationary coordinates) is identical for both, hence the phase variables x_{S1}, x_{S2}, and x_{S3} are introduced which need to be converted to a form $\vec{x}_{S123} = x_{S\alpha} + jx_{S\beta}$. Note that the space vector variables are identified by subscripts "Sα" and "Sβ," respectively. The conversion modules as shown in Fig. 4.18 are identical to those shown in Fig. 4.17. The zero sequence "output" and "input" lines are not shown in Fig. 4.18, given that the sum of the voltage and current phase variables is zero for this circuit configuration [see Eq. (4.1)].

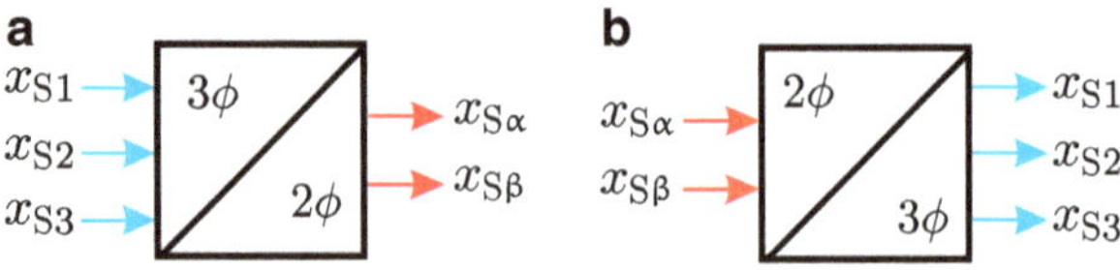

Fig. 4.18 Voltage/current conversion to space vector $\left(\vec{x}_{RST} = \vec{x}_{S123}\right)$ format: star connected. (**a**) Phase to space vector. (**b**) Space vector to phase

In some cases, a conversion is required where phase variables x_{S1}, x_{S2}, and x_{S3} or space vector variables $x_{S\alpha}$ and $x_{S\beta}$ need to be converted to supply (*RST*) based variables of the form $\vec{x}_{RST} = x_\alpha + jx_\beta$.

The starting point for this analysis is Eq. (4.17). The relationship between phase and supply variables for the star connected case can, according to Eqs. (4.3) and (4.6), be written as

$$x_R = x_{S1} + x_{S0} \tag{4.39a}$$

$$x_S = x_{S2} + x_{S0} \tag{4.39b}$$

$$x_T = x_{S3} + x_{S0} \tag{4.39c}$$

where x_{S0} will be zero in case the variable x represents the current i. Substitution of Eq. (4.39) into Eq. (4.17) leads to

$$\vec{x}_{RST} = \underbrace{C\left(x_{S1} + x_{S2}e^{j\gamma} + x_{S3}e^{j2\gamma}\right)}_{\vec{x}_{S123}} + Cx_{S0}\underbrace{\left(1 + e^{j\gamma} + e^{j2\gamma}\right)}_{0} \tag{4.40}$$

An important observation of Eq. (4.40) is that the presence of a zero sequence component in the supply variables will *not* have any impact on the conversion process. The reason for this is that the constant Cx_{S0} is multiplied by zero (the vector sum of the three terms is zero). A direct consequence of this conversion is that the inverse transformation, i.e., from space vector to supply variables, is only possible in case the zero sequence component x_{S0} is known, e.g., zero, or can be calculated from circuit analysis. A non-zero value x_{S0} is "lost" in the conversion $x_R, x_S, x_T \rightarrow \vec{x}_{RST}$. A further observation of Eq. (4.40) shows that the space vector representation in supply and phase format is same, hence

$$\vec{x}_{RST} = \vec{x}_{S123}. \tag{4.41}$$

Note that according to Eq. (4.41) the real and imaginary components of these vectors will be equal for the star connected circuit, hence, $x_\alpha = x_{S\alpha}$ and $x_\beta = x_{S\beta}$. This is *not* the case for a "delta" connected circuit, as will become apparent shortly.

4.6.2 Circuit Modeling Using Space Vectors: Star Connected

In this section we will demonstrate how we can use the space vector approach to build a dynamic generic module of this system according to Fig. 4.7. It is at this stage helpful to recall the differential equation set of the circuit in question, which is of the form

$$u_{S1} = i_{S1}R + L\frac{\mathrm{d}i_{S1}}{\mathrm{d}t} \tag{4.42a}$$

$$u_{S2} = i_{S2}R + L\frac{\mathrm{d}i_{S2}}{\mathrm{d}t} \tag{4.42b}$$

$$u_{S3} = i_{S3}R + L\frac{\mathrm{d}i_{S3}}{\mathrm{d}t} \tag{4.42c}$$

We can rewrite Eq. (4.42) in a space vector form by making use of, for example, Eq. (4.31). This equation tells us that we can build the space vector equation of this circuit by taking the following steps:

- multiply Eq. (4.42a) by a factor C.
- multiply Eq. (4.42b) by a factor $C\mathrm{e}^{\mathrm{j}\gamma}$.
- multiply Eq. (4.42c) by a factor $C\mathrm{e}^{\mathrm{j}2\gamma}$.
- Add the three previous terms together which in effect gives us the space vector form of the current and voltage space variables.

The resultant circuit equation in space vector form is then given as

$$\vec{u}_{S123} = \frac{\mathrm{d}\vec{\psi}_{S123}}{\mathrm{d}t} + \vec{i}_{S123}R \tag{4.43}$$

where $\vec{\psi}_{S123} = L\vec{i}_{S123}$. The development of the generic model proceeds along the lines discussed for the single phase R-L example. A possible generic implementation of the three-phase system is given in Fig. 4.19. The model according to Fig. 4.19 has as inputs the three-phase voltage variables which are then used as inputs to a "three- to two-phase" module, which produces the real and imaginary components of the voltage space vector $\vec{u}_{S123} = u_{S\alpha} + ju_{S\beta}$. A multiplexer function is used to

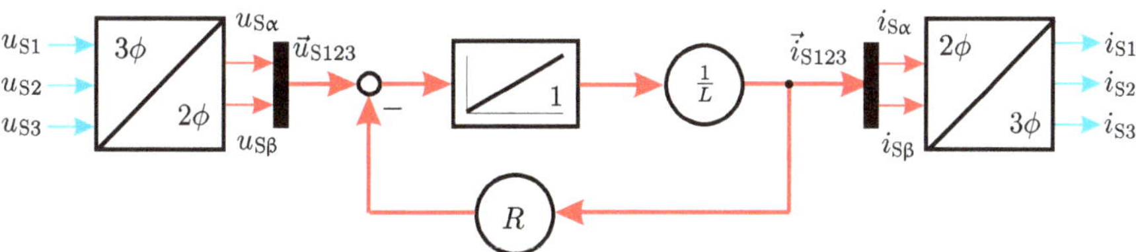

Fig. 4.19 Generic, space vector based, model of three-phase R-L circuit (star connected)

convert to a so-called vector line format, which simplifies the modeling process. The vector line format is represented as a "wide" line which in this case represents the variables $(u_{S\alpha}, u_{S\beta})$ in array format. The integrator shown has as input the space vector flux-linkage differential which is formed by the terms $\vec{u}_{S123}$ minus $R\vec{i}_{S123}$. The output of the integrator (with unity gain) is the flux-linkage space vector $\vec{\psi}_{S123}$, which is then multiplied by a gain $1/L$ to arrive at the current space vector $\vec{i}_{S123}$. Next, a de-multiplexer is used to convert from a vector line format in the form of the array variables $(i_{S\alpha}, i_{S\beta})$ to two scalar lines variables $(i_{S\alpha}, i_{S\beta})$, which are the inputs to the "two- to three-phase" conversion module. The outputs of this module represent the three phase currents i_{S1}, i_{S2}, and i_{S3} of this system. In conclusion, the use of space vectors allows us to model three-phase symmetrical circuits in the same way as single phase circuits, thus simplifying the process. The space vector based circuit model as discussed here (see Fig. 4.19) replaces the earlier circuit model (see Fig. 4.7 on page 88). This new approach allows us to model more complex circuits such as electrical machines and three-phase transformers.

4.6.3 Use of Space Vectors in Delta Connected Circuit

The conversion process from phase voltages and currents to a space vector form is identical for both, hence the phase variables x_{D1}, x_{D2}, and x_{D3} are introduced, which need to be converted to a form $\vec{x}_{D123} = x_{D\alpha} + jx_{D\beta}$. Note that the space vector variables are identified by subscripts "Dα" and "Dβ" ("D" for "delta"), respectively. The conversion modules as shown in Fig. 4.17 are directly applicable to the delta connected circuit as defined on page 89. The input and output variables are however tied to the "delta" configuration as is apparent from Fig. 4.20. The zero sequence "output" and "input" lines are not shown in Fig. 4.20. When considering this conversion for phase currents, the zero sequence connection between the two modules must be shown, given that the sum current can be non-zero.

In some cases a conversion is required where phase variables x_{D1}, x_{D2}, and x_{D3} or space vector variables $x_{D\alpha}$ and $x_{D\beta}$ must be converted to supply (RST) based variables of the form $\vec{x}_{RST} = x_\alpha + jx_\beta$. In this case, the voltage and current phase conversions need to be examined separately.

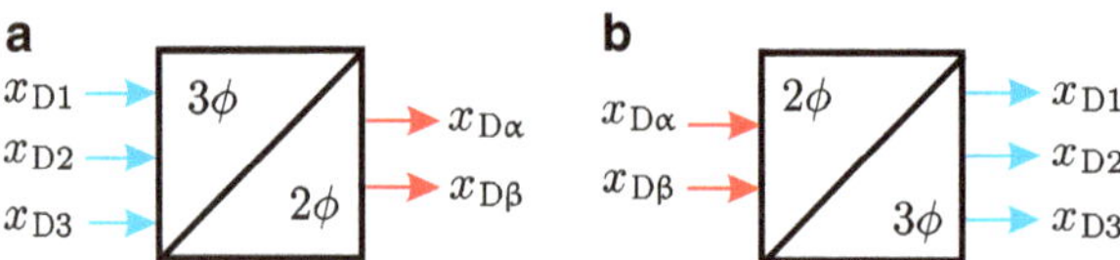

Fig. 4.20 Voltage/current conversion to space vector format $(\vec{x}_{D123})$: delta connected. (**a**) Phase to space vector. (**b**) Space vector to phase

A suitable starting point is again Eq. (4.17) which upon substitution of Eq. (4.16) may be written as

$$\vec{u}_{\mathrm{RST}} = \frac{C}{3}\left((u_{\mathrm{D1}} - u_{\mathrm{D2}}) + (u_{\mathrm{D2}} - u_{\mathrm{D3}})\,\mathrm{e}^{\mathrm{j}\gamma} + (u_{\mathrm{D3}} - u_{\mathrm{D1}})\,\mathrm{e}^{\mathrm{j}2\gamma}\right) + C u_{\mathrm{S0}} \underbrace{\left(1 + \mathrm{e}^{\mathrm{j}\gamma} + \mathrm{e}^{\mathrm{j}2\gamma}\right)}_{0} \tag{4.44}$$

Equation (4.44) may be rearranged by grouping the phase variables as shown in Eq. (4.45)

$$\vec{u}_{\mathrm{RST}} = \frac{C}{3}\left(u_{\mathrm{D1}} \underbrace{\left(1 - \mathrm{e}^{\mathrm{j}2\gamma}\right)}_{\sqrt{3}\,\mathrm{e}^{\mathrm{j}\frac{\gamma}{4}}} + u_{\mathrm{D2}} \underbrace{\left(-1 + \mathrm{e}^{\mathrm{j}\gamma}\right)}_{\sqrt{3}\,\mathrm{e}^{\mathrm{j}\left(\frac{\gamma}{4}+\gamma\right)}} + u_{\mathrm{D3}} \underbrace{\left(-\mathrm{e}^{\mathrm{j}\gamma} + \mathrm{e}^{\mathrm{j}2\gamma}\right)}_{\sqrt{3}\,\mathrm{e}^{\mathrm{j}\left(\frac{\gamma}{4}+2\gamma\right)}} \right) \tag{4.45}$$

The braced terms contain a common term $\sqrt{3}\,\mathrm{e}^{\mathrm{j}\frac{\gamma}{4}}$, which allows Eq. (4.45) to be written as

$$\vec{u}_{\mathrm{RST}} = \frac{1}{\sqrt{3}}\,\mathrm{e}^{\mathrm{j}\frac{\gamma}{4}} \underbrace{C\left(u_{\mathrm{D1}} u_{\mathrm{D1}} + u_{\mathrm{D2}}\mathrm{e}^{\mathrm{j}\gamma} + u_{\mathrm{D3}}\mathrm{e}^{\mathrm{j}2\gamma}\right)}_{\vec{u}_{\mathrm{D123}}} \tag{4.46}$$

Equation (4.46) is significant in that it tells us that the voltage space vector $\vec{u}_{\mathrm{RST}} = u_{\alpha} + \mathrm{j}u_{\beta}$ can be found by converting the three-phase voltages to the vector $\vec{u}_{\mathrm{D123}} = u_{\mathrm{D}\alpha} + \mathrm{j}u_{\mathrm{D}\beta}$, which needs to be rotated by an angle $\gamma/4$ (30°) and scaled by a factor $1/\sqrt{3}$. The three- to two-phase conversion is carried out with the conversion matrix according to Eq. (4.32). In the generic representation as given by Fig. 4.21a, the conversion as defined by Eq. (4.46) is clearly visible.

The second module shown with a "delta" and "star" symbol symbolizes the conversion $\vec{u}_{\mathrm{D123}} \rightarrow \vec{u}_{\mathrm{RST}}$, which takes place when a delta connected circuit is used. The transfer matrix linked to this conversion is given by equation

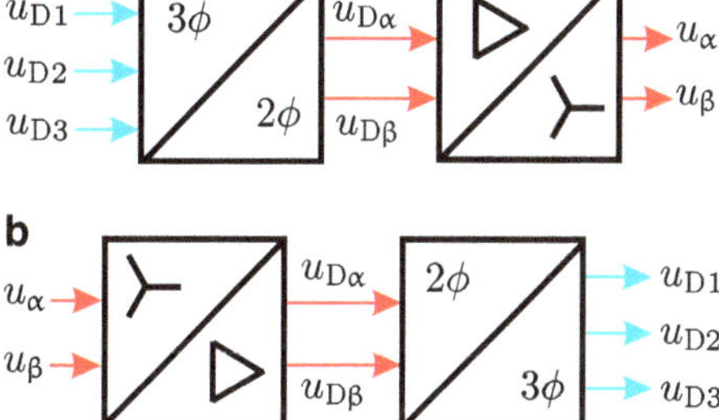

Fig. 4.21 Phase voltage to space vector ($\vec{u}_{\mathrm{RST}}$) conversions: delta connected. (**a**) Phase to space vector. (**b**) Space vector to phase

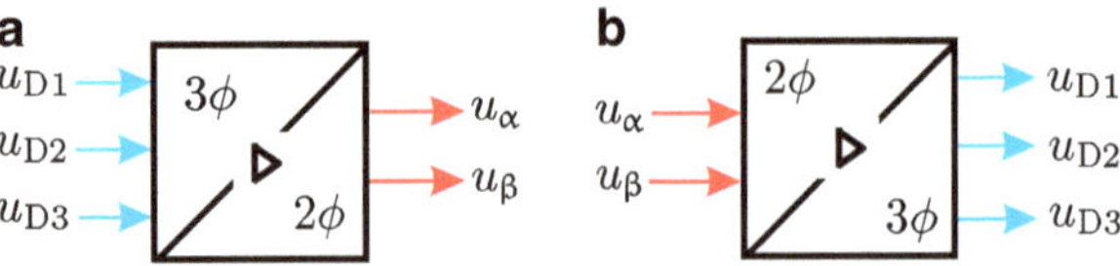

Fig. 4.22 Alternative phase voltage to space vector ($\vec{u}_{RST}$) conversions: delta connected. (**a**) Phase to space vector. (**b**) Space vector to phase

$$\begin{bmatrix} u_\alpha \\ u_\beta \end{bmatrix} = \frac{1}{\sqrt{3}} \begin{bmatrix} \cos\frac{\gamma}{4} & -\sin\frac{\gamma}{4} \\ \sin\frac{\gamma}{4} & \cos\frac{\gamma}{4} \end{bmatrix} \begin{bmatrix} u_{D\alpha} \\ u_{D\beta} \end{bmatrix} \tag{4.47}$$

The conversion module which converts the phase voltages to space vector format $u_{D1,D2,D3} \rightarrow \vec{u}_{RST}$ is shown in Fig. 4.22a. Its contents can either be according to the set of generic modules shown in Fig. 4.21a or the conversion matrix given by Eq. (4.48).

$$\begin{bmatrix} u_\alpha \\ u_\beta \end{bmatrix} = \begin{bmatrix} \frac{C}{2} & -\frac{C}{2} & 0 \\ \frac{C}{2\sqrt{3}} & \frac{C}{2\sqrt{3}} & -\frac{C}{\sqrt{3}} \end{bmatrix} \begin{bmatrix} u_{D1} \\ u_{D2} \\ u_{D3} \end{bmatrix} \tag{4.48}$$

This matrix is found by combining the matrices according to Eqs. (4.32) and (4.47). The inverse operation, namely, the conversion process from supply space vector variables to phase voltage variables, follows directly from Eq. (4.46). This expression may also be written as

$$\vec{u}_{D123} = \sqrt{3}\, e^{-j\frac{\gamma}{4}}\, \vec{u}_{RST} \tag{4.49}$$

Equation (4.49) states that the space vector $\vec{u}_{RST}$ must be rotated by an angle $-\gamma/4$ ($-30°$) and scaled by a factor $\sqrt{3}$ in order to arrive at the vector $\vec{u}_{D123}$, which can be converted to phase voltage variables using Eq. (4.38). The generic modules required for this conversion are shown in Fig. 4.21b. Included in this figure is a conversion module symbolized by the symbols "star" and "delta" and its transfer matrix is of the form given by Eq. (4.50).

$$\begin{bmatrix} u_{D\alpha} \\ u_{D\beta} \end{bmatrix} = \sqrt{3} \begin{bmatrix} \cos\frac{\gamma}{4} & \sin\frac{\gamma}{4} \\ -\sin\frac{\gamma}{4} & \cos\frac{\gamma}{4} \end{bmatrix} \begin{bmatrix} u_\alpha \\ u_\beta \end{bmatrix} \tag{4.50}$$

The conversion module which converts the supply space vector format to phase variables $\vec{u}_{RST} \rightarrow u_{D1,D2,D3}$ is shown in Fig. 4.22b. Its contents can either be according to the set of generic modules shown in Fig. 4.21b or the conversion matrix given by Eq. (4.51).

$$\begin{bmatrix} u_{D1} \\ u_{D2} \\ u_{D3} \end{bmatrix} = \begin{bmatrix} \frac{1}{C} & \frac{1}{C\sqrt{3}} \\ -\frac{1}{C} & \frac{1}{C\sqrt{3}} \\ 0 & -\frac{2\sqrt{3}}{3C} \end{bmatrix} \begin{bmatrix} u_\alpha \\ u_\beta \end{bmatrix} \tag{4.51}$$

This matrix is found by combining the matrices according to Eqs. (4.38) and (4.50), with $\cos{}^{\gamma}/_{4} = {}^{\sqrt{3}}/_{2}$ and $\sin{}^{\gamma}/_{4} = {}^{1}/_{2}$.

The starting point for determining the conversion modules for the phase currents is Eq. (4.17), which upon substitution of Eq. (4.9) may be written as

$$\vec{i}_{RST} = C\left((i_{D1} - i_{D2}) + (i_{D2} - i_{D3})\,e^{j\gamma} + (i_{D3} - i_{D1})\,e^{j2\gamma}\right) \tag{4.52}$$

Equation (4.52) can be rearranged by grouping the phase variables as shown in Eq. (4.53)

$$\vec{i}_{RST} = C\left(i_{D1}\underbrace{\left(1 - e^{j2\gamma}\right)}_{\sqrt{3}\,e^{j\frac{\gamma}{4}}} + i_{D2}\underbrace{\left(-1 + e^{j\gamma}\right)}_{\sqrt{3}\,e^{j\left(\frac{\gamma}{4}+\gamma\right)}} + i_{D3}\underbrace{\left(-e^{j\gamma} + e^{j2\gamma}\right)}_{\sqrt{3}\,e^{j\left(\frac{\gamma}{4}+2\gamma\right)}}\right) \tag{4.53}$$

The braced terms contain a common term $\sqrt{3}\,e^{j\frac{\gamma}{4}}$, which allows Eq. (4.53) to be written as

$$\vec{i}_{RST} = \sqrt{3}\,e^{j\frac{\gamma}{4}}\underbrace{C\left(i_{D1} + i_{D2}\,e^{j\gamma} + i_{D3}\,e^{j2\gamma}\right)}_{\vec{i}_{D123}} \tag{4.54}$$

Equation (4.54) is significant because it tells us that the current space vector $\vec{i}_{RST} = i_\alpha + j i_\beta$ can be found by converting the three-phase currents to a phase vector $\vec{i}_{D123} = i_{D\alpha} + j i_{D\beta}$, which needs to be rotated by an angle ${}^{\gamma}/_{4}$ and scaled by a factor $\sqrt{3}$. The required three- to two-phase conversion is carried out with the conversion matrix according to Eq. (4.32). In the generic representation, as given by Fig. 4.23a, the conversion steps as defined by Eq. (4.54) are clearly visible.

The second module, shown with a "delta" and "star" symbol, symbolizes the conversion $\vec{i}_{D123} \rightarrow \vec{i}_{RST}$ which takes place when a delta connected circuit (page 89) is used. The transfer matrix for this conversion is given by Eq. (4.55). Note that any zero sequence current component will not appear in these transformations.

$$\begin{bmatrix} i_\alpha \\ i_\beta \end{bmatrix} = \sqrt{3}\begin{bmatrix} \cos\frac{\gamma}{4} & -\sin\frac{\gamma}{4} \\ \sin\frac{\gamma}{4} & \cos\frac{\gamma}{4} \end{bmatrix}\begin{bmatrix} i_{D\alpha} \\ i_{D\beta} \end{bmatrix} \tag{4.55}$$

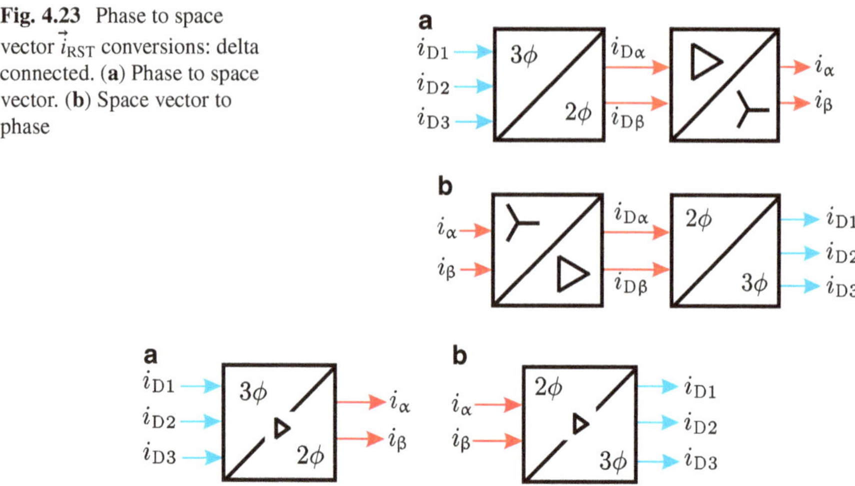

Fig. 4.23 Phase to space vector $\vec{i}_{RST}$ conversions: delta connected. (**a**) Phase to space vector. (**b**) Space vector to phase

Fig. 4.24 Alternative conversions phase current to space vector $\vec{i}_{RST}$: delta connected. (**a**) Phase to space vector. (**b**) Space vector to phase

An alternative conversion from phasor to space vector format can be made by combining the transfer matrices of the modules shown in Fig. 4.23a. The resultant transfer matrix found using Eqs. (4.32) and (4.55) is given by Eq. (4.56).

$$\begin{bmatrix} i_\alpha \\ i_\beta \end{bmatrix} = \begin{bmatrix} \frac{3C}{2} & -\frac{3C}{2} & 0 \\ \frac{\sqrt{3}C}{2} & \frac{\sqrt{3}C}{2} & -\sqrt{3}C \end{bmatrix} \begin{bmatrix} i_{D1} \\ i_{D2} \\ i_{D3} \end{bmatrix} \tag{4.56}$$

The corresponding generic diagram for this module is shown in Fig. 4.24a.

An example of the practical use of conversion modules is given in Fig. 4.25. This figure shows three measured currents i_{D1}, i_{D2}, and i_{D3}, respectively. An amplitude invariant ($C = 2/3$) conversion to space vector format (see Fig. 4.26) was made for the phase currents in this delta connected circuit. Hence, the model according to Fig. 4.24a was used. A zero sequence current i_{D0} is also shown, calculated using Eq. (4.8), also with $C = 2/3$.

The inverse operation, namely, the conversion process from phase space vector variables to supply current variables follows directly from Eq. (4.54). This expression may also be written as

$$\vec{i}_{D123} = \frac{1}{\sqrt{3}}\, e^{-j\frac{\gamma}{4}}\, \vec{i}_{RST} \tag{4.57}$$

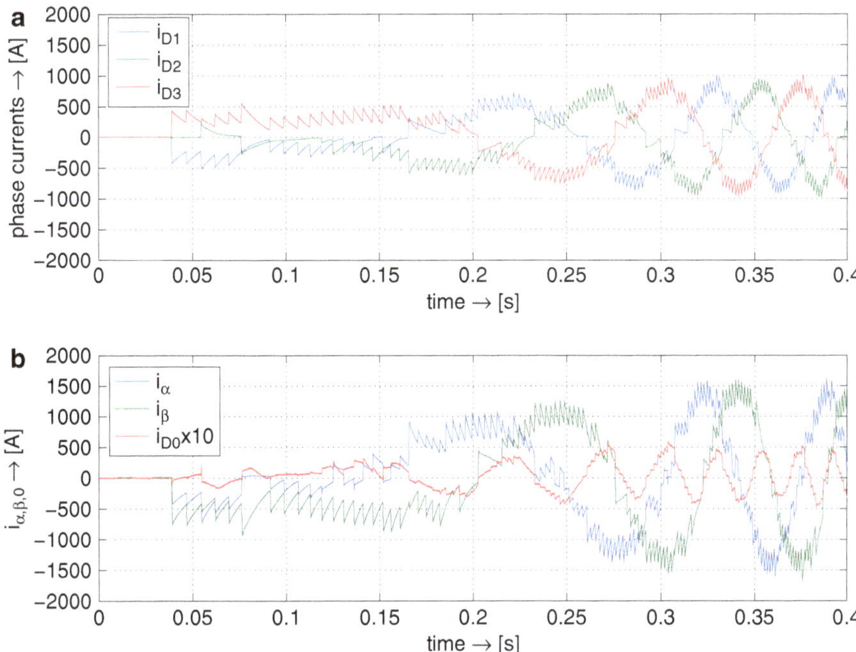

Fig. 4.25 Measured data from a linear synchronous machine (LSM) rated 2.4 MW in a roller-coaster launch application: conversion from phase currents to space vector format $\vec{i}_{\alpha,\beta,0}$ showing substantial third harmonic loop current (Courtesy of Electroproject, the Netherlands)

Equation (4.57) states that the space vector $\vec{i}_{\mathrm{RST}}$ must be rotated by an angle $-\gamma/4$ and scaled by a factor $1/\sqrt{3}$ in order to arrive at the phase vector $\vec{i}_{\mathrm{D123}}$, which can be converted to phase current variables using Eq. (4.38). The generic modules required for this conversion are shown in Fig. 4.23b. Included in this figure is a conversion module identified by the symbols *delta* and *star* which has a transfer matrix of the form given by Eq. (4.58).

$$\begin{bmatrix} i_{\mathrm{D}\alpha} \\ i_{\mathrm{D}\beta} \end{bmatrix} = \frac{1}{\sqrt{3}} \begin{bmatrix} \cos\frac{\gamma}{4} & \sin\frac{\gamma}{4} \\ -\sin\frac{\gamma}{4} & \cos\frac{\gamma}{4} \end{bmatrix} \begin{bmatrix} i_\alpha \\ i_\beta \end{bmatrix} \tag{4.58}$$

The two conversion modules shown in Fig. 4.23b can also be replaced by a single generic module as given in Fig. 4.24b. The corresponding transfer matrix as given by Eq. (4.59) is found by combining the matrices represented by Eqs. (4.38) and (4.58). It is emphasized that this conversion will not contain any zero sequence component, as it was not included.

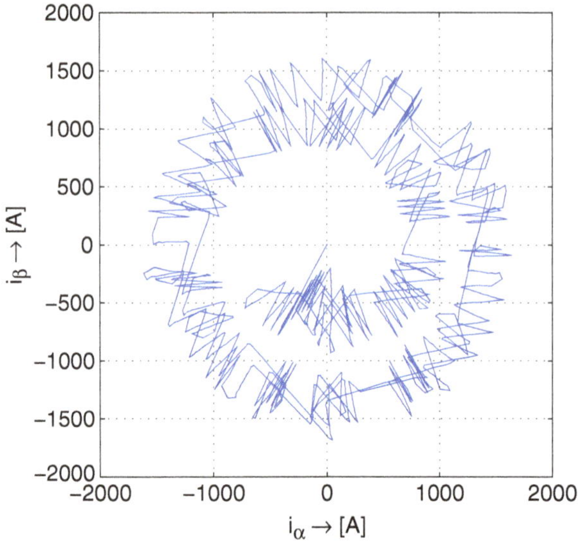

Fig. 4.26 Measured data: current vector locus diagram $(i_\alpha(t), i_\beta(t))$, the same data as in Fig. 4.25

$$\begin{bmatrix} i_{D1} \\ i_{D2} \\ i_{D3} \end{bmatrix} = \begin{bmatrix} \frac{1}{3C} & \frac{1}{C3\sqrt{3}} \\ -\frac{1}{3C} & \frac{1}{C3\sqrt{3}} \\ 0 & -\frac{2\sqrt{3}}{9C} \end{bmatrix} \begin{bmatrix} i_\alpha \\ i_\beta \end{bmatrix} \tag{4.59}$$

4.6.4 Circuit Modeling Using Space Vectors: Delta Connection

The space vector based circuit model, as given in Fig. 4.19, for the star connected three-phase system is directly applicable for the delta connected circuit. The only change lies with the use of *delta* input and output variables as shown in Fig. 4.27. The input variables of the three- to two-phase conversion unit are the phase voltages u_{D1}, u_{D2}, and u_{D3}. The output of the two- to three-phase conversion unit are the phase currents i_{D1}, i_{D2}, and i_{D3}. The model according to Fig. 4.27 replaces the component module shown in Fig. 4.13 which has as input the phase voltages u_{D1}, u_{D2}, and u_{D3} and as output the phase currents i_{D1}, i_{D2}, and i_{D3}.

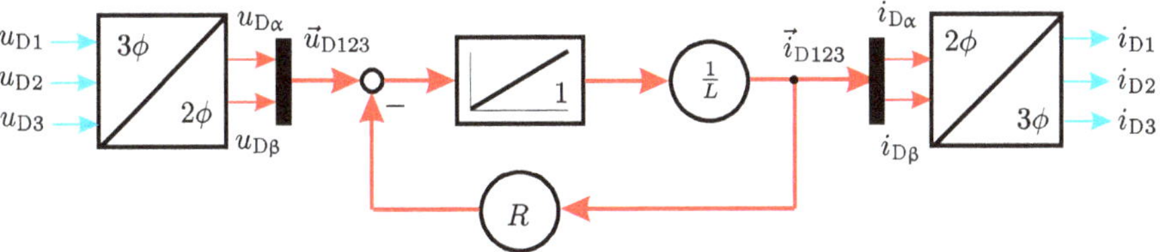

Fig. 4.27 Generic, space vector based, model of three-phase R-L circuit (delta connected)

4.7 Relationship Between Space Vectors and Phasors

The phasor in the form of a complex time independent variable has in an earlier chapter been introduced in the form $\underline{x}$. Basically, the phasor type analysis was and is used to study linear electrical circuits under steady-state AC conditions, i.e., carrying sinusoidally varying waveforms. Up till now, we have considered phasors for single phase applications. In this section, we will extend the use of phasors for three-phase systems.

In this context it is helpful to recall the three-phase example given in Sect. 4.5 where a three-phase sinusoidal supply voltage was assumed [see Eq. (4.19)], together with a set of sinusoidal supply currents. The three-phase variables were then transformed to a space vector form as given by Eq. (4.25). This equation set forms an excellent platform for establishing the link to phasors. In this context, it is helpful to rewrite Eq. (4.25) in the following form

$$\vec{u}_{\mathrm{RST}} = \underbrace{\hat{u}}_{\underline{u}_{\mathrm{RST}}} \mathrm{e}^{\mathrm{j}\omega t} \tag{4.60a}$$

$$\vec{i}_{\mathrm{RST}} = \underbrace{\hat{i}\,\mathrm{e}^{\mathrm{j}\rho}}_{\underline{i}_{\mathrm{RST}}} \mathrm{e}^{\mathrm{j}\omega t} \tag{4.60b}$$

An observation of Eq. (4.60) shows that the time dependent component of the space vectors has been written separately from the remaining term, which is precisely the phasor component of the space vector. Hence, for the example shown in Eq. (4.60) the voltage/current phasors are given as $\underline{u}_{\mathrm{RST}} = \hat{u}$ and $\underline{i}_{\mathrm{RST}} = \hat{i}\,\mathrm{e}^{\mathrm{j}\rho}$, respectively. Note that we introduced the subscript "RST" for these vectors and phasors to reinforce the fact that these are linked to the three-phase supply variables. The reader is reminded of the fact that the phasor concept was introduced in Chap. 2. Equation (2.11), as given in that chapter, was used to show the relationship that exists between phasors and sinusoidal time dependent waveforms. An observation of Eq. (2.11) shows the use of an amplitude invariant space vector representation. When dealing with phasors in single phase AC circuits, the real (for cosine

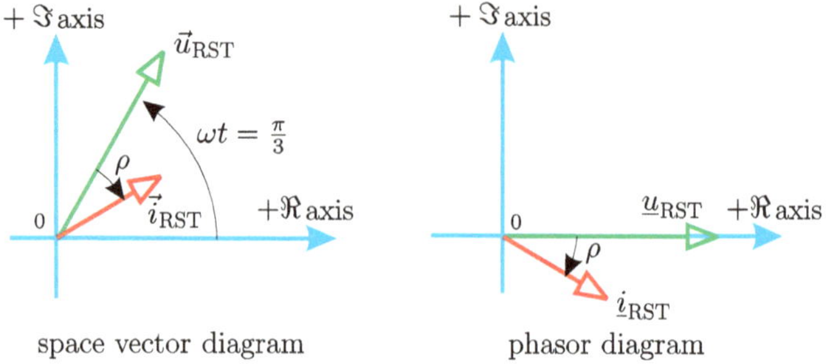

Fig. 4.28 Example showing "supply" space vector and corresponding phasor diagram

functions) or imaginary (for sine functions) space vector component is used in the transformation from time variables to phasors and vice versa.

An example of using space vectors in the form shown by Eq. (4.60) is given in Fig. 4.28 for $\omega t = \pi/3$ and $\rho = -\pi/6$, together with the corresponding phasor diagram.

4.7.1 Use of Phasors in Three-Phase Circuits

It is instructive to consider the transformation of a space vector differential equation to phasor format for the R-L type circuit considered in Sects. 4.6.2 and 4.6.4 for the star and delta connected configuration. The analysis is shown for variables of the star connected systems. However, the analysis for the delta case is identical. The only change lies with the change of subscripts from S123 to D123. The space vector load differential equation for the star connected case is given by Eq. (4.43). Substitution of $\vec{\psi}_{\mathrm{S123}} = L\vec{i}_{\mathrm{S123}}$ gives

$$\vec{u}_{\mathrm{S123}} = L\frac{\mathrm{d}\vec{i}_{\mathrm{S123}}}{\mathrm{d}t} + R\,\vec{i}_{\mathrm{S123}} \tag{4.61}$$

where $\vec{u}_{\mathrm{S123}}$ and $\vec{i}_{\mathrm{S123}}$ represent the voltage/current space vector linked to phase variables of the star configured circuit. The transformation of Eq. (4.61) to its phasor equivalent form is made using

$$\vec{u}_{\mathrm{S123}} = \underline{u}_{\mathrm{S123}}\,\mathrm{e}^{\mathrm{j}\omega t} \tag{4.62a}$$

$$\vec{i}_{\mathrm{S123}} = \underline{i}_{\mathrm{S123}}\,\mathrm{e}^{\mathrm{j}\omega t} \tag{4.62b}$$

$$\frac{\mathrm{d}\vec{i}_{\mathrm{S123}}}{\mathrm{d}t} = \mathrm{j}\omega\,\underline{i}_{\mathrm{S123}}\,\mathrm{e}^{\mathrm{j}\omega t} \tag{4.62c}$$

Equation (4.62) shows the conversion required as well as the current space vector differential, which in phasor terms leads to the addition of component $\mathrm{j}\omega$. The resultant phasor equation is found by eliminating the time dependent term $\mathrm{e}^{\mathrm{j}\omega t}$, which leads to

$$\underline{u}_{\mathrm{S123}} = \mathrm{j}\omega L\, \underline{i}_{\mathrm{S123}} + R\, \underline{i}_{\mathrm{S123}} \tag{4.63}$$

If, for example, the voltage phasor is known, then the current phasor is calculated using

$$\underline{i}_{\mathrm{S123}} = \frac{\underline{u}_{\mathrm{S123}}}{R + \mathrm{j}\omega L} \tag{4.64}$$

The peak phase current amplitude is then found using $\hat{i} = |\underline{i}|$. The phase angle with respect to the phase voltage is equal to $\rho = -\arctan {}^{\omega L}/_{R}$.

It is instructive to examine the process of calculating the supply phasor $\underline{i}_{\mathrm{RST}}$ on the basis of a given supply phasor according to (4.60a) for the star/delta connected circuit. In both cases, the voltage phasor $\underline{u}_{\mathrm{S123}}/\underline{u}_{\mathrm{D123}}$ must first be derived from the given supply phasor $\underline{u}_{\mathrm{RST}}$. Next, the current phasor $\underline{i}_{\mathrm{S123}}/\underline{i}_{\mathrm{D123}}$ needs to be calculated using Eq. (4.64). Finally, the conversion from $\underline{i}_{\mathrm{S123}}/\underline{i}_{\mathrm{D123}}$ to supply current phasor $\underline{i}_{\mathrm{RST}}$ needs to be made.

For the star connected case the relationship between phase and supply vectors (for currents and voltages) is given by Eq. (4.41). Consequently, the relationship between the phasors in a star connected circuit is of the form

$$\underline{u}_{\mathrm{S123}} = \underline{u}_{\mathrm{RST}} \tag{4.65a}$$

$$\underline{i}_{\mathrm{S123}} = \underline{i}_{\mathrm{RST}} \tag{4.65b}$$

This means that the calculation of the current phasor from a given voltage phasor is as discussed above, see Eq. (4.64). A phasor diagram example with $\rho = -{}^{\pi}/_{3}$ is given in Fig. 4.29.

For the delta connected case the relationship between phase and supply vectors is given by Eqs. (4.46) and (4.57). The corresponding phasor relationships between the phase and supply based phasors are of the form

$$\underline{u}_{\mathrm{D123}} = \sqrt{3}\, \mathrm{e}^{-\mathrm{j}\frac{\gamma}{4}}\, \underline{u}_{\mathrm{RST}} \tag{4.66a}$$

$$\underline{i}_{\mathrm{D123}} = \frac{1}{\sqrt{3}}\, \mathrm{e}^{-\mathrm{j}\frac{\gamma}{4}}\, \underline{i}_{\mathrm{RST}} \tag{4.66b}$$

For the calculation of the current phasor $\underline{i}_{\mathrm{D123}}$ we make use of Eq. (4.64) (with subscript D123), in which $\underline{u}_{\mathrm{D123}}$ is calculated using Eq. (4.66a). Once the phasor $\underline{i}_{\mathrm{D123}}$ is found, Eq. (4.66b) can be used to find the supply current phasor. The phasor diagram as given by Fig. 4.29 shows the conversion process with the same circuit model as used for the star connected example.

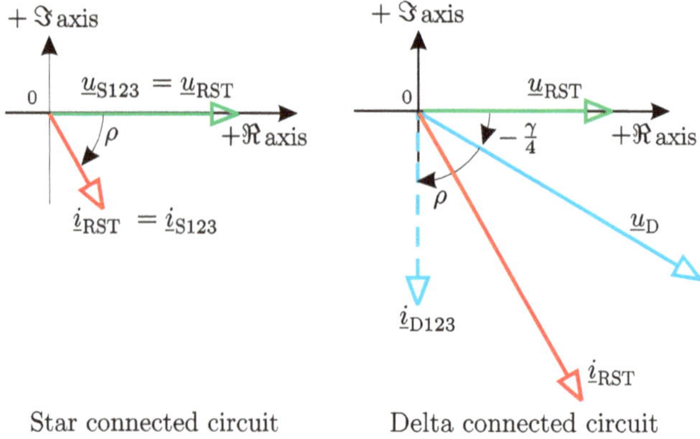

Fig. 4.29 Phasor diagram for star and delta R-L circuit

An important conclusion to make is that the supply current level, which will appear for the delta connected case, is three times larger than that which will appear in the star connected configuration (provided that the impedances remain the same). This is clearly apparent from Fig. 4.29, where the currents and voltage for the delta connected configuration are scaled as required in comparison to the star diagram. It is for this reason that grid connected three-phase induction machines are often star configured during the initial start-up sequence. Once past the start up phase, the machine phase configuration is changed to delta. This approach reduces the initial peak starting currents in the supply lines. However, significant transient peaks can still occur just after the Y-Δ reconfiguration. As explained earlier, delta connected machines are commonplace in applications with high phase currents (low supply voltage, high power). Inverter connected machines are usually star configured to avoid circulating phase currents.

It is noted that the process of modeling a delta connected circuit can be avoided if we simply take the delta circuit parameter values, divide these by a factor of three, and virtually re-configure the circuit in "star." However, we loose access to the "delta phase variables" under these circumstances. In applications where the delta connected voltages/currents are to be measured for control purposes, the conversion process as described above would need to be implemented. In any event, it is considered important to realize that there is a substantial difference between the modeling processes linked to the "star" and "delta" connected circuit.

4.8 Tutorials

4.8.1 Tutorial 1: PLECS Based Model of Conversion Modules for Star Connected Circuits

In this tutorial the conversion process (*supply→phase* and *phase→supply*) for a star connected circuit will be considered. The PLECS model according to Fig. 4.30 forms the basis for this tutorial. Shown in this figure is a `3ph_pulse` module which provides the supply voltages u_R, u_S, and u_T for this simulation. The "run time" for this simulation is set to 20.01 ms, which is slightly higher than the 20 ms of the generated pulse waveforms in order to include the voltage transition at $t = 20$ ms. Details of the waveform versus time functions to be implemented are given in Fig. 4.31. It is noted that the set of waveforms used in this tutorial are in fact those which are typically generated by a power electronic converter when operating in a so-called block or six-step mode. As a second step in this tutorial, provide an implementation of the conversion module "*RST→(star)123*" according to the matrix defined by Eq. (4.4). An example of the output of this conversion is shown in Fig. 4.32, where the supply voltage waveform u_R is also added for reference purposes. Clearly observable from Fig. 4.32 is the presence of a non-zero sequence voltage component u_{S0} as calculated with the aid of Eq. (4.5). The second module to be implemented is the "three to two" phase conversion module identified in Fig. 4.30 as "*123→vector*." The matrix according to Eq. (4.32) must be implemented in this case with $C = {}^2/_3$, given that an amplitude invariant notation is assumed. In PLECS a "standard abc → αβ" is already available which implements said matrix. The only difference is that in PLECS the phase notation abc is used instead of 123 as used in this book. The real and imaginary components of the space vector $\vec{u}_{S123}$ which should appear as output variables of this module are shown in Fig. 4.33. A further cross check on the output of this module is given in Fig. 4.34, which shows the locus of the space vector end point $\vec{u}_{S123}$ versus time for the 20.01 ms duration of

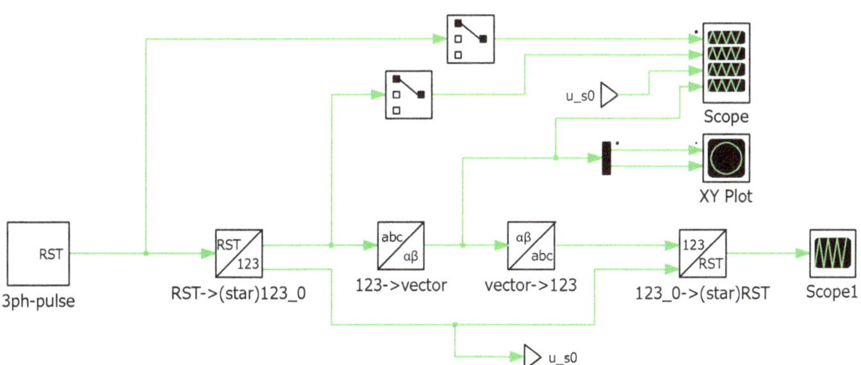

Fig. 4.30 PLECS model: conversion modules for star connected circuits

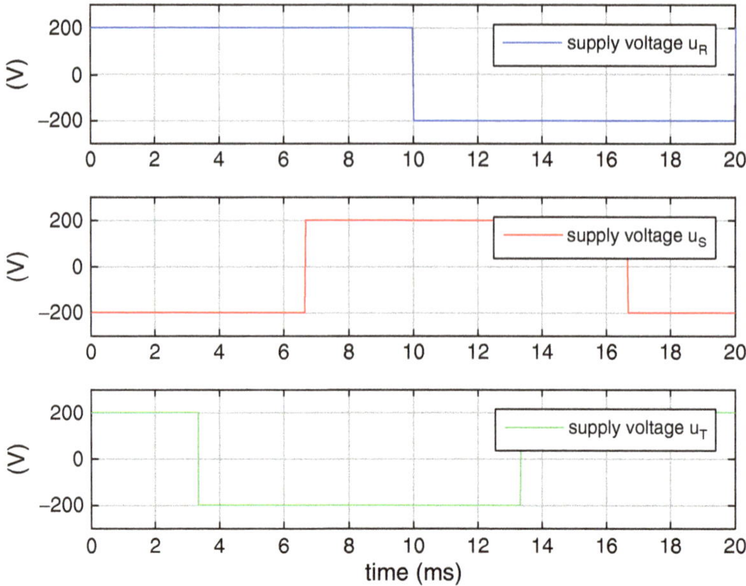

Fig. 4.31 Supply voltage waveforms: u_{R}, u_{S}, and u_{T}

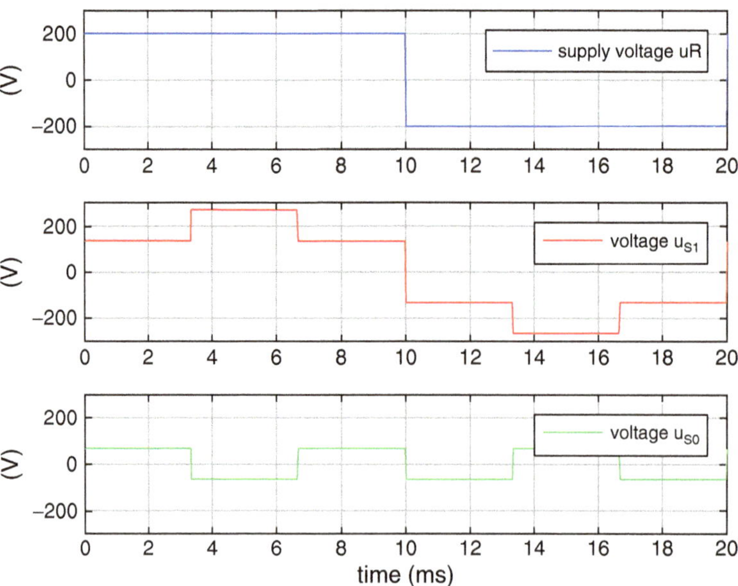

Fig. 4.32 Waveforms: u_{R}, u_{S1}, and u_{S0}

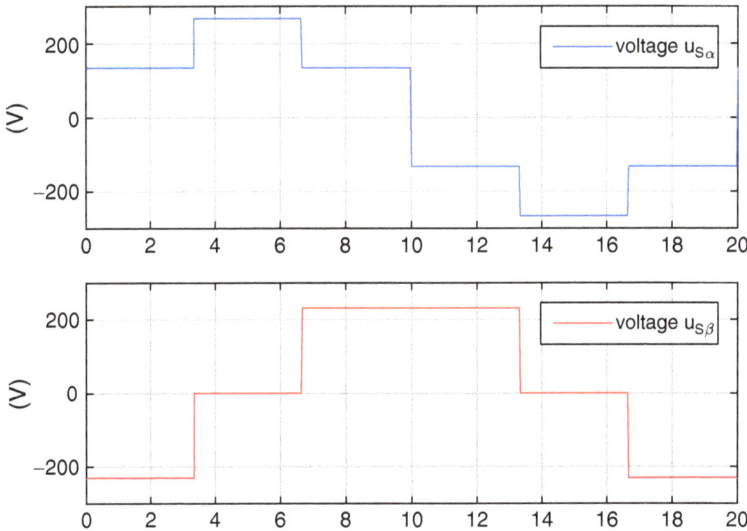

Fig. 4.33 Waveforms: $u_{S\alpha}$ and $u_{S\beta}$

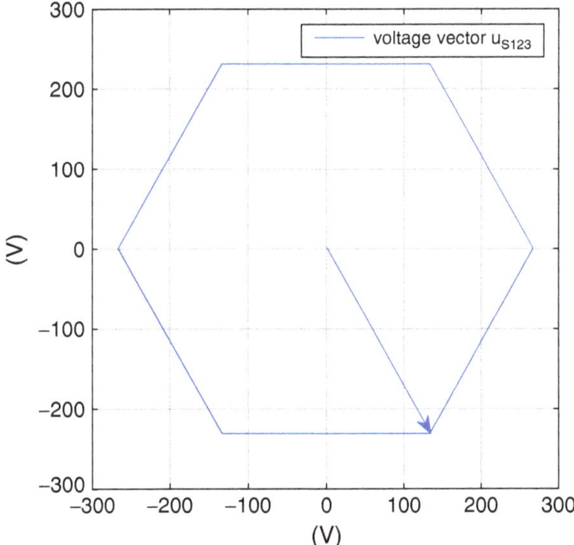

Fig. 4.34 Waveforms: $\vec{u}_{S123}$, vector locus diagram

this simulation generated by the "XY plot" module. Note that the shape is that of a hexagon where the vector end point moves from corner to corner which is why this mode of operation is referred to as "six step." It is instructive to carefully consider how the vector end point moves in Fig. 4.34 in relation to the simulation time. From such an analysis it may be observed that the vector end point changes from hexagon corner to corner whenever a waveform voltage transition in either $u_{S\alpha}(t)$ or $u_{S\beta}(t)$ occurs. Finally, implement the two remaining conversion modules "*vector→123*" and "*123→(star)RST*" according to Eqs. (4.38) and (4.6), respectively. The output waveforms from these modules can be directly checked against those obtained from the output of the "*RST→(star)123*" module and `3ph_pulse` module which provides the supply voltage waveforms.

4.8.2 Tutorial 2: Symbolic Representation for Star Connected Circuits Conversion Modules

This tutorial considers an alternative approach to that shown in the previous case. In this tutorial use is to be made of "electrical" PLECS models, where possible to implement the previous conversion process. The star connected load is to be represented by three 100 Ω resistors and the star connected three-supply as used in the previous example should also be used here.

A possible implementation of said problem as given in Fig. 4.35 shows three star connected voltage sources which in turn are controlled by the `3ph-pulse` unit used in the previous example. The star connected load formed by the resistors `R1`, `R2`, and `R3` has a "star point" which can be connected via a "manual switch" `Switch` to the "star point" of the three-phase supply. A set of voltage measurement modules have been added which measure the phase supply voltage u_R, load phase voltages u_{S1}, u_{S2}, and u_{S3}, and zero sequence voltage u_{S0}. A standard 123 → vector module is used to generate the vector variables $u_{S\alpha}$ and $u_{S\beta}$ from the load phase voltage variables. When the manual switch is open the results on the "Scope" and "XY plot" modules will be identical to those shown in Figs. 4.32, 4.33, and 4.34. If, however, the manual switch is closed prior to running the simulation, the load "star point" is then connected to the supply "star point" in which case the load phase voltage waveforms will be equal to the supply waveforms given in Fig. 4.31. The vector representation according to Figs. 4.33 and 4.34 is NOT affected by the switch change as the presence of a zero sequence voltage component u_{S0} does not affect this transformation. Note that opening the manual switch in this example is equivalent to removing the u_{S0} link in Fig. 4.30 between conversion modules "*RST→(star)123*" and "*123→(star)RST*."

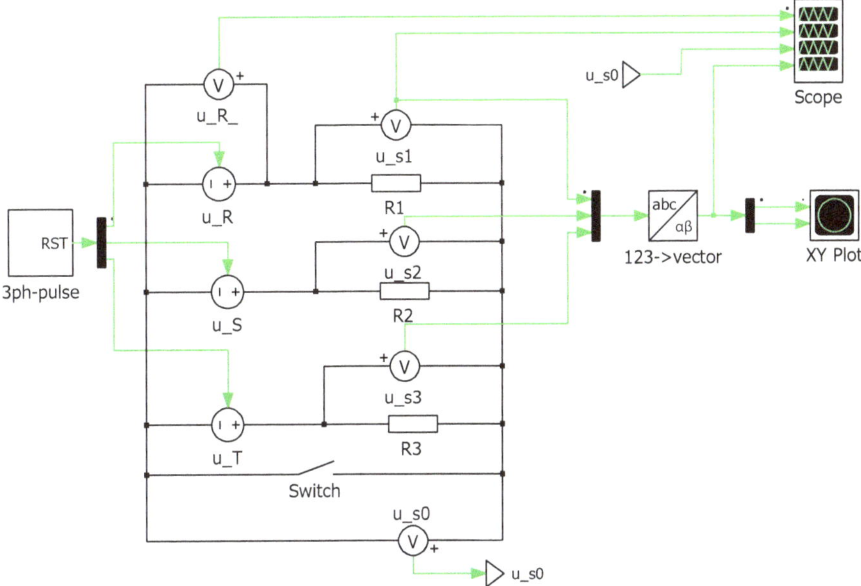

Fig. 4.35 PLECS model: conversion modules for star connected circuits using an electrical (symbolic) representation

4.8.3 Tutorial 3: PLECS Based Model of Star Connected Circuit Example

An implementation of the generic diagram shown in Fig. 4.19 is considered in this tutorial. The load is represented in this figure by the module "R-L-e-3ph." This module will be used more extensively in Sect. 4.8.8. In this tutorial this model is simplified by setting the resistance value to zero and the inductance value to $L = 100\,\mathrm{mH}$. Furthermore, the "sinus" modules shown in this figure are not implemented. The supply waveforms as used in tutorial 1 (see Sect. 4.8.1) remain unchanged. Add a "three to three" phase conversion module as shown in Fig. 4.6a to convert the supply voltages to phase voltages. Build this simulation and examine the current waveforms $i_{S\alpha}$, $i_{S\beta}$, and i_{S1}. Repeat this exercise after removing the "three-phase to three-phase" module used to convert the supply voltages to phase voltages. Explain if there is likely to be a difference in the current output waveforms. Note that for the star connected circuit the phase currents are equal to the supply currents. An example of the current waveforms, which should appear in your simulation, is given in Fig. 4.36. Shown are the real and imaginary current space vector components together with the phase/supply current i_{S1}. The phase voltage u_{S1} is also added for reference purposes. Removal of the module "*RST→(star)123*" will not change the current waveforms despite the presence of a zero sequence

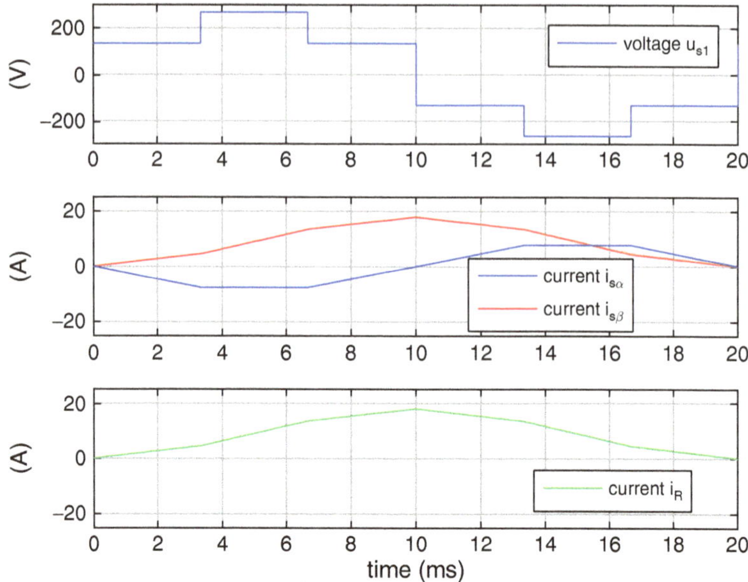

Fig. 4.36 Waveforms: u_{S1}, $(i_{S\alpha}, i_{S\beta})$, and i_{S1}/i_R

component u_{S0} in the supply voltages. The reason for this is that the space vector components $u_{S\alpha}$ and $u_{S\beta}$ are not affected by the presence of a component u_{S0} [see Eq. (4.40)].

4.8.4 Tutorial 4: PLECS Based Symbolic Model of Star Connected Circuit Example

This tutorial considers an alternative approach to that shown in the previous case. In this tutorial use is to be made of "electrical" PLECS models, where possible to implement the previous conversion process. The star connected load is represented by three 100 mH inductors and the star connected three-supply as used in the previous examples should also be used here.

A possible implementation of said problem as given in Fig. 4.37 shows three star connected voltage sources which in turn are controlled by the `3ph-pulse` unit used in the previous examples. The star connected load formed by the inductances `L1`, `L2`, and `L3` has a "star point" which is NOT connected to the "star point" of the three-phase supply (as is usually the case for drive applications). A set of voltage measurement modules have been added which measure the phase supply voltage u_R and load phase voltage u_{S1}. In addition a set of current sensors have been added, which measure the phase currents i_{S1}, i_{S2}, and i_{S3}. A standard 123 → vector

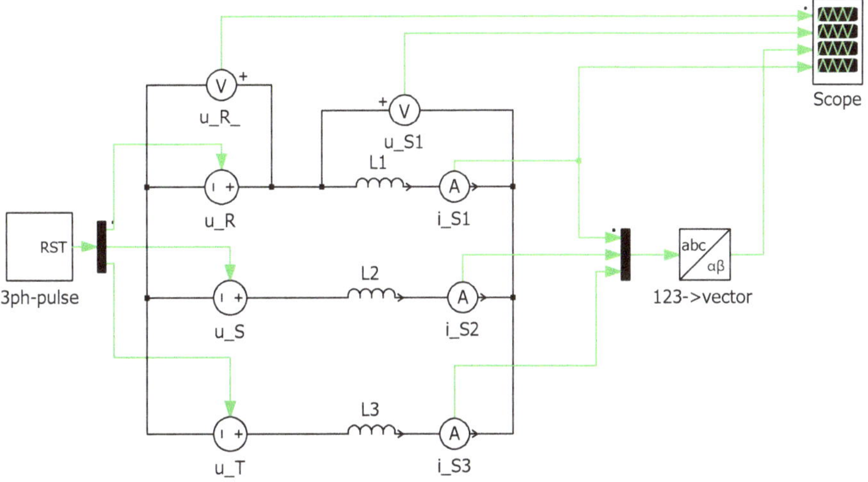

Fig. 4.37 PLECS symbolic model: star connected circuit

module is use to generate the vector variables $i_{S\alpha}$ and $i_{S\beta}$ from the load phase current variables. When running this simulation the results as shown in Fig. 4.36 should appear.

4.8.5 *Tutorial 5: PLECS Based Model of Conversion Modules for Delta Connected Circuits*

Tutorial 1 as discussed in Sect. 4.8.1 for the "star" configured three-phase circuit is repeated here for the "delta" connected case, i.e., load connected in "delta," supply in "star" configuration. This means that the module "*RST→(star)123*" (see Fig. 4.30) must be replaced by a new module "*URST→(delta)123*" as shown in Fig. 4.38, which has a conversion matrix according to Eq. (4.13). Note that the conversion matrices for supply to phase variables and vice versa are different for voltages and currents. An example of the phase voltage u_{D1}, which should appear from this module, is given in Fig. 4.39. The waveforms u_R and u_{S0} which also appear in this figure remain unchanged. The module "*123→vector*" is unchanged but the output of this module will be different given the new phase voltage inputs. The space vector $\vec{u}_{D123}$, formed by the variables $u_{D\alpha}$ and $u_{D\beta}$, should be of the form given by Fig. 4.40. A further cross check on the output of this module can also be made for this case. Figure 4.41 shows the locus of the space vector end point $\vec{u}_{D123}$ versus time for the delta connected case. Note that the shape is again that of a hexagon. However, the hexagon is now rotated clockwise by an angle of $-\pi/6$. Furthermore, the hexagon is larger by a factor of $\sqrt{3}$ when compared to the previous example (see Fig. 4.34). The conversion from space vector to phase variables remains unchanged.

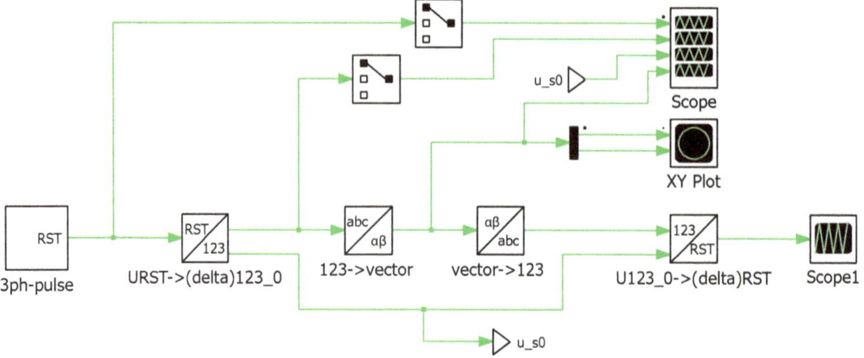

Fig. 4.38 PLECS model: conversion modules for delta connected circuit

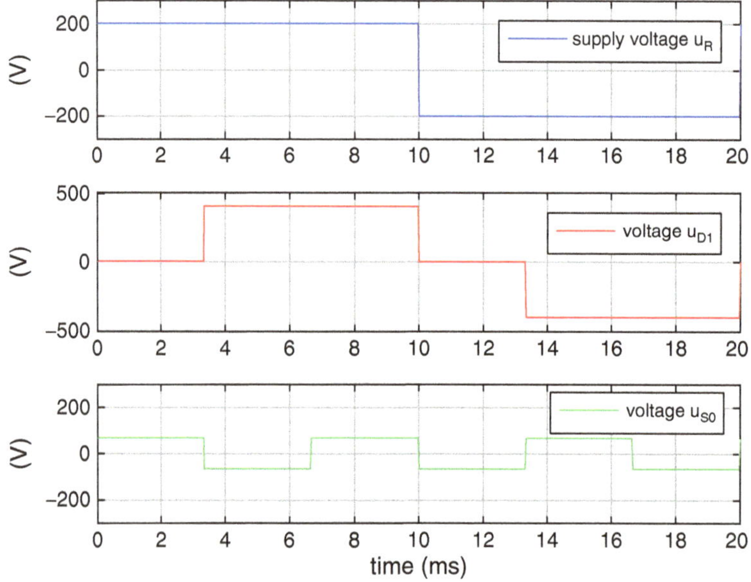

Fig. 4.39 Waveforms: u_R, u_{D1}, and u_{S0}

However, the voltage module "*123→(star)RST*" must be replaced by a new module "*U123→(delta)RST*" which has a conversion matrix given by Eq. (4.16). The output waveforms from this module should be the supply voltage waveforms according to Fig. 4.31.

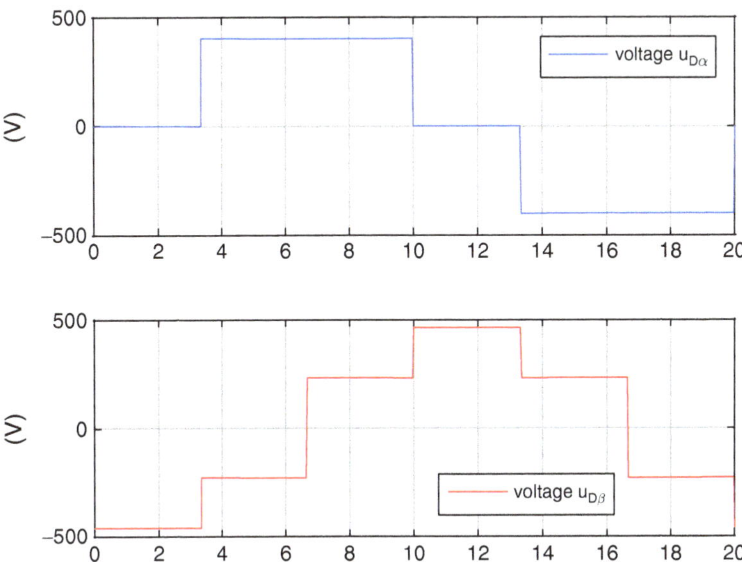

Fig. 4.40 Waveforms: $u_{D\alpha}$ and $u_{D\beta}$

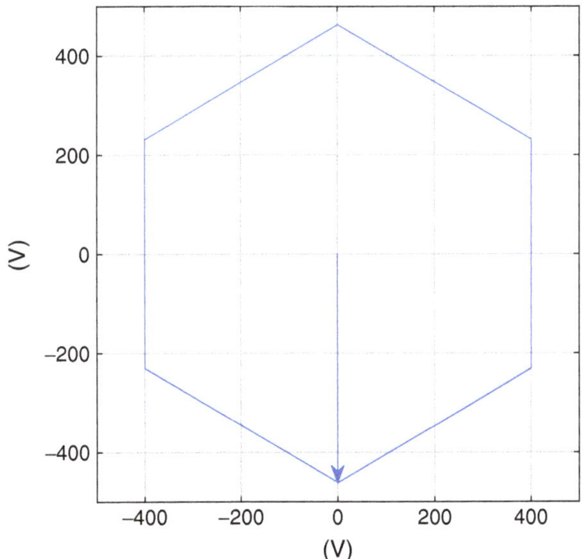

Fig. 4.41 Waveforms: $u_{D\alpha}$ and $u_{D\beta}$, vector locus diagram

4.8.6 Tutorial 6: Symbolic Representation for Delta Connected Circuits Conversion Modules

This tutorial considers a symbolic implementation of the previous tutorial. The excitation and parameters are identical to those used in the previous examples. Furthermore three single phase pulse generator modules are used to generate the three supply voltages u_R, u_S, and u_T as shown in Fig. 4.32. The aim is to reproduce the same waveforms shown in the previous example, with exception being the zero sequence voltage waveform u_{S0}, given that there is no load "star point" present in this example hence access to this variable is not directly possible. A set of 100 Ω resistors may be used to represent the delta connected load. A possible implementation example of this problem as given in Fig. 4.42 shows the star connected three-phase supply together with a delta connected load which is represented by the resistors `R1`, `R2`, and `R3`. A set of voltage measurement modules are used to obtain the supply phase voltage u_R and phase load voltages u_{D1}, u_{D2}, and u_{D3}. The latter are connected to a PLECS "standard" 123 → vector module, which generates the real and imaginary variables $u_{D\alpha}$ and $u_{D\beta}$. The results shown on the "Scope" and "XY plot" are identical to those shown in Figs. 4.39 (waveform u_{S0} not shown), 4.40, and 4.41.

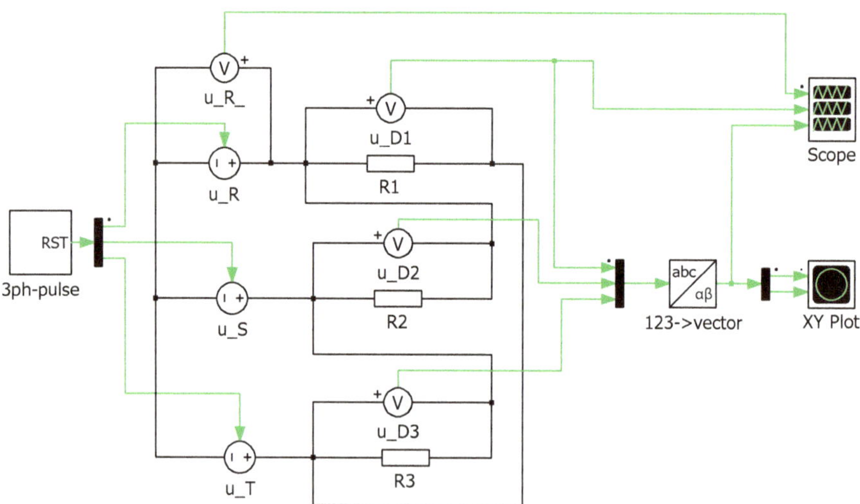

Fig. 4.42 PLECS symbolic model: delta connected circuit

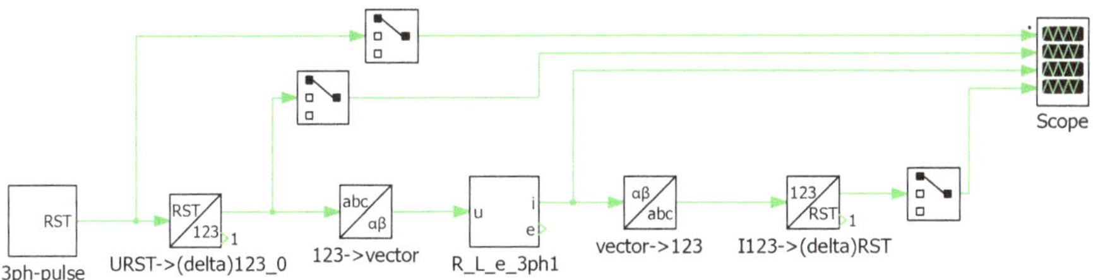

Fig. 4.43 PLECS model: delta connected circuit

4.8.7 Tutorial 7: PLECS Based Model of a Delta Connected Circuit Example

This exercise is concerned with the implementation of the generic diagram according to Fig. 4.27. The delta connected load is in the form of an ideal inductance value with value $L = 100\,\text{mH}$, as discussed in Sect. 4.8.3. The revised circuit model as given by Fig. 4.43 shows the three-phase pulse generator module `3ph_pulse` and conversion modules (as discussed in the previous tutorials) needed to arrive at the voltage space vector $\vec{u}_{\text{D123}}$ for the delta connected case. The output from the "R-L-e-3ph" module is the current vector $\vec{i}_{\text{D123}} = i_{\text{D}\alpha} + \text{j}\, i_{\text{D}\beta}$ which must be converted to three-phase currents using a PLECS standard vector to three-phase conversion module. This module is unchanged, but the phase current to RST current module "*I123→(delta)RST*" as shown in Fig. 4.43 must be build with a conversion matrix as defined by Eq. (4.9). An example of the current waveforms, which should appear (with a zero resistance coil) in your simulation, is given in Fig. 4.44. Shown are the real and imaginary current space vector components $i_{\text{D}\alpha}$ and $i_{\text{D}\beta}$ together with the supply current i_{R}. The phase voltage u_{D1} is also added for reference purposes. When comparing the supply current i_{R} waveform according to Fig. 4.44 against the result obtained with the star connected circuit (see Fig. 4.36) we see that the latter is three times smaller (as expected). The waveform shape remains unchanged. It is again noted that the process of modeling a delta connected circuit could be avoided if we simply take the delta circuit parameter values, divide these by a factor of three, and re-configure the circuit in "star."

It is left to the reader to redo this tutorial example using an electrical circuit model approach.

4.8.8 Tutorial 8: PLECS Based Model of a Star Connected Circuit Example with Sinusoidal Excitation

This tutorial is concerned with a phasor analysis of a star configured three-phase circuit connected to a sinusoidal supply. The simulation model as illustrated by

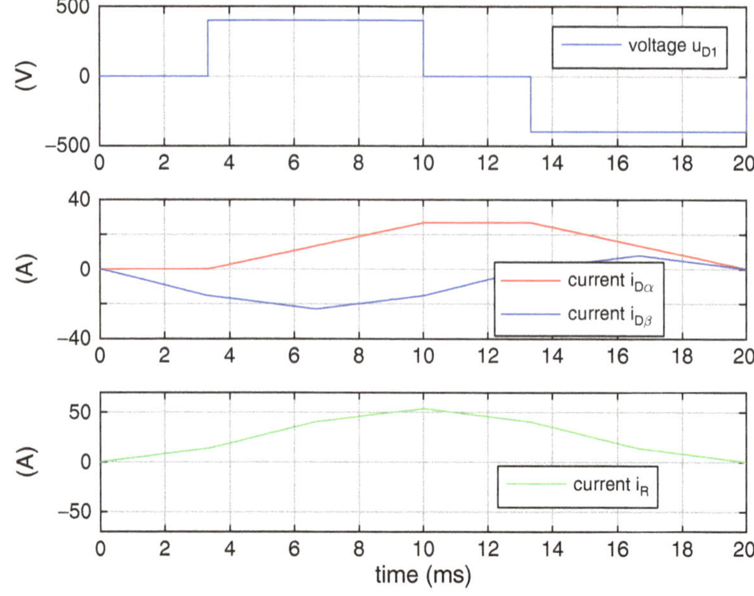

Fig. 4.44 Waveforms: u_{D1}, $(i_{D\alpha}, i_{D\beta})$, and i_R

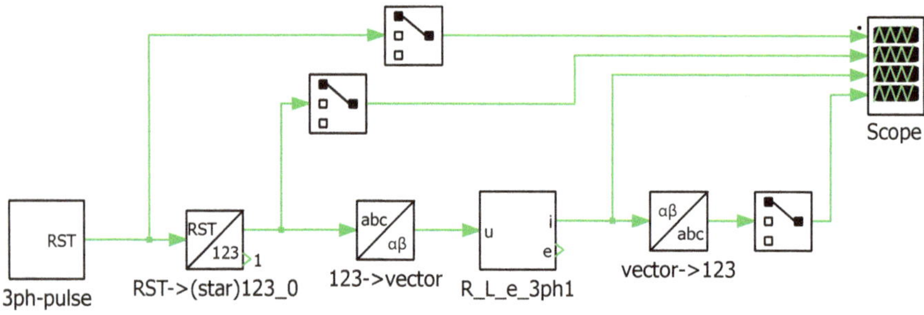

Fig. 4.45 PLECS model: star connected circuit

Fig. 4.45 needs to be modified by replacing the `3ph_pulse` module with a three-phase sinusoidal excitation module `3ph_Sine` as shown in Fig. 4.46.

The three-phase supply unit needs to be built and an implementation example of this module is given in Fig. 4.47a. The output waveforms should be according to Eq. (4.19). The sub-module shown is made with three "sine" modules which represent the "RST" supply voltages. The sub-module should have the numerical inputs as given by Table 4.1.

Note that the "phase" entry is set to $\pi/2$ given that our supply voltages are chosen as cosine functions. If we set the ρ parameter to zero, the output will be a sine function, i.e., at $t = 0$ the value of u_R will be zero. In our case its value should be $U_R\sqrt{2}$.

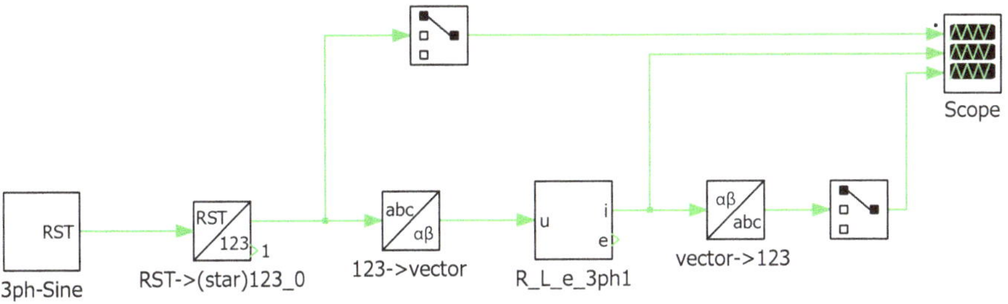

Fig. 4.46 PLECS model: star connected circuit, sinusoidal excitation

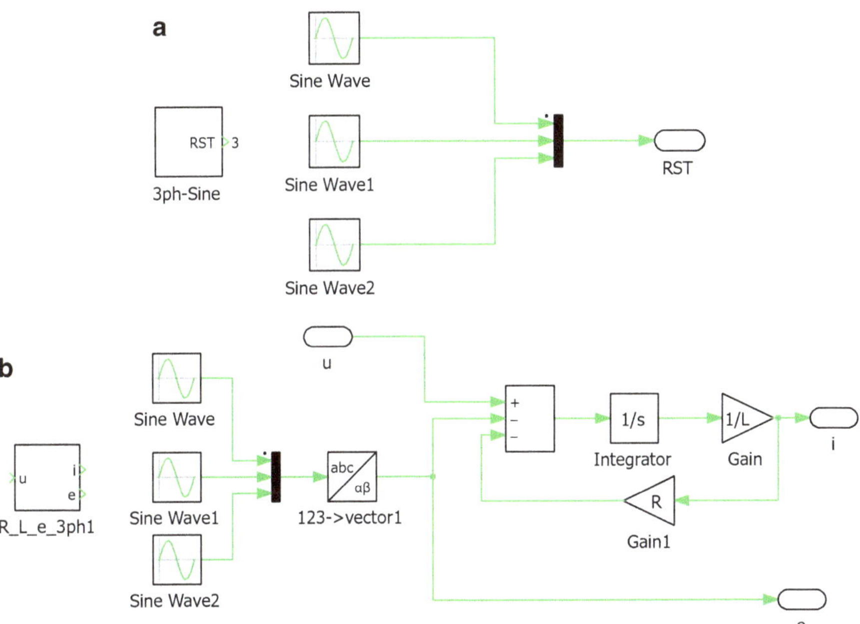

Fig. 4.47 PLECS: supply and branch circuit modules. (**a**) Three-phase supply. (**b**) Branch circuit

Table 4.1 Parameters for three-phase supply unit

Parameters		Value
RMS supply voltage	U_R	220 V
RMS supply voltage	U_S	220 V
RMS supply voltage	U_T	220 V
Supply frequency	f	50 Hz
Phase	ρ	$\pi/2$ rad

The branch circuit model as shown in Fig. 4.47b (in space vector form) is extended to include an “EMF” three-phase voltage source defined according to Eq. (4.67).

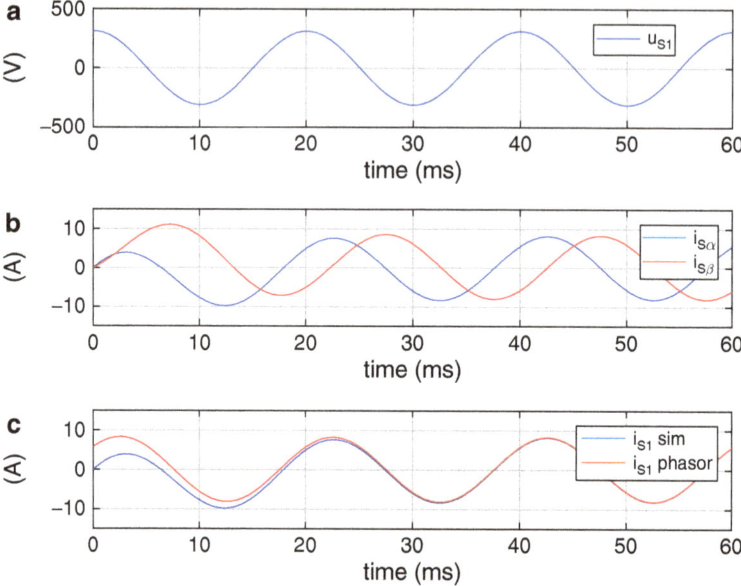

Fig. 4.48 Waveforms: u_{S1}, $(i_{S\alpha}, i_{S\beta})$, and i_{S1}/i_R

$$e_1 = E\sqrt{2}\cos(\omega t + \rho_e) \tag{4.67a}$$

$$e_2 = E\sqrt{2}\cos(\omega t + \rho_e - \gamma) \tag{4.67b}$$

$$e_3 = E\sqrt{2}\cos(\omega t + \rho_e - 2\gamma) \tag{4.67c}$$

where the parameters E and ρ_e are set to 100 V and $-\pi/3$ rad, respectively. Note that this equation is again built with "sine" modules, which means that an additional phase shift of $\pi/2$ must be added, given that we have chosen cosine functions in this example. Furthermore, the supply angular frequency is set to $\omega = 2\pi f$, with $f = 50$ Hz. In addition, a "three to two-phase" conversion module as discussed in Sect. 4.8.1 must be added in order to arrive at the vector $\vec{e}_{123}$. The remaining parameters for this module are taken to be $R = 10\,\Omega$ and $L = 100$ mH. The results of the simulation, which should be run over a period of 20 ms, are shown on the `Scope` module and subsequently exported to MATLAB via a `.csv` file. An example of the results obtained is given in Fig. 4.48. Shown in Fig. 4.48 are the following waveforms: phase voltage u_{S1} (which in this case is also the supply voltage u_R), current vector components $i_{S\alpha}$ and $i_{S\beta}$, and phase/supply current i_{S1} (which is also i_R). The phase/supply current obtained with the simulation "i_{S1} *sim*" is shown together with the "steady-state" phase/supply current "i_{S1} *phasor*," as obtained via a phasor analysis of the problem at hand. An observation of Fig. 4.48 shows that the two waveforms converge towards the end of the simulation. This implies that the "transient" components present in our dynamic simulation become less significant after approximately one cycle of operation.

On the basis of the approach outlined in Sect. 4.7, build an M-file which allows you to calculate the steady-state current waveform i_{S1}, as shown in Fig. 4.48 with the notation "i_{S1} *phasor*." An example of this phasor calculation is shown in the M-file below.

M-file Code

```
%tutorial 8, chapter 4
close all
%phasor analysis
C=2/3;                              %space vector amplitude invariant
U=220;                              %supply  phase voltage RMS
E=100;                              %EMF  phase voltage RMS
rho_e=-pi/3;                        %EMF phase angle
u_123=3/2*C*U*sqrt(2);              %supply phasor
e_p=3/2*C*E*sqrt(2);                %emf peak voltage
e_123=e_p*cos(rho_e)+j*e_p*sin(rho_e);%EMF phasor
R=10;                               %phase resistance
L=100e-3;                           %phase inductance(H)
w=100*pi;                           %supply frequency (rad/s)
X=w*L;                              %load reactance
i_123=(u_123-e_123)/(R+j*X);        %current phasor
ip=abs(i_123);                      %peak value phasor
rho_i=angle(i_123);                 %phase angle current phasor
i1p=2/(3*C)*ip;                     %phase/supply current amplitude
%plot data
dat = csvread('tut8ch4_data.csv',1,0) %read in data from PLECS
subplot(3,1,1)
plot(dat(:,1)*1e3,dat(:,2))         %load  phase voltage
grid on
legend('u_{S1}');
xlabel('(a) time (ms)')
ylabel('(V)')
subplot(3,1,2)
plot(dat(:,1)*1e3,dat(:,3))         %real current
 hold on
plot(dat(:,1)*1e3,dat(:,4))         %imaginary current
grid
legend('i_{S\alpha}','i_{S\beta}');
xlabel('(b) time (ms)')
ylabel('(A)')
ylim([-15 15])
%%%%%%plot phase current i1 in sub-plot with waveform from simulation
subplot(3,1,3)
plot(dat(:,1)*1e3,dat(:,5))
hold on
t=[0:.1e-3:60e-3];
i1vt=i1p*cos(w*t+rho_i);            %phase current versus time
plot(t*1e3,i1vt)
grid
legend('i_{S1} sim','i_{S1} phasor')
xlabel('(c) time (ms)')
ylabel('(A)')
ylim([-15 15])
```

Chapter 5
Concept of Real and Reactive Power

5.1 Introduction

In this chapter the meaning of "real" and "reactive" power is explored for sinusoidal systems. Initially, single phase (so-called two-wire) circuits are discussed to gain an understanding of the energy flow within a circuit configuration that is representative for electrical machines. We will then extend this analysis to three-phase (three-wire) circuits. In the final part of this chapter a set of tutorials is introduced to reinforce the concepts discussed.

5.2 Power in Single Phase Systems

The concept of power is introduced with the aid of Fig. 5.1, which shows a load in the form of an inductor L, resistance R, and voltage source u_e in series connection. The voltage source u_e is generally known as the induced voltage or back-emf. Hence, the circuit configuration as described above is representative for electrical machines. A current source $i(t)$ is connected to this network. The reason for using a sinusoidal supply current source instead of a supply voltage source is to simplify the mathematical analysis. Application of Kirchhoff's voltage laws to this circuit shows that the voltage across the current source can be written as

$$u = u_R + u_L + u_e \tag{5.1}$$

where $u_R, u_L, and u_e$ represent the instantaneous voltages across the resistance, inductor, and voltage source u_e, respectively. If we multiply Eq. (5.1) with the

---

Electronic supplementary material The online version of this chapter (doi: 10.1007/978-3-319-29409-4_5) contains supplementary material, which is available to authorized users.

© Springer International Publishing Switzerland 2016

A. Veltman et al., *Fundamentals of Electrical Drives*, Power Systems,
DOI 10.1007/978-3-319-29409-4_5

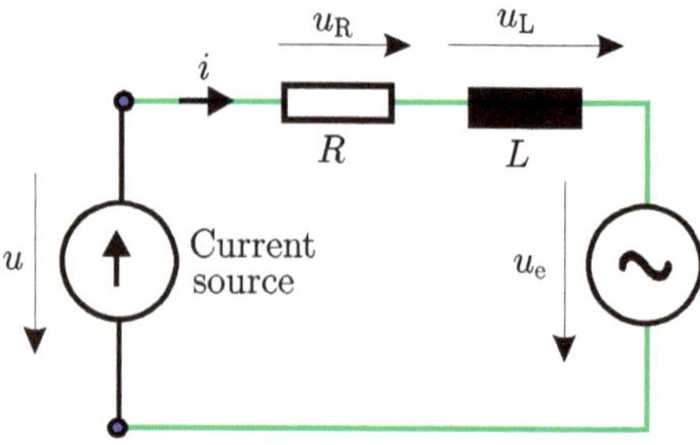

Fig. 5.1 R-L-u_e load connected to current source

instantaneous current produced by the current source, the so-called power balance equation appears as shown in Eq. (5.2).

$$\underbrace{u\,i}_{p_{in}} = \underbrace{i\,u_R}_{p_R} + \underbrace{i\,u_L}_{p_L} + \underbrace{i\,u_e}_{p_e} \tag{5.2}$$

The instantaneous power, which is a physical quantity, is simply the product of the instantaneous voltage and instantaneous current. Note that the voltage and current variables of the current source are shown in the so-called generator arrow system, while the voltage and current variables of the R-L-u_e load follow the "motor arrow system." Hence, if the value of the instantaneous power is positive then power flows out of the current source into the circuit element and internal voltage source u_e. With reference to Eq. (5.2), the term p_{in} refers to the power supplied to the network. For the resistance and inductor the instantaneous power is given as p_R and p_L, respectively. The same definition in terms of energy/power flow also applies to the voltage source u_e. It will be shown later that this type of circuit element is also present in electrical machines where it is instrumental in the electrical to mechanical energy conversion process. It is noted that for a resistance the instantaneous power is always positive as it is dissipated, which means that its energy is given as heat to the environment. Note that energy (Ws = Joule) is defined as $\Delta W_e = \int_0^t p(\tau)\,d\tau$, i.e. in a time-diagram the area underneath the respective power function and the horizontal time line.

It is instructive to discuss these concepts with the aid of an example where we assume a sinusoidal current time function of the form

$$i(t) = \hat{i}\cos\omega t \tag{5.3}$$

The "steady-state" voltage across the current source will be of the form

$$u(t) = \hat{u}\cos(\omega) \tag{5.4}$$

In addition we will assume that the voltage across the induced voltage source u_{e} can be written as

$$u_{\mathrm{e}}(t) = \hat{e}\cos(\omega) \tag{5.5}$$

It is convenient at this stage to introduce a phasor representation of the variables $u(t), andi(t)$ according to the approach discussed in Sect. 2.5. The variables according to Eqs. (5.3) and (5.4) may also be written as

$$i(t) = \Re\left\{\underbrace{\hat{i}}_{\underline{i}}\ \mathrm{e}^{\mathrm{j}\omega t}\right\} \tag{5.6a}$$

$$u(t) = \Re\left\{\underbrace{\hat{u}\,\mathrm{e}^{\mathrm{j}\rho}}_{\underline{u}}\ \mathrm{e}^{\mathrm{j}\omega t}\right\} \tag{5.6b}$$

An observation of Fig. 5.2 shows that we can also represent the voltage phasor in terms of the vector sum of the two phasors $\underline{u}_{\mathrm{re}} = \hat{u}\cos\rho$ and $\underline{u}_{\mathrm{im}} = \mathrm{j}\hat{u}\sin\rho$, i.e. $\underline{u} = \underline{u}_{\mathrm{re}} + \underline{u}_{\mathrm{im}}$. The phasor $\underline{u}_{\mathrm{re}}$ is aligned (in phase) with the current phasor, the other $\underline{u}_{\mathrm{im}}$ is orthogonal, i.e. at right angles to $\underline{i}$. If the nature of the circuit is "inductive" (as shown) the angle ρ, which is conventionally measured from the current vector to the voltage vector, will be greater than zero. Alternatively, circuits which exhibit a negative ρ are referred to as being "capacitive." This definition ties in with the fact that the phasor diagram of, for example, an inductor corresponds to the case $\rho = {}^{\pi}/_{2}\,\mathrm{rad}$, whereas for a capacitor the angle equals $\rho = -{}^{\pi}/_{2}\,\mathrm{rad}$.

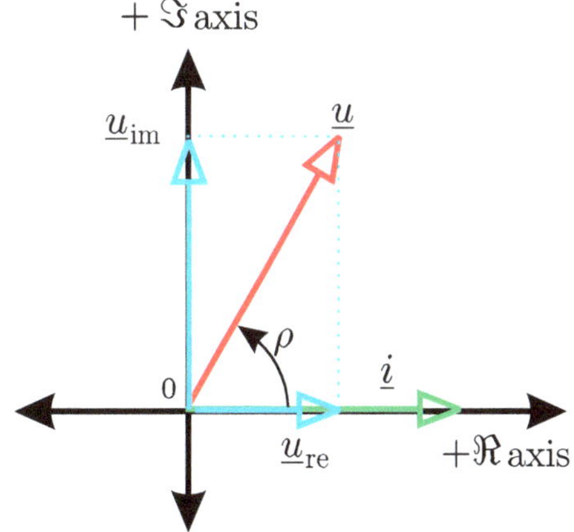

Fig. 5.2 Phasor diagram of phasors $\underline{i}$, $\underline{u}$

The voltage phasor components $\underline{u}_{\mathrm{re}}$ and $\underline{u}_{\mathrm{im}}$ can also be converted to variables as function of time (by using the transformations given in Eq. (5.6)) namely

$$u_{\mathrm{re}}(t) = \hat{u}\cos\rho\cos\omega t \tag{5.7a}$$

$$u_{\mathrm{im}}(t) = -\hat{u}\sin\rho\sin\omega t \tag{5.7b}$$

Note that expression (5.7) may also be written in space vector form as

$$\vec{u}(t) = u_{\mathrm{re}}(t) + \mathrm{j}u_{\mathrm{im}}(t) \tag{5.8}$$

The introduction of phasor components $\underline{u}_{\mathrm{re}}$, and $\underline{u}_{\mathrm{im}}$ allows us to rewrite the input power equation $p_{\mathrm{in}} = i\,u$ as

$$p_{\mathrm{in}} = \underbrace{i\,u_{\mathrm{re}}}_{p_{\mathrm{re}}} + \underbrace{i\,u_{\mathrm{im}}}_{p_{\mathrm{im}}} \tag{5.9}$$

which shows that the instantaneous input power expression is now defined in terms of two components which with the aid of Eqs. (5.3) and (5.7) can also be written as

$$p_{\mathrm{re}} = \hat{u}\,\hat{i}\cos\rho\cos^2\omega t \tag{5.10a}$$

$$p_{\mathrm{im}} = -\hat{u}\,\hat{i}\sin\rho\cos\omega t\sin\omega t \tag{5.10b}$$

Equation (5.10) can also be rewritten in the form given below

$$p_{\mathrm{re}} = \underbrace{\frac{\hat{u}\,\hat{i}}{2}\cos\rho}_{P}\,(1+\cos 2\omega t) \tag{5.11a}$$

$$p_{\mathrm{im}} = \underbrace{\frac{\hat{u}\,\hat{i}}{2}\sin\rho}_{Q}\,(-\sin 2\omega t) \tag{5.11b}$$

An analysis of Eq. (5.11a) shows that there is a time-independent term, known as the "real power," P, with units in "Watts," which represents the average power level P of the function p_{re}. In other words, the average power level corresponds to the total amount of energy supplied to the circuit for one cycle period T (the area enclosed by the power function p_{re} and time line) divided by T. Note that the average power level is also present in the function p_{in} given that the average value of the variable p_{im} is zero. The so-called reactive power value, Q, expressed in volt-ampere reactive VAr (to differentiate from Apparent power S in VA) is tied to the energy flow associated with expression (5.11b). Note that this power expression is based on the use of the voltage phasor component, which is at right angles to the current phasor (see Fig. 5.2). An observation of expression (5.11b) shows that the average power level of p_{im} is zero. The amplitude of the power function p_{im} is known as the reactive power value "Q."

The real, reactive, and apparent power of the circuit may be written as

$$P = UI\cos\rho \qquad \mathrm{W} \tag{5.12a}$$

$$Q = UI\sin\rho \qquad \mathrm{VAr} \tag{5.12b}$$

$$S = UI \qquad \mathrm{VA} \tag{5.12c}$$

where $U = \hat{u}/\sqrt{2}$ and $I = \hat{i}/\sqrt{2}$ are the respective RMS values of the voltage/current waveforms of the source connected to the R-L-u_{e} circuit. The term $\cos\rho$ is referred to as the *displacement factor*. When no harmonics are present this factor equals the *power factor*, which is defined as the ratio of real and reactive power.

It is at this stage helpful to consider a simple numerical example where we assume the current to be of the form $i = \cos\omega t$, where $\omega = 100\,\pi$ rad/s. The circuit elements are chosen purposely as to arrive at a voltage across the current source which is of the form $u = 2\cos(\omega t + \rho)$, with $\rho = \pi/3$. Hence, the circuit is "inductive" given that the voltage/time function leads the current/time function. For this example, the values of P and Q are according to Eq. (5.12), equal to $P = 0.5\,\mathrm{W}$ and $Q = 0.866\,\mathrm{VAr}$, respectively. The input power versus time plot together with its components p_{re} and p_{im} are shown in Fig. 5.3 for one 20 ms cycle of operation. An observation of Fig. 5.3 shows that the energy flow is towards the circuit for some parts of the cycle (shown in "green") and back to the current source for other parts (shown in "red"). There is, however, an average energy flow (the difference between the "green" and "red" areas) and the power associated with this net energy is known as the "real" power P given in watts as was discussed above.

Also shown in Fig. 5.3 are the two components p_{re} and p_{im} of the input power function. An observation of this example confirms that the energy linked with the power function p_{re} is precisely the energy supplied to the circuit over one period $T = 20\,\mathrm{ms}$. The amount of energy supplied (colored "green" in the power function p_{re}) is equal to P times the cycle time T. The energy linked to the power waveform p_{im} represents the energy which is temporarily stored in the circuit (in either the inductive, capacitive elements or the internal voltage source). This energy oscillates between circuit and supply source (as shown by the "green" and "red" areas which identify the energy direction).

The reactive power value Q is equal to the amplitude of the waveform p_{im}. Note that the value of Q can be positive or negative, in both cases its value remains linked to the amplitude of the energy fluctuations. It is interesting to consider the changes to Fig. 5.3 for the case $\rho = 0$. Under these circumstances the reactive power level Q is zero, hence there are no energy fluctuations linked to this term. The energy level still fluctuates, but the energy flow is unidirectional, i.e. from supply to circuit. On the other hand, if we choose $\rho = \frac{1}{2}\pi$ the real power P will be zero in which case the average amount of energy transferred from supply to the circuit is zero. The energy

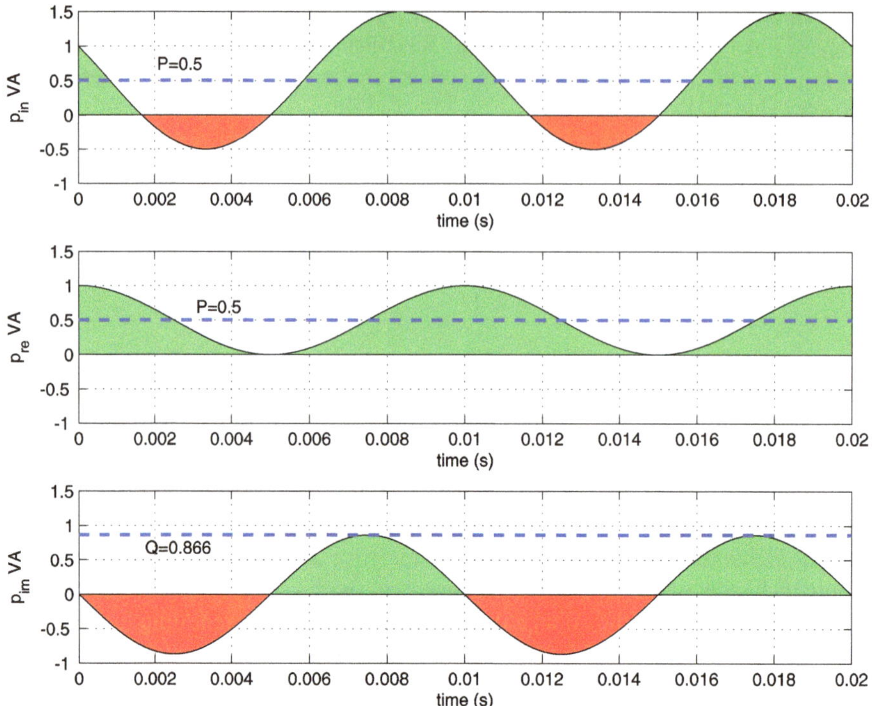

Fig. 5.3 Power plot for R-L-u_{e} circuit, supply side

under these circumstances is stored and recovered from the circuit, i.e. for parts of the cycle it flows from source to the circuit and for the other (equal amount) it flows in the opposite direction.

At this point we have considered the power and energy situation when viewed from outside the circuit, i.e. from the source connected to the circuit. We will now consider the energy flow within the circuit itself.

Firstly, we examine the resistive component where the voltage across this element is given as $u_{\mathrm{R}} = i\,R$, which can also be linked to its phasor representation namely

$$u_{\mathrm{R}}(t) = \Re\left\{\underbrace{\hat{i}R}_{\underline{u}_{\mathrm{R}}}\, \mathrm{e}^{\mathrm{j}\omega t}\right\} \tag{5.13}$$

The phasor $\underline{u}_{\mathrm{R}}$ is in phase with the current phasor $\underline{i}$ as shown in Fig. 5.4. The instantaneous power linked to this component is equal to $p_{\mathrm{R}} = i^2 R$, which after substitution of Eq. (5.3) can (after some manipulation) be written as

$$p_{\mathrm{R}}(t) = \frac{\hat{i}^2 R}{2}\left(1 + \cos(2\omega t)\right) \tag{5.14}$$

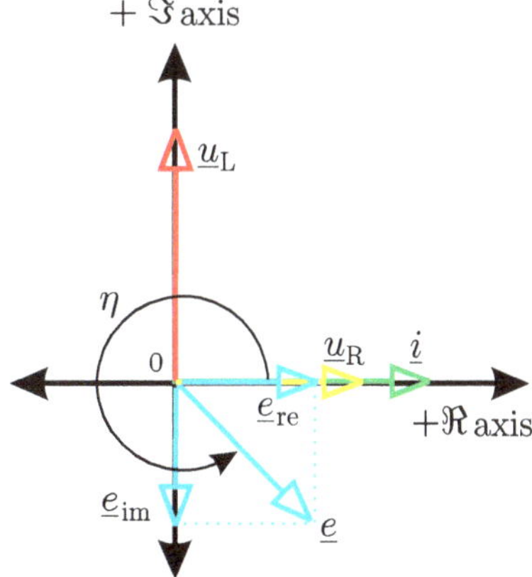

Fig. 5.4 Phasor diagram for R-L-u_e circuit, component side

An important observation from Eq. (5.14) is that the power p_R is greater than or equal to zero, as mentioned previously. Furthermore, the real power P_R (which is defined as the average value of $p_R(t)$) is equal to $P_R = I^2 R = {U^2}/{R}$, where $I = {\hat{i}}/{\sqrt{2}}$ and $U = {\hat{u}}/{\sqrt{2}}$ are equal to the RMS value of the current and voltage, respectively.

We will now consider the inductance L of the circuit. The voltage across this element is given as $u_L = L\,{di}/{dt}$. Substitution of the current expression (5.3) leads to $u_L = -\hat{i}\,\omega L \sin \omega t$, which can also be written in terms of the phasor $\underline{u}_L$ linked to this function, namely

$$u_L(t) = \Re\left\{\underbrace{\mathrm{j}\,\hat{i}\,\omega L}_{\underline{u}_L}\ \mathrm{e}^{\mathrm{j}\omega t}\right\} \tag{5.15}$$

Equation (5.15) shows that the voltage phasor $\underline{u}_L$ is orthogonal (at right angles) to the current phasor as shown in Fig. 5.4. The power linked with this component is given as $p_L = u_L i$, which can also be written as

$$p_L(t) = -\frac{\omega L \hat{i}^2}{2} \sin(2\omega t) \tag{5.16}$$

The first observation to be made from Eq. (5.16) is that its average value is zero, which is to be expected given that an inductor cannot dissipate energy. Hence energy taken into the device is stored and must (at a later interval) be recovered. The peak value of p_L represents the reactive power Q_L of this component and is equal to $Q = I^2 \omega L$. Note that a similar analysis of this type can also be done for the capacitor in which case the reactive power is taken to be negative and of the form $I^2\,{1}/{\omega C}$.

Finally, the circuit component u_e is considered, which is of the form given in Eq. (5.5). The phasor $\underline{e}$ linked to this voltage function is of the form

$$u_e(t) = \Re\left\{\underbrace{\hat{e}\,\mathrm{e}^{\mathrm{j}\eta}}_{\underline{e}}\ \mathrm{e}^{\mathrm{j}\omega t}\right\} \tag{5.17}$$

The phasor $\underline{e}$ shown in Fig. 5.4 (with $\eta = -\pi/4$) can be defined in terms of two phasors $\underline{e}_{re}$ and $\underline{e}_{im}$ (also given in Fig. 5.4 according to the approach outlined for the voltage phasor (see Eq. (5.7)). The corresponding voltages are of the form

$$e_{re}(t) = \hat{e}\cos\eta\cos\omega t \tag{5.18a}$$

$$e_{im}(t) = -\hat{e}\sin\eta\sin\omega t \tag{5.18b}$$

The introduction of these two voltage components allows us to rewrite the power equation $p^{e} = i\,u_{e}$ as

$$p^{e} = \underbrace{i\,e_{re}}_{p^{e}_{re}} + \underbrace{i\,e_{im}}_{p^{e}_{im}} \tag{5.19}$$

which shows that the instantaneous power expression is again defined by two terms, which with the aid of Eqs. (5.3) and (5.18) can also be written as

$$p^{e}_{re} = \hat{e}\,\hat{i}\cos\eta\cos^{2}\omega t \tag{5.20a}$$

$$p^{e}_{im} = -\hat{e}\,\hat{i}\sin\eta\cos\omega t\sin\omega t \tag{5.20b}$$

Equation (5.20) can also be rewritten in the form given below

$$p^{e}_{re} = \underbrace{\frac{\hat{e}\,\hat{i}}{2}\cos\eta}_{P_{e}} + \frac{\hat{e}\,\hat{i}}{2}\cos\eta\cos 2\omega t \tag{5.21a}$$

$$p^{e}_{im} = \underbrace{\frac{\hat{e}\,\hat{i}}{2}\sin\eta}_{Q_{e}}\,(-\sin 2\omega t) \tag{5.21b}$$

Expression (5.21) clearly shows the real and reactive power contributions P_{e} and Q_{e}, respectively, which are associated with the voltage source u_{e}.

It is at this stage helpful to return to the numerical example given for the supply side. For example, if we set $\hat{e} = \sqrt{2}/2\,\mathrm{V}$, $\eta = -\pi/4$ rad, $R = 0.5\,\Omega$, $\omega L = \left(\sqrt{3} + 0.5\right)\Omega$, and $\hat{i} = 1\,\mathrm{A}$, then a simple phasor analysis of this circuit shows that the voltage across the current supply source equals $u = 2\cos(\omega t + \pi/3)$, which is the waveform used for the supply example that corresponds to the power/energy Fig. 5.3. Use of these circuit parameters with Eqs. (5.14), (5.16), and (5.21) leads to the power waveforms and "energy" surfaces as given in Fig. 5.5. The axis scaling has purposely been chosen to match that of Fig. 5.3. The energy levels linked with this circuit are again shown, where "green" implies an energy flow into the circuit element and "red" out of an element. For example, in the resistance, energy flow is always into this component, given that this energy is dissipated as heat. The average value of the waveform p_{R} represents the real power $P_{R} = 0.25\,\mathrm{W}$ dissipated in the resistance given our choice of parameter values. For the inductor

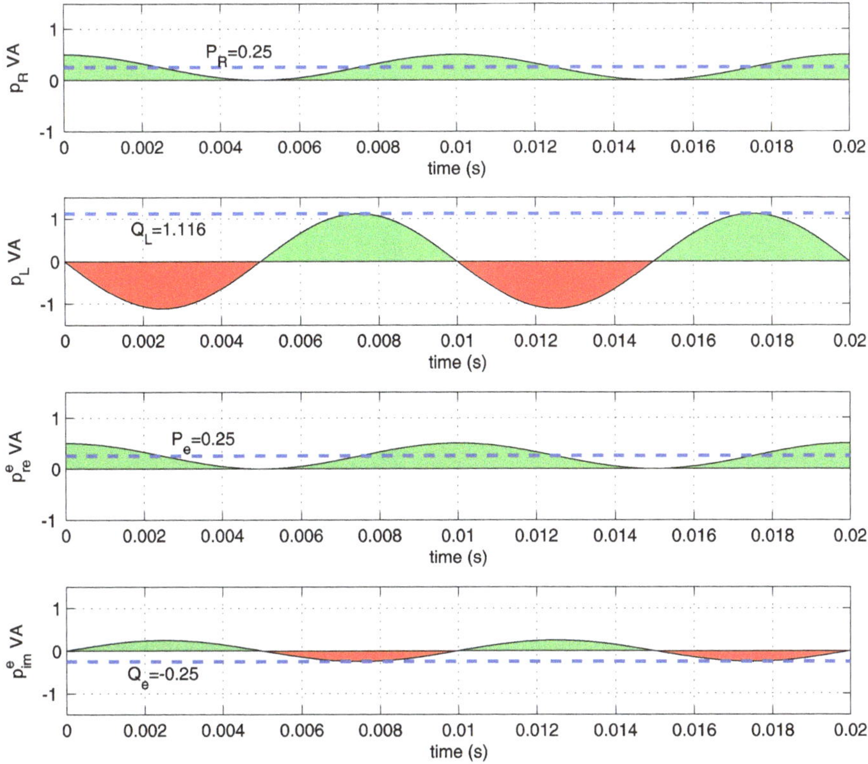

Fig. 5.5 Power plot for *L-R-e* circuit, component side

we observe a pulsating energy flow, which corresponds to a reactive power value of $Q_L = 1.116$ VAr. Finally, we need to consider the voltage source u_e that has two power components, of which the first p^e_{re} has a unidirectional energy flow (for this example) into this element. The average power level with the present parameter values is equal to $P_e = 0.25$ W. A reactive power term is also evident, which in this case was arbitrarily chosen to be capacitive, which corresponds with a negative reactive value of $Q_e = -0.25$ VAr. The total real power which is absorbed by the circuit is equal to $P_R + P_e = 0.25 + 0.25 = 0.5$ W, which corresponds to the power $P = 0.5$ W (supplied by the current source), as shown in Fig. 5.3. The reactive component sum is equal to $Q_L + Q_e = 1.116 - 0.25 = 0.866$ VAr, which corresponds to the reactive power level of the circuit as shown in Fig. 5.3. From this analysis we can also observe where the real and reactive power ends up in the circuit. In this example, real power supplied by the source is equally divided between the resistance and voltage source u_e. Furthermore, the analysis shows that the total reactive power $Q = 0.866$ VAr is predominantly linked to the inductor. However, a small capacitive component equal to -0.25 VAr is also linked to the voltage source u_e, as shown in Fig. 5.5.

5.3 Power in Three-Phase Systems

The approach used to explain the concept of power in single phase systems is extended in this section to three-phase systems, using the material presented in Chap. 4, in particular with respect to the use of space vectors.

It is helpful to assume a three-phase supply current source of the form

$$i_R = \hat{i} \cos(\omega t) \tag{5.22a}$$

$$i_S = \hat{i} \cos(\omega t - \gamma) \tag{5.22b}$$

$$i_T = \hat{i} \cos(\omega t - 2\gamma) \tag{5.22c}$$

with $\gamma = {}^{2\pi}/_{3}$, as mentioned earlier. The phase voltages which appear across the respective supply current sources may be written as

$$u_R = \hat{u} \cos(\omega t + \rho) \tag{5.23a}$$

$$u_S = \hat{u} \cos(\omega t + \rho - \gamma) \tag{5.23b}$$

$$u_T = \hat{u} \cos(\omega t + \rho - 2\gamma) \tag{5.23c}$$

The corresponding space vector representation (amplitude invariant notation, see Eq. (4.30)) of the supply waveforms is in this case given as

$$\vec{u} = \hat{u}\, e^{j\rho} e^{j\omega t} \tag{5.24a}$$

$$\vec{i} = \hat{i}\, e^{j\omega t} \tag{5.24b}$$

where $\hat{u}$ and $\hat{i}$ represent the peak values of the three-phase sinusoidal variables.

The three-phase circuit configuration as described for the single phase is again used here, which means that each phase consists of a series network in the form of a resistance R, inductor L, and voltage source u_e which for the three phases 1, 2, 3 is now of the form

$$u_{e1} = \hat{e} \cos(\omega t + \eta) \tag{5.25a}$$

$$u_{e2} = \hat{e} \cos(\omega t + \eta - \gamma) \tag{5.25b}$$

$$u_{e3} = \hat{e} \cos(\omega t + \eta - 2\gamma) \tag{5.25c}$$

which can also be presented in its space vector form

$$\vec{u}_e = \hat{e}\, e^{j\eta} e^{j\omega t} \tag{5.26}$$

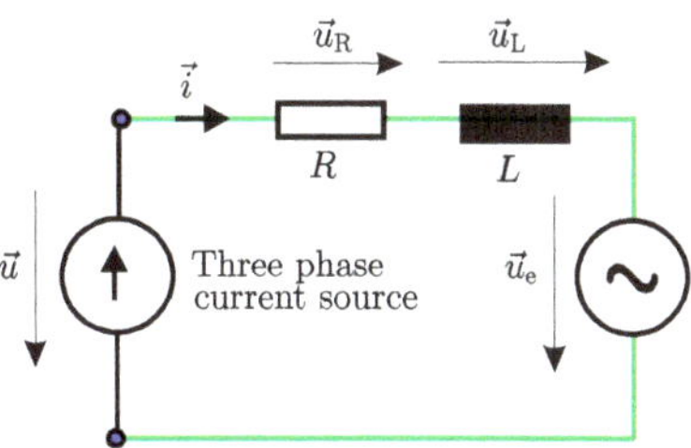

Fig. 5.6 Space vector representation of the three-phase circuit

It is instructive to represent the three-phase network as considered here in its space vector form according to the approach outlined in Chap. 4. We will assume for simplicity a "star" connected circuit, given that phase variables are under these conditions equal to supply variables (there is no zero sequence voltage component in this case $u_0 = 0$). The resultant space vector based circuit diagram is of the form shown in Fig. 5.6. On the basis of Fig. 5.6 we will examine the power/energy concepts. A suitable starting point is the total instantaneous input power which is now of the form

$$p_{\text{in}} = \underbrace{u_R\, i_R}_{p_R} + \underbrace{u_S\, i_S}_{p_S} + \underbrace{u_T\, i_T}_{p_T} \tag{5.27}$$

in which the instantaneous power for each phase is also shown. The real power in, for example, the "R" phase is according to Eq. (5.12a) equal to $P_R = \frac{\hat{u}\,\hat{i}}{2}\cos\rho$. The total real power for this system will be three times this amount, on the grounds that the circuit configuration for all three phases is identical, hence the total real power is equal to

$$P = \frac{3\hat{u}\,\hat{i}}{2}\cos\rho \tag{5.28}$$

Note that this expression can also be written in terms of RMS values, namely

$$P = 3\,UI\cos\rho \tag{5.29}$$

In this case it is helpful to use the form according to Eq. (5.28), given that it provides a relative easy transition to the space vector form of the power, which is of the form

$$P = \frac{3}{2}\Re\left\{\vec{u}\,\left(\vec{i}\right)^{*}\right\} \tag{5.30}$$

Note that the superscript $*$ indicates the conjugate of a vector. The validity of this expression is readily verified upon substitution of Eq. (5.24), in which case the end result must conform with Eq. (5.28).

Equation (5.30) can also be given in terms of its space vector components $\vec{u} = u_\alpha + j\,u_\beta$ and $\vec{i} = i_\alpha + j\,i_\beta$, which leads to Eq. (5.31).

$$P = \frac{3}{2}\left(u_\alpha\, i_\alpha + u_\beta\, i_\beta\right) \tag{5.31}$$

The space vector components can also, with the aid of the conversion matrix (4.32), be rewritten in terms of phase variables, which (with sinusoidal variables and star connected circuit) also correspond to the supply variables u_R, u_S, u_T, i_R, i_S, and i_T. After some mathematical manipulation Eq. (5.32) appears.

$$P = \underbrace{u_R\, i_R}_{p_R} + \underbrace{u_S\, i_S}_{p_S} + \underbrace{u_T\, i_T}_{p_T} \tag{5.32}$$

A comparison between Eqs. (5.32) and (5.27) shows that they are identical. The significant conclusion is therefore that the total instantaneous input power of the circuit is equal to the real power, hence

$$P = p_{in} \tag{5.33}$$

Equation (5.33) also states that there is no energy fluctuation present in the waveform, p_{in}, given that the average value (as defined by the value P) corresponds to the instantaneous value. This observation is of fundamental importance as it means that the energy flow in a three-phase system from supply to source is fully utilized in terms of transport efficiency. This is the fundamental reason why three-wire (three-phase) circuits are used for energy conversion processes. A two-wire circuit (single phase system) cannot realize such an efficient energy transport as can be observed from Fig. 5.3, waveform $p_{in}\,(t)$.

For simulation purposes it is helpful to introduce a new building block as given in Fig. 5.7, which calculates the real power P according to Eq. (5.31). The input variables are in this case "vector" lines, which means that the voltage and current input lines are given by the variables u_α, u_β and i_α, i_β, respectively. The output of this module is the power P. Prior to discussing the concept of reactive power in three-phase systems it is instructive to extend the single phase example according to Fig. 5.3 for the three-phase case. The same circuit parameters and phase current/voltage parameters are used here, which means that $\hat{i} = 1\,\mathrm{A}$, $\hat{u} = 2\,\mathrm{V}$,

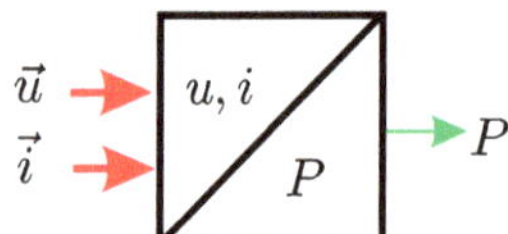

Fig. 5.7 Real power module

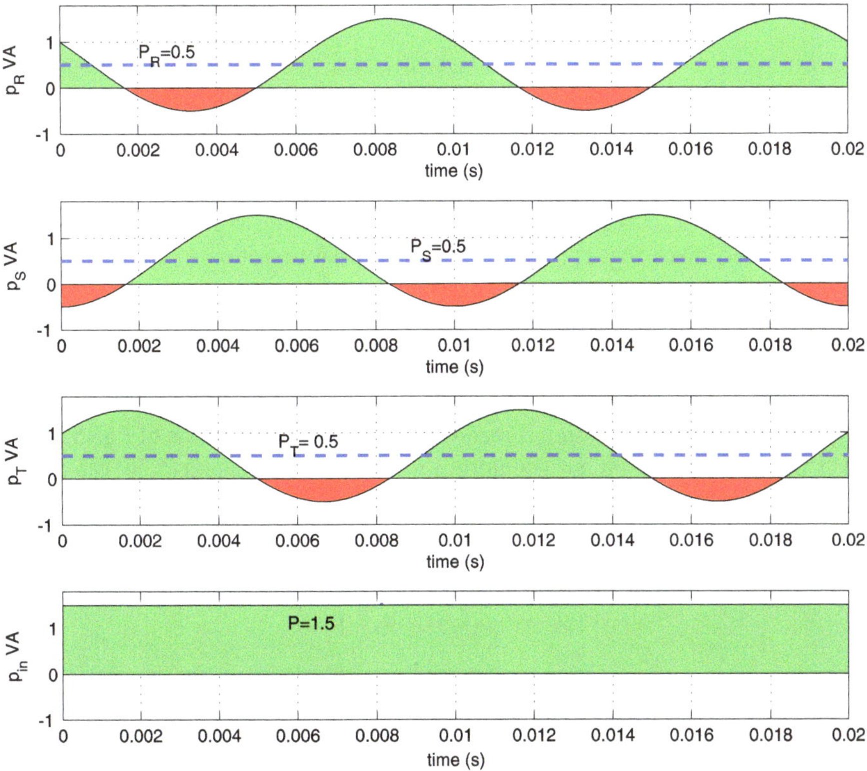

Fig. 5.8 Three-phase "real" power plot

$\rho = \pi/3$ rad, $\omega = 100$ rad/s. On the basis of this data the instantaneous power for the three phases can be plotted using Eqs. (5.22), (5.23), and (5.27). The results as given in Fig. 5.8 also show the energy linked with these waveforms, where "green" is used to identify power into the circuit, whereas "red" depicts an outgoing energy flow.

As expected, Fig. 5.8 has a constant instantaneous power value, whereas the phase power waveforms have an average value, which corresponds to the phase power, P_{RST}, calculated using Eq. (5.12a). The reader could perhaps at this point come to the erroneous conclusion that the reactive power Q for the three-phase circuit is zero given that the input power level is constant. This is *not* the case as will become apparent in the following discussion.

The reactive power in, for example, the "R" phase is according to Eq. (5.12b) equal to $Q_{\text{R}} = \frac{\hat{u}\hat{i}}{2} \sin \rho$. The total reactive power for this system will be three times this amount on the grounds that the circuit configuration for all three phases is identical (as mentioned earlier), hence the total reactive power is equal to

$$Q = \frac{3\hat{u}\,\hat{i}}{2} \sin \rho \tag{5.34}$$

Note that this expression can also be written in terms of RMS values, namely

$$Q = 3\,UI\sin\rho \tag{5.35}$$

In this case it is helpful to use the form according to Eq. (5.34) for further development, given that it provides a relative easy transition to the space vector format of the reactive power equation, namely

$$Q = \frac{3}{2}\Im\left\{\vec{u}\,\left(\vec{i}\right)^{*}\right\} \tag{5.36}$$

The validity of this expression is readily verified upon substitution of Eq. (5.24), in which case the end result must conform with Eq. (5.34). Equation (5.36) can also be given in terms of its space vector components $\vec{u}$ and $\vec{i}$ which leads to Eq. (5.37).

$$Q = \frac{3}{2}\left(u_{\beta}\,i_{\alpha} - u_{\alpha}\,i_{\beta}\right) \tag{5.37}$$

At this point it is convenient to introduce a simulation building block which has as output the reactive power value as defined in Eq. (5.37). The input variables are in this case "vector" lines which means that the voltage and current input lines are given by the variables u_{α}, u_{β} and i_{α}, i_{β}, respectively (Fig. 5.9). Prior to discussing the power/energy flow within the three phase circuit it is instructive to return to our numerical example used to plot the waveforms given in Fig. 5.8. The reactive power plot (for one cycle $T = 20\,\text{ms}$) which corresponds with this example is given in Fig. 5.10. The parameters used to obtain these results are $\hat{i} = 1\,\text{A}$, $\hat{u} = 2\,\text{V}$, $\rho = \pi/3\,\text{rad}$, $\omega = 100\,\text{rad/s}$.

The reactive power per phase corresponds to the amplitude of the p^{R}_{im}, p^{S}_{im}, p^{T}_{im} waveforms. Note that the instantaneous sum of these three waveforms is zero. The total reactive power of the circuit is defined as the sum of the reactive power per phase contributions, hence this sum is NOT zero but three times the amplitude of the phase contributions.

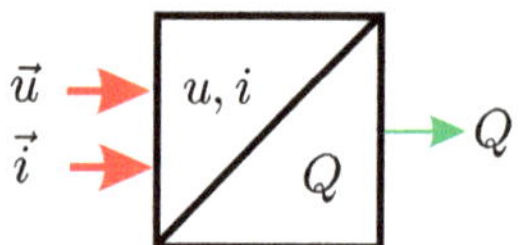

Fig. 5.9 Reactive power module

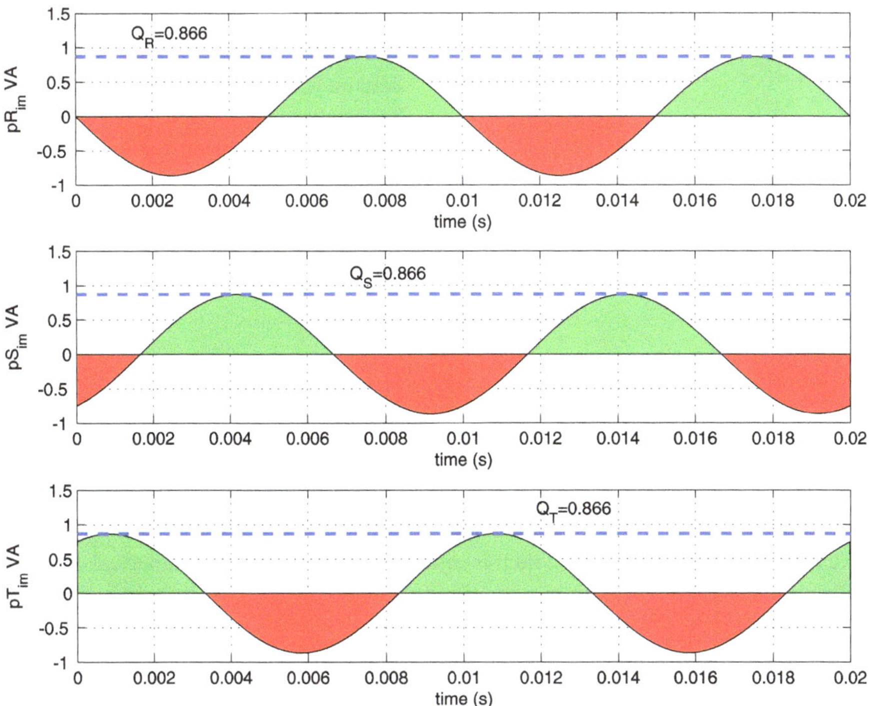

Fig. 5.10 Three-phase "reactive" power plot

The approach used to examine the power contributions (for each element of the circuit) is based on the real and reactive power definitions as given by Eqs. (5.30) and (5.36), respectively. The vector $\vec{u}$ must be replaced by the vector which corresponds to the circuit element under discussion.

For the resistive element the voltage vector is of the form $\vec{u}_{\mathrm{R}} = \vec{i}R$ in which case the total power dissipated is given as

$$P_{\mathrm{R}} = \frac{3}{2}\vec{i}\,\vec{i}^{*}R \tag{5.38}$$

The reactive power for the inductor is found by use of the voltage vector $\vec{u}_{\mathrm{L}} = L\frac{\mathrm{d}\vec{i}}{\mathrm{d}t}$ with Eq. (5.36), which leads to the reactive power Q_{L} expression defined as

$$Q_{\mathrm{L}} = \frac{3}{2}\vec{i}\,\vec{i}^{*}\omega L \tag{5.39}$$

The real and reactive power contributions linked to the three phase voltage source u_{e} are found by using Eq. (5.26) with Eqs. (5.30) and (5.36) which lead to

Table 5.1 Real and reactive value for circuit example

Variable		Value
Real power resistance	P_R	0.75 W
Reactive power inductor	Q_L	3.348 VAr
Real power source u_e	P_e	0.75 W
Reactive power source u_e	Q_e	−0.75 VAr

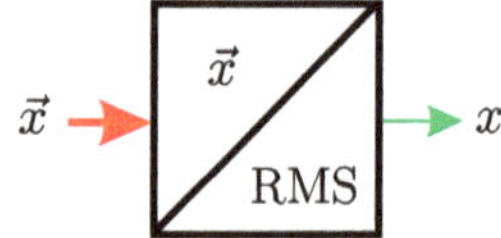

Fig. 5.11 Vector to RMS module

$$P_e = \frac{3}{2}\Re\left\{\vec{e}\,\vec{i}^*\right\} \tag{5.40a}$$

$$Q_e = \frac{3}{2}\Im\left\{\vec{e}\,\vec{i}^*\right\} \tag{5.40b}$$

As with the single phase example it is instructive to return to the numerical example given for the three-phase supply side (Figs. 5.8 and 5.10). If we again set $\hat{e} = \sqrt{2}/2\,\mathrm{V}$, $\eta = -\pi/4\,\mathrm{rad}$, $R = 0.5\,\Omega$, $\omega L = \left(\sqrt{3} + 0.5\right)\Omega$, and $\hat{i} = 1\,\mathrm{A}$ then a space vector analysis of this circuit shows that the voltage across the current supply source will be given by Eq. (5.24a) with $\hat{u} = 2\,\mathrm{V}$, $\rho = \pi/3\,\mathrm{rad}$. The real and reactive power contributions as calculated using Eqs. (5.38)–(5.40) for these circuit elements are given in Table 5.1.

According to Table 5.1 the total real power utilized by the circuit elements is equal to $P_R + P_e = 0.75 + 0.75 = 1.5\,\mathrm{W}$ which is precisely the power level shown in Fig. 5.8. The reactive power of the circuit is according to Table 5.1 equal to $Q_L + Q_e = 3.348 - 0.75 = 2.598\,\mathrm{VAr}$. This value divided by three gives the reactive power contribution per phase as shown in Fig. 5.10.

At the conclusion of this section it is helpful to introduce a simulation module which has as input a space vector $\vec{x} = x_\alpha + j\,x_\beta$ and as output the "RMS" value "x." This module is useful for simulations with three-phase sinusoidal systems, where the RMS value linked to a space vector is often of interest. The building block which represents this conversion process is given in Fig. 5.11.

The relationship between input and output is found by realizing that the amplitude of the vector is given as $|\vec{x}| = \sqrt{(x_\alpha)^2 + (x_\beta)^2}$. Furthermore, for amplitude invariant type transformations (and sinusoidal waveforms) the *RMS* phase value is given as $x = \hat{x}/\sqrt{2}$, where $\hat{x} = |\vec{x}|$. The resultant conversion may also be written as

$$x = \frac{1}{\sqrt{2}}\sqrt{(x_\alpha)^2 + (x_\beta)^2} \tag{5.41}$$

5.4 Phasor Representation of Real and Reactive Power

In Sect. 4.7.1, the relationship between phasors and space vectors was discussed. In this section the use of phasors with real and power concepts is considered. According to Eq. (4.60) the phasor representation of a voltage/current based power invariant space vectors may be written as

$$\vec{u} = \underbrace{\hat{u}\,\mathrm{e}^{\mathrm{j}\rho}}_{\underline{u}}\,\mathrm{e}^{\mathrm{j}\omega t} \tag{5.42a}$$

$$\vec{i} = \underbrace{\hat{i}}_{\underline{i}}\,\mathrm{e}^{\mathrm{j}\omega t} \tag{5.42b}$$

The calculation of the real power of, for example, the R-L-u_{e} circuit is defined in Eq. (5.30). Substitution of Eq. (5.42) leads to the phasor based form, namely

$$P = \frac{3}{2}\Re\left\{\underline{u}\,(\underline{i})^{*}\right\} \tag{5.43}$$

A similar approach for conversion from space vector to phasor format can also be carried out for the reactive power equation (5.36) which gives

$$Q = \frac{3}{2}\Im\left\{\underline{u}\,(\underline{i})^{*}\right\} \tag{5.44}$$

Similarly, the real and reactive power phasor based equations for the circuit components may be obtained. Use of Eqs. (5.38)–(5.40) leads to

$$P_{\mathrm{R}} = \frac{3}{2}\underline{i}\,(\underline{i})^{*}\,R \tag{5.45}$$

$$Q_{\mathrm{L}} = \frac{3}{2}\underline{i}\,(\underline{i})^{*}\,\omega L \tag{5.46}$$

$$P_{\mathrm{e}} = \frac{3}{2}\Re\left\{\underline{e}\,(\underline{i})^{*}\right\} \tag{5.47}$$

$$Q_{\mathrm{e}} = \frac{3}{2}\Im\left\{\underline{e}\,(\underline{i})^{*}\right\} \tag{5.48}$$

5.5 Tutorials

5.5.1 *Tutorial 1: PLECS Based Model of a Single Phase Resonant Circuit*

This tutorial is concerned with the energy flow within a resonant circuit connected to a battery source $u_\mathrm{b} = 10\,\mathrm{V}$ via a switch S which is closed at time $t = 0$. The resonant period of the circuit, as shown in Fig. 5.12 is given as $T = 2\pi\sqrt{LC}$. In this example, the period is set to $T = 20\,\mathrm{ms}$. By choosing $L = 100\,\mathrm{mH}$, the C value can be found using $T = 2\,\pi\sqrt{LC}$ to give the selected T. A simple computation shows that the capacitor value must be set to $C = 101.32\,\mu\mathrm{F}$. Furthermore, we will assume that the capacitor is fully discharged at $t = 0$, hence $u_\mathrm{C}(0) = 0$. A PLECS "control block" representation of Fig. 5.12 is to be made based on the equation set which is linked with this circuit. An example of a PLECS representation of this circuit is given in Fig. 5.13. Note the significance of the integrator outputs namely: integration of the inductance voltage u_L leads to the flux-linkage ψ and the current is then found using $i = {}^{\psi}/_{L}$ (which assumes a linear inductance model). For a capacitor an analogous reasoning is possible, where the input of the integrator is the current i and the output the electrical charge q, on the basis of which the voltage across the capacitor is calculated using $u_\mathrm{C} = {}^{q}/_{C}$. The objective of this tutorial is to examine the voltage, current, and instantaneous power waveforms of this circuit for the time interval $0 \rightarrow 10\,\mathrm{ms}$, i.e. for one-half period cycle ${}^{T}/_{2}$ after the switch

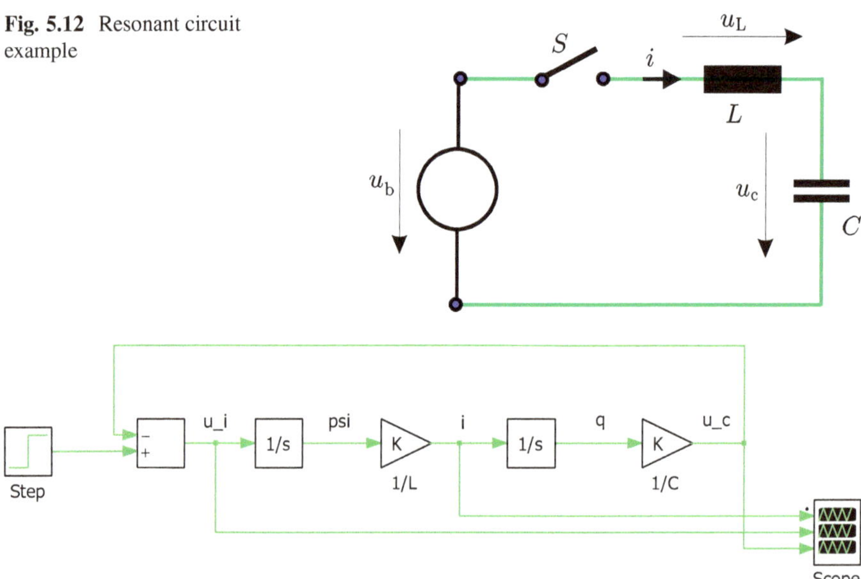

Fig. 5.12 Resonant circuit example

Fig. 5.13 PLECS control block model: resonant circuit model

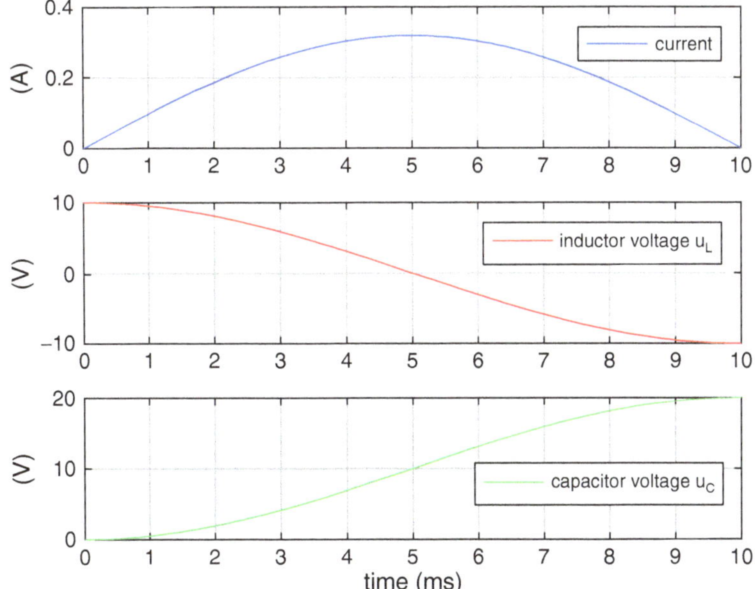

Fig. 5.14 Resonant circuit model: results $i\,(t)$, $u_{\mathrm{L}}\,(t)$, $u_{\mathrm{C}}\,(t)$

is closed. This means that the "run" time of your simulation must be set to 10 ms. An example of the results which should appear on the `Scope` module in terms of the current i, voltage across the capacitor u_{C}, and inductor u_{L} is given in Fig. 5.14. The energy flow for this circuit can be studied with the aid of the instantaneous power for each of the elements. The power for the battery, inductor, and capacitor are of the form $p_{\mathrm{s}} = u_{\mathrm{b}}\, i$, $p_{\mathrm{L}} = u_{\mathrm{L}}\, i$, and $p_{\mathrm{C}} = u_{\mathrm{C}}\, i$, respectively. The power plots for this example are given in Fig. 5.15 and these show the energy supplied to and from the circuit elements. Areas shaded "green" correspond to energy supplied to an element, whereas "red" relates to energy which is recovered. An observation of Fig. 5.15 shows that during the first quarter (0 → 5 ms) of the cycle, energy from the battery source is supplied to the capacitor and inductor. During the second part of the cycle (5 → 10 ms) the capacitor receives energy from the battery as well as the inductor.

5.5.2 Tutorial 2: PLECS Circuit Model Representation of a Single Phase Resonant Circuit

In this tutorial, a circuit model implementation of Fig. 5.12 is considered. The parameters for this example correspond to those given in the previous tutorial. The simulation model as given in Fig. 5.16 on page 148 is in this case constructed with the aid of "circuit" modules instead of generic (control block) modules as

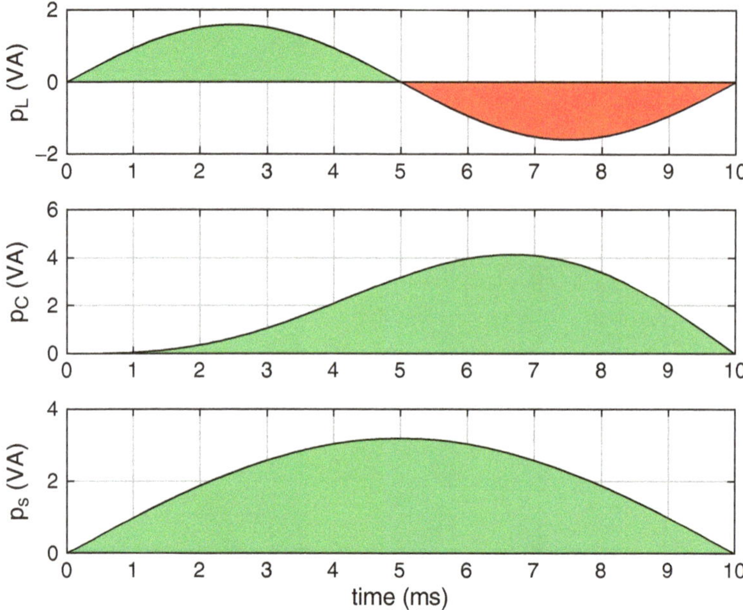

Fig. 5.15 PLECS: results $p_L(t)$, $p_C(t)$, $p_s(t)$

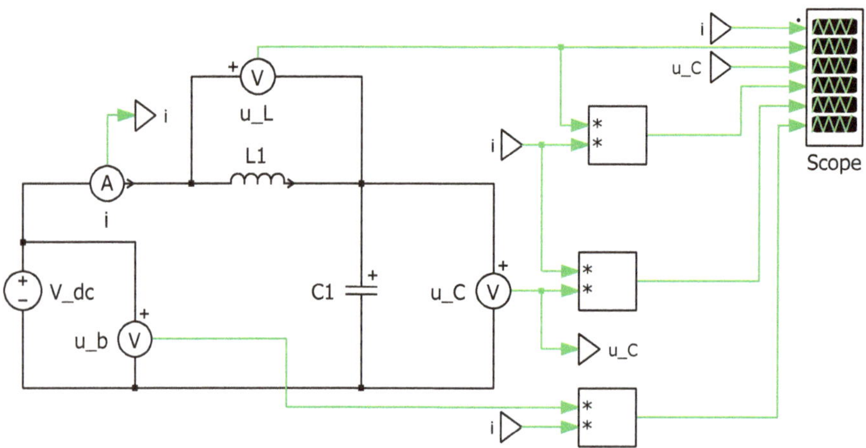

Fig. 5.16 Circuit model representation: resonant circuit

shown in the previous tutorial. The run time of 10 ms remains unchanged. A `Scope` module is used to monitor the circuit current i, inductor voltage u_L, capacitor voltage u_C, and DC supply voltage u_b. In addition, a set of multiplier control blocks have been introduced to measure the instantaneous power of the DC supply, inductance, and capacitor. The results shown on the `Scope` module are identical to those shown in Figs. 5.13 and 5.15.

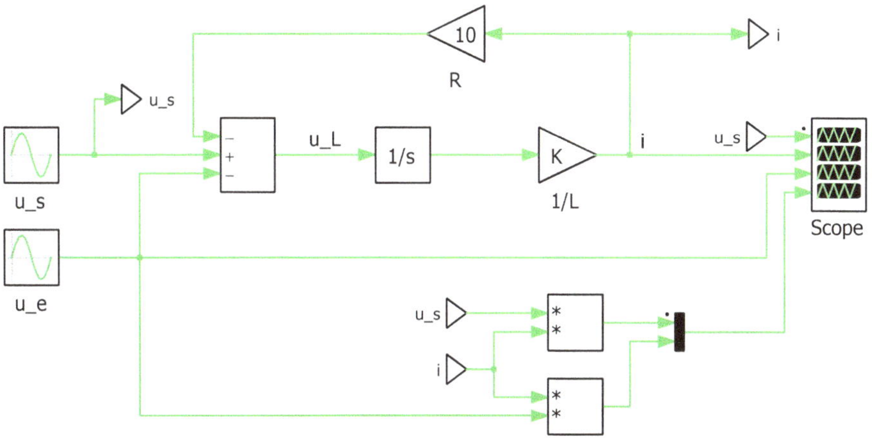

Fig. 5.17 PLECS: single phase R-L-u_e circuit example

5.5.3 *Tutorial 3: PLECS Based Model of a Single Phase Circuit Used for Power Analysis*

This tutorial is concerned with an R-L-u_e series circuit as given in Fig. 5.1. However, in this case a voltage source is connected to the load. A PLECS implementation of this circuit configuration, as given in Fig. 5.17, has an input voltage function $u = \hat{u}\cos(\omega t)$, where $\hat{u} = 220\sqrt{2}\,\mathrm{V}$, $\omega = 100\pi$ rad/s. The voltage source u_e is given as $e = \hat{e}\cos(\omega t + \zeta)$, where $\hat{e} = 100\sqrt{2}$, $\zeta = -{}^{7\pi}/_{6}$. The resistance and inductor value are taken to be $R = 10\,\Omega$ and $L = 100\,\mathrm{mH}$, respectively.

Run the simulation for a period of 60 ms and plot the voltage/current waveforms u, e, and i. An example of the results after running this simulation is given in Fig. 5.18. The instantaneous power waveforms $p_{\mathrm{in}} = u\,i$ for the circuit and for the voltage source u_e, $p_e = e\,i$ are given in Fig. 5.19. The objective of this tutorial is to examine the power linked with the voltage source u_e on the basis that the latter is an unknown quantity. A problem of this type is typical for machines, hence its inclusion here. The input to this problem are the "measured" steady-state current (as observed from the simulation, see Fig. 5.18), which is equal to $I = 9.29\,\mathrm{A}$ and the "measured" real power level (taken from Fig. 5.19) $P = 293\,\mathrm{W}$.

On the basis of the data provided calculate the reactive power Q for the circuit, inductor Q_L and u_e source Q_e. In addition, calculate the real power associated with the resistance P_R and u_e source P_e. Finally, calculate on the basis of the Q_e and P_e values the amplitude and phase angle of the u_e source. An example of this calculation together with the M-file used to generate the plots for this tutorial is given at the end of this tutorial. The real and reactive results obtained after running the simulation and corresponding M-file are given in Table 5.2.

The amplitude and phase of the u_e source are found by taking the ratio of the terms Q_e and P_e which gives the angle $\eta = \arctan\left({}^{Q_e}/_{P_e}\right)$. Substitution of the values

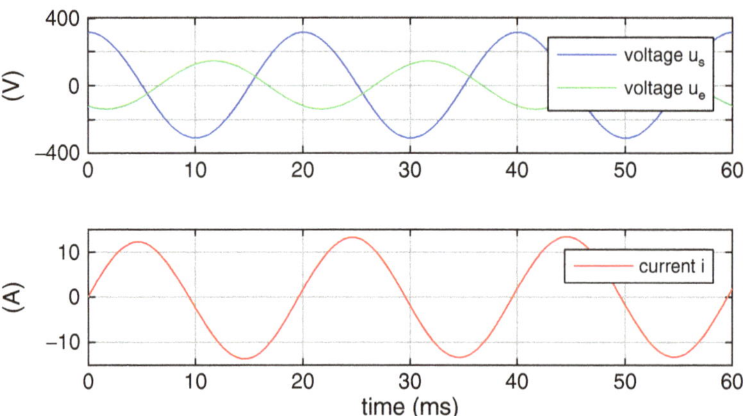

Fig. 5.18 Results u, i, u_e

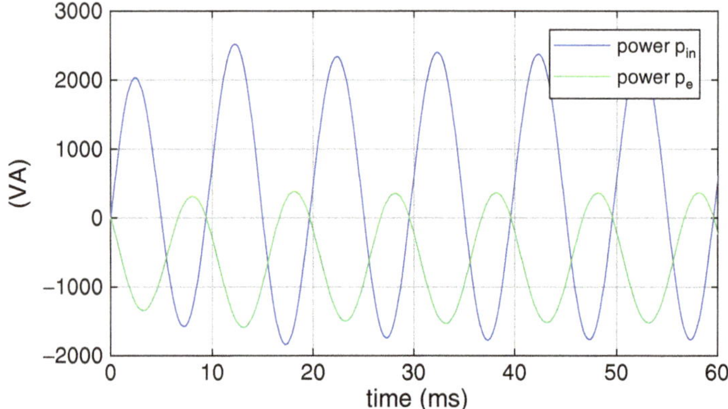

Fig. 5.19 Results p_{in}, p_e

Table 5.2 Real and reactive values for tutorial 2 example

Variable		Value
Real power resistance	P_R	863.0 W
Reactive power inductor	Q_L	2711.3 VAr
Real power source u_e	P_e	−570.0 W
Reactive power source u_e	Q_e	−688.6 VAr

Q_e and P_e as given in Table 5.2, gives $\eta = -129°$ which is the angle between the phasors $\underline{e}$, $\underline{i}$ as shown in the phasor diagram (see Fig. 5.20). Once the angle η is known, the amplitude of the u_e source can be calculated using Eq. (5.21) which gives $\hat{e} \simeq 96.22 \cdot \sqrt{2}$. The value calculated using this approach compares favorably with the actual value of $100 \cdot \sqrt{2}$ as used for the simulation. The phase angle between the voltage and current phasors was found to be $\rho_i = -81.7°$ and this angle is also shown in the phasor diagram given in Fig. 5.20.

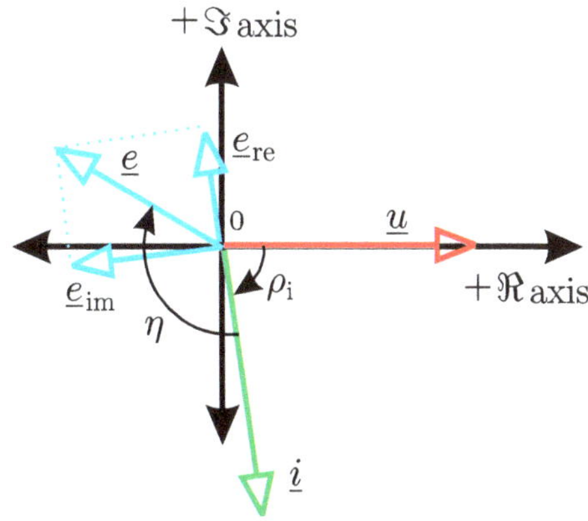

Fig. 5.20 Phasor diagram for tutorial 3

Also shown in the phasor diagram are the phasor components $\underline{e}_{re}$ and $\underline{e}_{im}$ used to derive the real and reactive power components P_e and Q_e for the u_e source. Note that these components are in phase and orthogonal to the current phasor, *not* to the voltage phasor.

M-file Code

```
%Tutorial 3, chapter 5
%we set E=100; eta=-7*pi/6 assume we don't know these
   values
%determine real reactive components on the basis of
   measured
%current
close all
R=10;                                % resistance
L=100e-3;                            % inductor
w=100*pi;                            % frequency in rad/s
%%%%%%%%%%%%%%%%
t=dat(:,4); u=dat(:,3); i=dat(:,1);
e=dat(:,2);                          % used to cross check
                                        results calculation

%%%%%%%%%%%%%%%%%%5
plot(t,u);
grid
hold on
plot(t,i*10,'r')
plot(t,e,'g')
legend('u','i*10','e')
ylabel('u (V), i (A), e (V)')
xlabel('time(s)')
%%%%analysis
%measured current
I=9.29;%measured RMS current
U=220; %RMS voltage
```

```
%watt meter,
figure
plot(t,u.*i);
grid
hold on
plot(t,e.*i,'g')
legend('p_{in}','p_e')
ylabel('p_{in} (VA), p_e (VA)')
xlabel('time (s)')
P=293;                          % measured average power
rho_i=acos(P/(U*I));            % phase angle  u and i
                                  (rad)
%Current lags voltage, i.e., inductive circuit
P_R=I^2*R;                      % dissipated power
P_e=P-P_R;                      % calculated power in
                                  $u_e$
Q=U*I*sin(rho_i);
Q_L=I^2*w*L;                    % reactive power in
                                  inductor
Q_e=Q-Q_L;                      % calculated reactive
                                  power in $u_e$
eta=atan(Q_e/P_e);              % phase angle $u_e$,
                                  between e,i ,
                                % solution in third
                                  quadrant
eta3=(pi-eta)*180/pi;           % result in degrees
E=1/I*sqrt(P_e^2+Q_e^2);        % amplitude e source
```

5.5.4 Tutorial 4: Three-Phase Star/Delta Circuit Model Used for Power Analysis

The objective of this tutorial is to modify the simulation model given in Sect. 4.8.8 in such a manner as to give the user the option of choosing a star or delta configured R-L-u_e branch circuit. This means that we should be able to examine how the supply power and RMS current changes when the phase configuration is changed. Furthermore, the real and reactive power modules and RMS conversion modules are to be added as shown in Fig. 5.21. The circuit shown in Fig. 5.21 appears on the surface to be very similar to that given in Fig. 4.47b. The difference is that the power and RMS conversion modules according to Figs. 5.7, 5.9, and 5.11 have been added together with “display” modules to observe the respective values. A set of modules needs to be added which will allow the R-L-u_e space vector based phase circuit to be used in a star or delta configuration. The key is to introduce a menu variable for this sub-module which is linked to a variable `SD` shown in Fig. 5.22 within a “constant” module. The output of this module must change with the configuration selected. For example, `SD=0` for star, `SD=1` for delta, in which case a switch module can be used for the conversion as shown in Fig. 5.23. In addition to the use of a switch, a set of conversion modules must be added for the delta connected case, as discussed

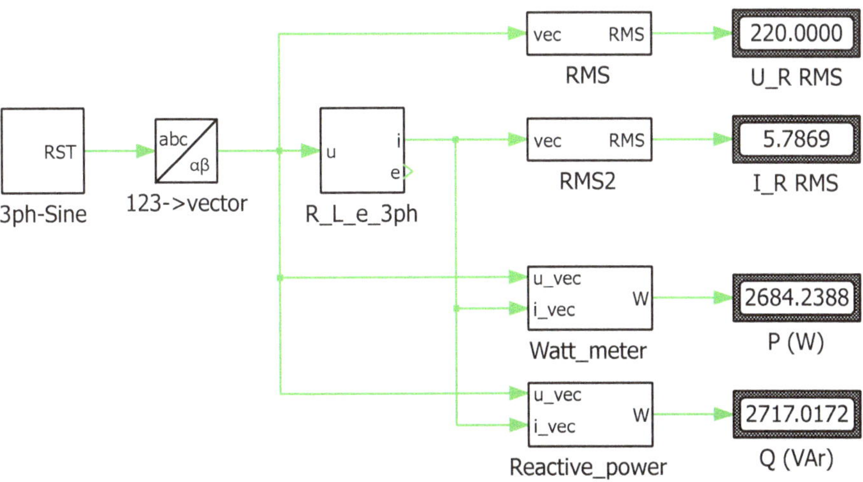

Fig. 5.21 Three-phase L-R-u_e circuit example, with star/delta selection

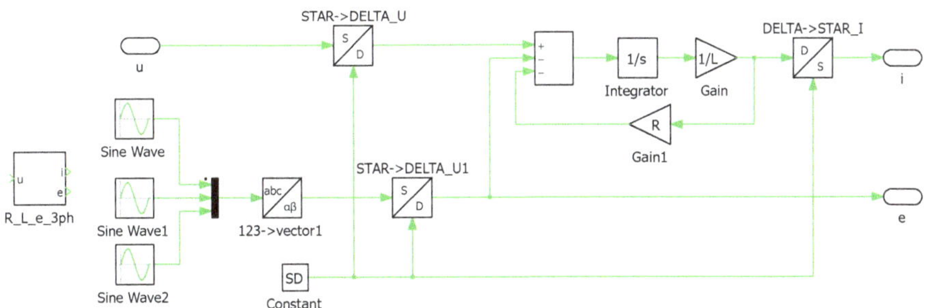

Fig. 5.22 Star/delta circuit model

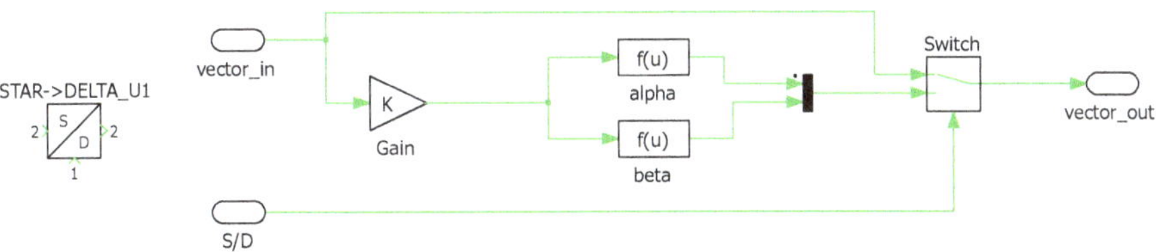

Fig. 5.23 Star/delta conversion

in Sect. 4.6.3. The conversion modules for conversions from $\vec{u}_{\mathrm{RST}} \rightarrow \vec{u}_{\mathrm{D123}}$ and $\vec{i}_{\mathrm{D123}} \rightarrow \vec{i}_{\mathrm{RST}}$ need to be implemented for the delta connected case, as shown in Fig. 5.23. The "function" modules shown in Fig. 5.23 are used to realize the vector rotation $\pm\gamma/4$ and $\sqrt{3}$ scaling as required for the current and voltage space vectors. The supply voltage and circuit components remain unchanged when compared to those given in Sect. 4.8.8. Run your simulation with a "run" time of 600 ms

Table 5.3 Simulation results tutorial 4

Parameters		Value
RMS supply current (star)	I_S	5.79 A
RMS supply current (delta)	I_D	17.36 A
Real power (star)	P_S	2684.2 W
Real power (delta)	P_D	8052.7 W
Reactive power (star)	Q_S	2717.0 VAr
Reactive power (delta)	Q_D	8151.0 VAr

and observe the results on the numerical display modules for the star and delta configured circuit case. If your circuit is implemented correctly the results according to Table 5.3 should appear. The sequel to this tutorial is concerned with verifying the results shown in Table 5.3 by way of a phasor analysis (in the form of an M-file) of this problem for the star and delta connected case. The results of this analysis should be the variables and values as given in Table 5.3. Note that the values for the delta connected circuit configuration are three times higher than those shown for the star case. An example of an M-file for this problem is given below:

M-file Code

```
%Tutorial 4, chapter 5
%%%%eta=-pi/3
R=10;                                    % phase resistance
L=100e-3;                                % phase inductance(H)
w=100*pi;                                % supply frequency
                                           (rad/s)
X=w*L;                                   % load reactance
C=2/3;                                   % space vector ampl
                                           invariant
U=220;                                   % supply  phase
                                           voltage RMS
E=100;                                   % EMF  phase voltage
                                           RMS
rho_e=-pi/3;                             % EMF phase angle
u_RST=3/2*C*U*sqrt(2);                   % supply phasor
U=u_RST/sqrt(3);                         % RMS suply
e_p=3/2*C*E*sqrt(2);                     % emf peak voltage
e_RST=e_p*cos(rho_e)+j*e_p*sin(rho_e);   % EMF phasor
gamma=2*pi/3;
% Star/delta choice in circuit module
% phasor analysis-star connected
u_123s=u_RST;% phase vector
e_123s=e_RST;%e phase vector
i_123s=(u_123s-e_123s)/(R+j*X);          % current phasor
ips=abs(i_123s);                         % peak value phasor
rho_is=angle(i_123s);                    % phase angle current
                                           phasor

% RMS  supply current
```

```
Is=ips/sqrt(2);                          % RMS current
Ps=real(u_123s*conj(i_123s))*3/2;        % real power W
Qs=imag(u_123s*conj(i_123s))*3/2;        % reactive power VAr
%%%%%%%%%%%%%%%%%%%%%%%5
%Delta connected
u_123d=u_RST*sqrt(3)*(cos(-gamma/2)+j*sin(-gamma/2));
                                         % U_phase  phasor
e_123d=e_RST*sqrt(3)*(cos(-gamma/2)+j*sin(-gamma/2));
                                         % e_phase  phasor
i_123d=(u_123d-e_123d)/(R+j*X);          % current phasor
ipd=abs(i_123d);                         % peak value phasor
rho_id=angle(i_123d);                    % phase angle current
                                           phasor
%RMS  supply current
Id=ipd/sqrt(2)*sqrt(3);                  % RMS current
Pd=real(u_123d*conj(i_123d))*3/2;        % real power W
Qd=imag(u_123d*conj(i_123d))*3/2;        % reactive power VAr
```

Chapter 6
Space Vector Based Transformer Models

6.1 Introduction

This chapter considers an extension of the (single phase) ideal transformer (ITF) model to a two-phase space vector based version. The introduction of a two- phase (ITF) model is instructive as a tool for moving towards the so-called ideal rotating transformer "IRTF" concept, which forms the basis of machine models for this book. The reader is reminded of the fact that a two-phase model is a convenient method of representing three-phase systems as discussed in Sect. 4.6. The development from ITF to a generalized two-inductance model, as discussed for the single phase model (see Chap. 3), is almost identical for the two-phase model. Consequently, it is not instructive to repeat this process here. Instead, emphasis is placed in this chapter on the development of a two-phase space vector based ITF symbolic and generic model.

6.2 Development of a Space Vector Based ITF Model

The process of moving from a single phase ITF model to a space vector based version is readily done by making use of Fig. 3.1, which is modified to a two-phase configuration as shown in Fig. 6.1.

A comparison between the single phase (Fig. 3.1) and two-phase (Fig. 6.1) shows that there are now two windings (or coils) on the primary and two on the secondary side of the transformer. The primary and secondary "alpha" winding

---

Electronic supplementary material The online version of this chapter (doi: 10.1007/978-3-319-29409-4_6) contains supplementary material, which is available to authorized users.

© Springer International Publishing Switzerland 2016

A. Veltman et al., *Fundamentals of Electrical Drives*, Power Systems,
DOI 10.1007/978-3-319-29409-4_6

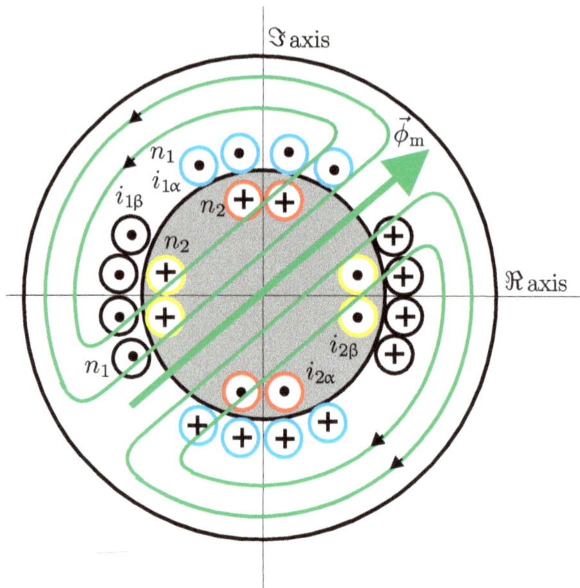

Fig. 6.1 Two-phase physical transformer model

pair are orthogonal to the "beta" winding pair. The number of "effective" primary and secondary turns in both winding sets is equal to n_1 and n_2, respectively. Note that the windings are shown in symbolic form to show how the windings are positioned. The primary and secondary phase windings are assumed to be sinusoidally distributed. A discussion on the concept of sinusoidally distributed windings and "effective" number of turns is given in Appendix A. The currents in the primary and secondary windings are defined as $i_{1\alpha}$, $i_{1\beta}$ and $i_{2\alpha}$, $i_{2\beta}$, respectively. A complex plane with a real and imaginary axis is also introduced in Fig. 6.1. Also shown in this figure is the circuit flux distribution which is linked to a flux space vector $\vec{\psi}_{\rm m} = \psi_{{\rm m}\alpha} + {\rm j}\psi_{{\rm m}\beta}$. The complex plane is purposely tied to the orientation of the primary windings of the transformer, as can be explained by considering the two-phase model in a single phase form. If we ignore, for the purpose of this discussion, the windings which carry the currents ($i_{1\beta}$, $i_{2\beta}$), then a primary current $i_{1\alpha}$ (formerly i_1 in the single phase model) leads to a primary MMF $n_1 i_{1\alpha}$. In an ITF, this primary MMF must, for reasons discussed in Chap. 3, correspond to a secondary MMF $n_2 i_{2\alpha}$, where $i_{2\alpha}$ is now used instead of i_2 (as used in the single phase model). Furthermore, the circuit flux vector $\vec{\psi}_{\rm m}$ is under these circumstances oriented along the horizontal axis, which is precisely the chosen direction for the "real" axis of the new complex plane, in which case $\vec{\psi}_{\rm m} = \psi_{{\rm m}\alpha}$.

The relationship between currents and flux-linkages for the two-phase ITF model proceeds along similar lines as discussed for the single phase ITF model. The relationship between the primary and secondary currents is given as

$$n_1 i_{1\alpha} - n_2 i_{2\alpha} = 0 \tag{6.1a}$$

$$n_1 i_{1\beta} - n_2 i_{2\beta} = 0 \tag{6.1b}$$

The primary and secondary flux-linkages are defined as

$$\psi_{1\alpha} = n_1 \phi_{m\alpha} \tag{6.2a}$$

$$\psi_{2\alpha} = n_2 \phi_{m\alpha} \tag{6.2b}$$

$$\psi_{1\beta} = n_1 \phi_{m\beta} \tag{6.2c}$$

$$\psi_{2\beta} = n_2 \phi_{m\beta} \tag{6.2d}$$

Expressions (6.1) and (6.2) correspond to the space vector form given in Sect. 4.4 where the general notational form $\vec{x} = x_\alpha + jx_\beta$ was introduced. The resultant space vector based ITF equation set is given by (6.3).

$$\vec{u}_1 = \frac{d\vec{\psi}_1}{dt} \tag{6.3a}$$

$$\vec{u}_2 = \frac{d\vec{\psi}_2}{dt} \tag{6.3b}$$

$$\vec{\psi}_2 = \left(\frac{n_2}{n_1}\right) \vec{\psi}_1 \tag{6.3c}$$

$$\vec{i}_1 = \left(\frac{n_2}{n_1}\right) \vec{i}_2 \tag{6.3d}$$

The corresponding symbolic model and generic model of the space vector based ITF are given in Fig. 6.2. Note that the generic model shown in Fig. 6.2b represents the so-called ITF-flux version, which is one of two possible model configurations available (see Fig. 3.3). One configuration is shown here to demonstrate the transition from single phase phasor representation to space vector form. However, both are equally applicable for the space vector based ITF model. To illustrate the difference between the single and two-phase ITF models, we use "vector lines" (lines which are drawn wider when compared to single phase), which now represent the real and imaginary components. For example, the "vector line" $i_{1\alpha}$, $i_{1\beta}$ represents the vector $\vec{i}_1$. The symbolic and generic models as discussed for the single phase ITF based transformer remain unchanged in terms of configuration. Consequently,

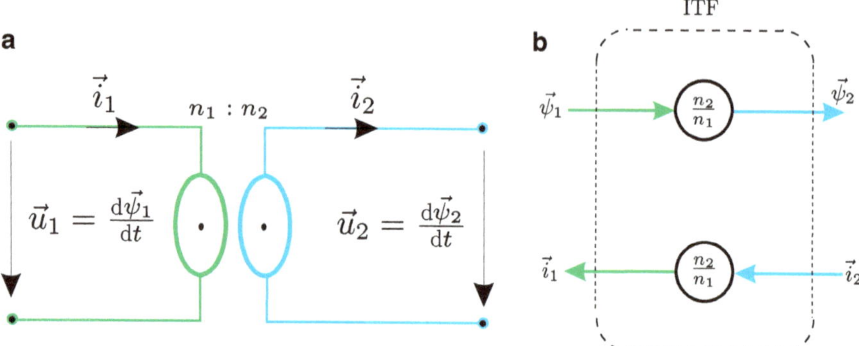

Fig. 6.2 Symbolic and Generic space vector based ITF models. (**a**) Symbolic model. (**b**) Generic model

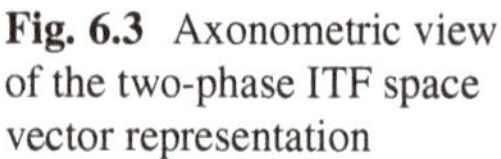

Fig. 6.3 Axonometric view of the two-phase ITF space vector representation

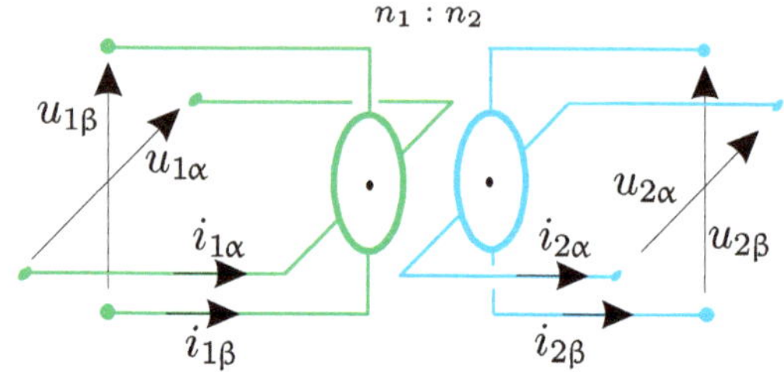

the ITF based single phase models with its extensions as discussed in Chap. 3 are directly applicable here. An alternative symbolic way to represent the space vector ITF model of Fig. 6.2a is given in Fig. 6.3. In this figure, the ITF is shown in terms of its primary and secondary α, β components. This representation will be particularly instructive when discussing the so-called IRTF module.

6.2.1 *Simplified ITF Based Transformer Example*

In this section an application example is given which demonstrates the use of the space vector based ITF model (see Fig. 6.2). The ITF model is extended by the introduction of a finite magnetizing inductance L_m as discussed in Sect. 3.4 for the single phase case. Leakage inductance and winding resistance are ignored in this model. A series configured resistive/inductive load is connected to the secondary winding. The symbolic model representation of this system is given in Fig. 6.4.

The aim is to build a generic model of this system which will be transformed to a PLECS model (see the tutorial's at the end of this chapter). As input to the generic model we will assume the primary flux-linkage space vector $\vec{\psi}_1$ rather than the primary supply voltage vector $\vec{u}_1$. The reason for doing this is to emphasize the fact that it is the flux-linkage vector, which is central to the behavior of this type of

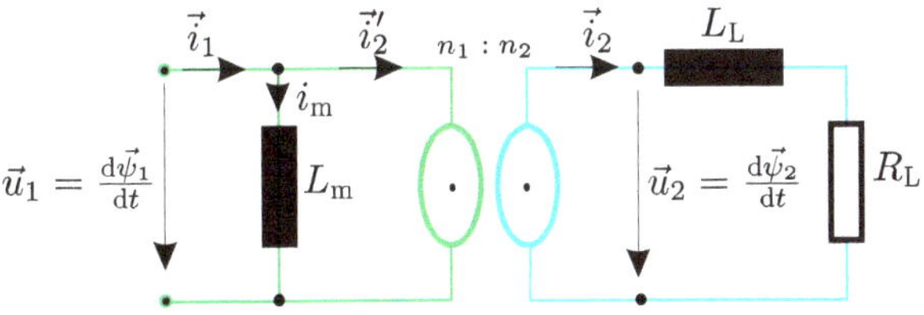

Fig. 6.4 Two-phase transformer example with R, L load (space vector model)

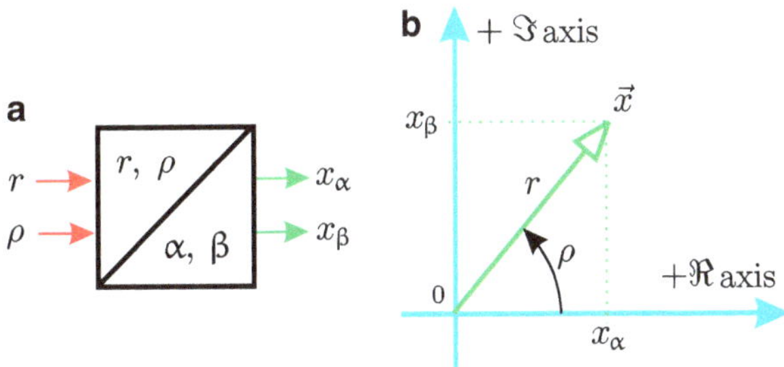

Fig. 6.5 Module and vector diagram from polar to cartesian conversion. (**a**) Module. (**b**) Vector diagram

system. It is noted that generally we are not able to use the flux-linkage as an input vector given that its amplitude will change when a more complicated transformer model is used (as discussed earlier, see single phase transformer Sect. 3.6).

We will assume that the primary flux-linkage vector is given by

$$\vec{\psi}_1 = \hat{\psi}_1 \, e^{j\omega t} \tag{6.4}$$

The implementation of Eq. (6.4) in generic form calls for the introduction of a new building block, namely a "polar to cartesian" conversion module, as indicated in Fig. 6.5a. The conversion equation set is derived with the aid of Fig. 6.5b, which tells us that a vector $\vec{x}$ can be either written in its polar form $\vec{x} = r\,e^{j\rho}$ or cartesian form $\vec{x} = x_\alpha + jx_\beta$. A comparison of these notation forms and observation of Fig. 6.5b gives

$$x_\alpha = r \cos \rho \tag{6.5a}$$

$$x_\beta = r \sin \rho \tag{6.5b}$$

For this problem it is helpful to reconsider the module according to Fig. 6.5a, as this can be directly used to build the generic model shown in Fig. 6.6. The model according to Fig. 6.6 shows the polar to cartesian conversion unit which has as input the amplitude and argument of the space vector $\vec{\psi}_1$. The output represents the real and imaginary components of the flux-linkage vector. These components

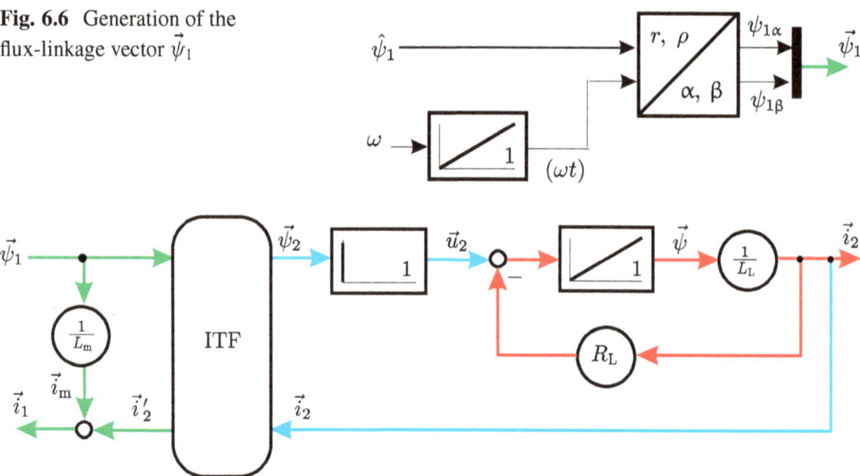

Fig. 6.6 Generation of the flux-linkage vector $\vec{\psi}_1$

Fig. 6.7 Generic ITF based transformer model with load: space vector form

are then combined via a "multiplexer" to give a single "vector." In the following the multiplexer will be placed inside the conversion module. The resultant vector $\vec{\psi}_1$ serves as an input to the generic model of the transformer.

A suitable generic model of the ITF based transformer model with load can found by making use of the single phase model shown in Fig. 3.9. In the latter case the load was a resistance, which in this case needs to be replaced by a resistance-inductance combination. Furthermore, the reader is reminded of the fact that the generic model is now used in its space vector form, i.e., "vectors" are now used in the diagram. The generic model of the load is directly taken from the earlier example shown in Fig. 4.19 given that it represents precisely the resistance-inductance network. In this case, the resistance and inductance values are defined as R_L and L_L.

The model according to Fig. 6.7 shows a differentiator module, which we can implement in different ways to avoid potential simulation problems. The reader is reminded that the use of differentiator modules in simulations can be problematic and should therefore be handled with care. For the tutorial exercise linked with this chapter, we need access to the primary voltage vector which is defined in Eq. (6.3a). Furthermore, the differentiator module shown in Fig. 6.7 is there to implement Eq. (6.3b). Both equations differentiate a flux vector which (in this section) in its general form is given as

$$\vec{\psi} = \hat{\psi}\, e^{j\omega t} \tag{6.6}$$

where $\hat{\psi}$ is (in this example) *not* a function of time. If we differentiate Eq. (6.6) according to $\vec{u} = {d\vec{\psi}}/{dt}$, we find a very simple representation, namely $\vec{u} = j\omega\vec{\psi}$. The generic implementation of this alternative "differentiator" module is shown in Fig. 6.8.

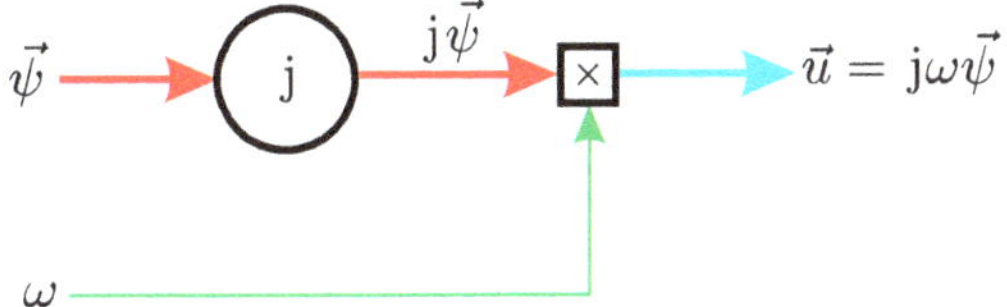

Fig. 6.8 Alternative implementation of differentiator module for sinusoidal flux-linkage time functions

The generic module according to Fig. 6.8 is *only* usable when the flux amplitude and frequency are constant. These conditions are valid here and consequently the module can be used to implement Eqs. (6.3a) and (6.3b) with the appropriate flux vector.

The gain module shown in Fig. 6.8 requires some further attention in terms of modeling such a unit. Essentially the gain $\mathrm{j} = e^{\mathrm{j}\pi/2}$ rotates an input vector $\vec{x} = x_\alpha + \mathrm{j}x_\beta$ by $\pi/2$ rad (90°). Hence the relation between input and output (for the gain module) vector $\vec{y} = y_\alpha + \mathrm{j}y_\beta$ is of the form $\vec{y} = \mathrm{j}\,\vec{x}$, which may also be written in the form given in Eq. (6.7).

$$\begin{bmatrix} y_\alpha \\ y_\beta \end{bmatrix} = \begin{bmatrix} 0 & -1 \\ 1 & 0 \end{bmatrix} \begin{bmatrix} x_\alpha \\ x_\beta \end{bmatrix} \tag{6.7}$$

In the Simulink environment, Eq. (6.7) is directly usable with a "Matrix gain" type element.

6.2.2 *Phasor Analysis of Simplified Model*

The analysis shown here is in fact very similar to that carried out for the three-phase R, L model (see Sect. 4.7.1). Its inclusion here can therefore be seen as a revision exercise applied to the transformer system.

The flux-linkage vector (see Eq. (6.4)) is the input vector which corresponds with the phasor $\underline{\psi}_1 = \hat{\psi}_1$. The remaining phasors are found using the phasor based equation set of this system which is of the form

$$\underline{u}_1 = \mathrm{j}\omega\,\underline{\psi}_1 \tag{6.8a}$$

$$\underline{u}_2 = \mathrm{j}\omega\,\underline{\psi}_2 \tag{6.8b}$$

$$\underline{\psi}_2 = \left(\frac{n_2}{n_1}\right)\underline{\psi}_1 \tag{6.8c}$$

$$\underline{i}_2' = \left(\frac{n_2}{n_1}\right)\underline{i}_2 \tag{6.8d}$$

$$\underline{i}_1 = \underline{i}_2' + \underline{i}_\mathrm{m} \tag{6.8e}$$

$$\underline{i}_\mathrm{m} = \frac{\underline{\psi}_1}{L_\mathrm{m}} \tag{6.8f}$$

$$\underline{u}_2 = (R_\mathrm{L} + \mathrm{j}\omega L_\mathrm{L})\,\underline{i}_2 \tag{6.8g}$$

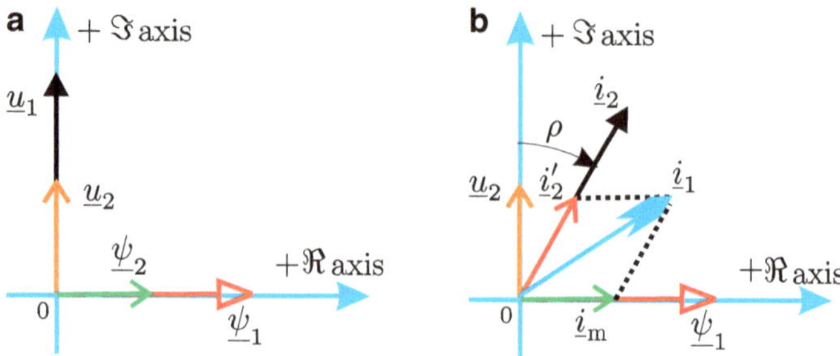

Fig. 6.9 Phasor diagrams for transformer with *R-L* load. (**a**) Voltage/flux. (**b**) Current/flux/voltage

The reader is advised to look carefully at Eq. (6.8) in terms of correlating with the generic block diagram (Fig. 6.7), which in fact represents the space vector based equation set for the system under consideration.

We will now proceed with the analysis to find the unknown phasors. The input voltage phasor is found using Eq. (6.8a) with $\underline{\psi}_1 = \hat{\psi}_1$, which gives $\underline{u}_1 = j\omega\hat{\psi}_1$. The secondary flux-linkage vector is found using (6.8c), which gives $\underline{\psi}_2 = (n_2/n_1)\,\hat{\psi}_1$. This vector in turn allows us to find (with the aid of Eq. (6.8b)) the secondary voltage phasor, namely $\underline{u}_2 = j\omega\frac{n_2}{n_1}\hat{\psi}_1$. We are now able to find the load current phasor with the aid of Eq. (6.8g) which yields

$$\underline{i}_2 = \frac{\underline{u}_2}{R_L + j\omega L_L} \tag{6.9}$$

The load current phasor on the secondary side corresponds with a current phasor (known as the primary referred secondary current phasor) $\underline{i}'_2$ on the primary side of the ITF, which is calculated using Eqs. (6.9) and (6.8d). Finally, the primary current is calculated by making use of Eqs. (6.8e) and (6.8f) with $\underline{\psi}_1 = \hat{\psi}_1$. Two phasor diagrams are given in Fig. 6.9 for the case $n_2/n_1 = 0.5$.

An observation of Fig. 6.9a shows that the primary flux-linkage is aligned with the real axis. This can be expected given that the primary flux-linkage has no imaginary component. The secondary flux-linkage phasor must be aligned with the primary phasor and its amplitude is reduced by a factor 0.5, which corresponds to the chosen winding ratio. The voltages are $\pi/2$ rad rotated forward with respect to their flux phasors. The secondary current phasor $\underline{i}_2$, as shown in Fig. 6.9b, lags the secondary voltage phasor by an angle $\rho = -\arctan(\omega L_L/R_L)$ as may be deduced from Eq. (6.9). The primary referred secondary current must be in phase with the secondary current phasor and its value is reduced by a factor 0.5, given our choice of winding ratio. Adding, in vector terms, the primary referred secondary current $\underline{i}'_2$ and the magnetizing current phasor (which must be in phase with the primary flux-linkage phasor) yields the primary current phasor $\underline{i}_1$, as may be observed from Fig. 6.9b.

6.3 Two-Phase ITF Based Universal Transformer Model

The use of a space vector type notation allows us to take any single phase symbolic or generic model (as developed in Chap. 3) and use it in a two-phase system context.

The single phase transformer development to include magnetizing inductance and leakage is therefore equally applicable to two-phase systems. Likewise, the universal model as given in Fig. 3.16 is readily converted to a space vector form as indicated in Fig. 6.10. Note that the lines shown in Fig. 6.10 now represent two variables. Furthermore, the primary and secondary winding resistances are added to this model. The equation set which corresponds with Fig. 6.10 is as follows:

$$\vec{u}_1 = \vec{i}_1 R_1 + \frac{\mathrm{d}\vec{\psi}_1}{\mathrm{d}t} \tag{6.10a}$$

$$\vec{\psi}_1 = \vec{i}_1 L_\sigma^{\mathrm{pr}} + \vec{\psi}_\mathrm{M} \tag{6.10b}$$

$$\vec{\psi}_\mathrm{M} = L_\mathrm{M} \vec{i}_\mathrm{M} \tag{6.10c}$$

$$\vec{\psi}\,'_2 = \vec{\psi}_\mathrm{M} - L_\sigma^{\mathrm{se}} \vec{i}\,'_2 \tag{6.10d}$$

$$\frac{\mathrm{d}\vec{\psi}_2}{\mathrm{d}t} = \vec{u}_2 + \vec{i}_2 R_2 \tag{6.10e}$$

$$\vec{\psi}\,'_2 = k\,\vec{\psi}_2 \tag{6.10f}$$

$$\vec{i}_2 = k\,\vec{i}\,'_2 \tag{6.10g}$$

where the transformer ratio k and inductance parameters L_σ^{pr}, L_σ^{se}, and L_M are a function of the transformation factor a as may be deduced from Eqs. (3.26) and (3.27).

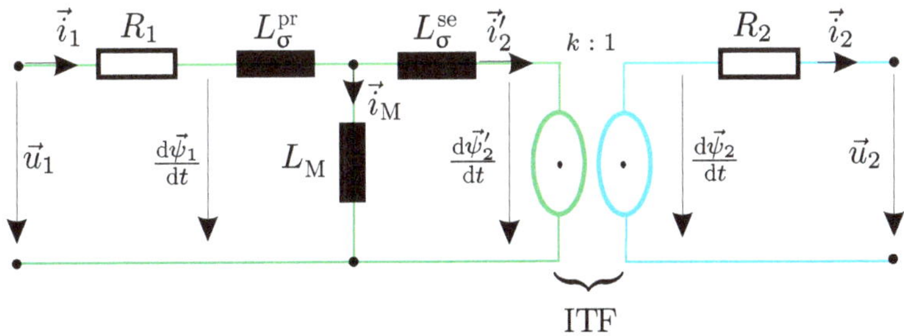

Fig. 6.10 Space vector based universal transformer model

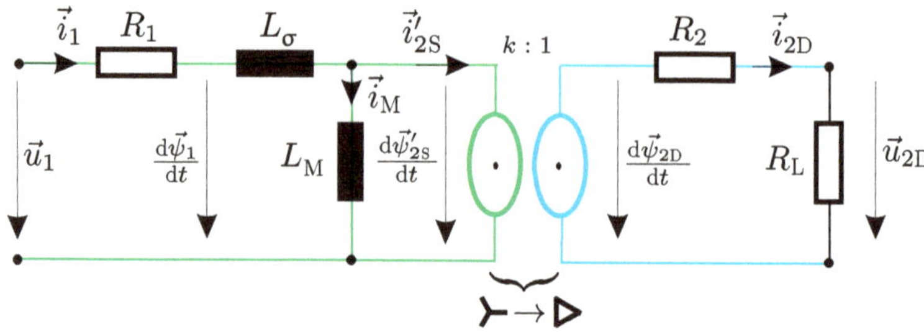

Fig. 6.11 Symbolic model, transformer example

6.3.1 Primary Leakage Inductance Based Transformer Example

A transformer application example is considered in this section which is based on the symbolic model given in Fig. 6.10 for the case $L_\sigma^{se} = 0$. A two-inductor model representation thus appears which is similar to that discussed in Sect. 3.7.1 (for a single phase model). The resultant model, shown in Fig. 6.11, consists of a leakage inductance L_σ^{pr}, which for the sake of readability is renamed as L_σ, and a magnetizing inductance L_M. Observation of Fig. 6.11 shows that a load resistance R_L is connected to the secondary winding. Furthermore, the secondary phase windings are assumed to be connected in delta while the primary windings are star configured. The load resistance R_L as shown in Fig. 6.10 represents in three-phase terms a delta connected symmetrical load where each load phase consists of a resistance R_L. The *star→delta* symbol underneath the ITF module identifies the presence of a winding connection change between primary and secondary. No such notation is normally shown with the ITF module if the winding configuration between secondary and primary is unchanged. Equation set (6.10) is applicable to this example where the subscripts S and D need to be added (as given in Fig. 6.11) to identify secondary star/delta space vectors. Furthermore, an additional equation must be added to equation set (6.10), namely

$$\vec{u}_{2D} = \vec{i}_{2D} R_L \tag{6.11}$$

where R_L represents the load resistance. The primary windings are taken to be connected to a three-phase sinusoidal grid with angular frequency ω, which in space vector form corresponds to a vector $\vec{u}_1 = \hat{u}_1 e^{j\omega t}$.

An example of a generic diagram based on Eqs. (6.10) and (6.11) which can be used for dynamic simulation purposes is given in Fig. 6.12. The diagram according to Fig. 6.12 uses an *ITF-current* module (because the primary current is designated as an input). In addition, if the internal delta variables are required, a star/delta and delta/star conversion module are explicitly shown in this example. The first

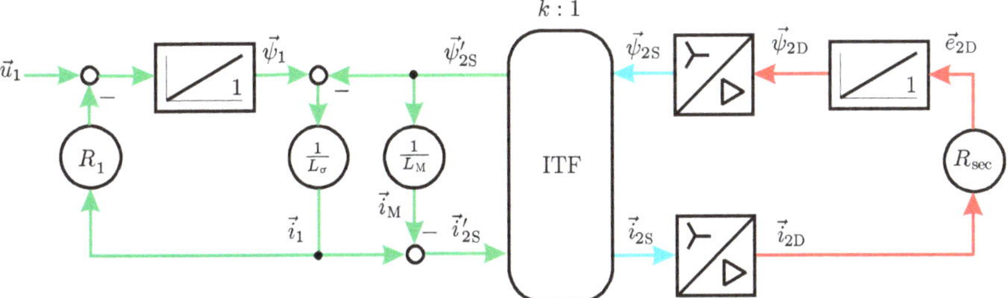

Fig. 6.12 Generic model, transformer example

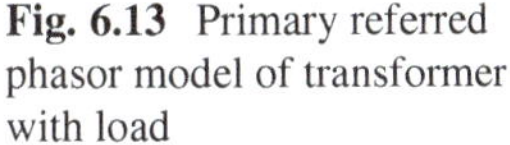

Fig. 6.13 Primary referred phasor model of transformer with load

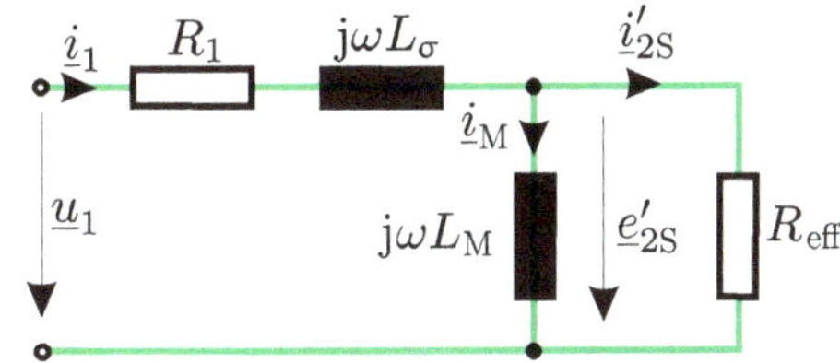

conversion $\vec{\psi}_{2D} \rightarrow \vec{\psi}_{2S}$ uses the "voltage" conversion equation (4.49), while the second conversion $\vec{i}_{2S} \rightarrow \vec{i}_{2D}$ uses Eq. (4.57). A voltage vector $\vec{e}_{2D} = {}^{d\vec{\psi}_{2D}}/_{dt}$ is introduced on the secondary side of model, given that the input to the integrator (on the secondary side) is equal to $\vec{e}_{2D} = R_{sec}\vec{i}_{2D}$, where $R_{sec} = R_2 + R_L$.

6.3.2 *Phasor Analysis of Primary Leakage Inductance Based Model Example*

The model according to Fig. 6.11 can also use phasors. Such an analysis may, in some cases, be useful if it is sufficient to consider the model under steady-state conditions. In other instances a phasor analysis is useful as means of verifying the steady-state results obtained with a dynamic simulation model.

In this example, the aim is to calculate the phasors $|\underline{i}_1|$, $|\underline{i}_{2D}|$, $|\underline{\psi}_1|$, and $|\underline{\psi}_{2D}|$ which represent the output variables from the dynamic simulation. The excitation to the model is the phasor $\underline{u}_1$ which is linked with the space vector representation $\vec{u}_1 = \underline{u}_1 e^{j\omega t}$. The supply vector amplitude is $\hat{u}_1$, hence $\underline{u}_1 = \hat{u}_1$. The calculation of the primary current phasor $\underline{i}_1$ is similar to the approach discussed for the single phase transformer. This approach makes use of a primary referred model as shown in Fig. 6.13.

The components which represent the primary resistance, leakage inductance, and magnetizing inductance are readily identifiable. Also shown is a component R_{eff}, which is the effective "primary referred" load resistance. Its value is of the form given in Eq. (6.12).

$$R_{\mathrm{eff}} = \frac{1}{3}k^2 \underbrace{(R_{\mathrm{L}} + R_2)}_{R} \tag{6.12}$$

Equation (6.12) can be made plausible by realizing that the total secondary resistance R must be transformed from a delta to a star equivalent form, which, according to Sect. 4.7.1, is possible by multiplying the impedance by a factor $\frac{1}{3}$. Furthermore, Eq. (6.12) shows a factor k^2, where k represents the ITF winding ratio. This factor as introduced for the single phase models is required to "refer" a secondary impedance to the primary side.

The primary current phasor $\underline{i}_1$ is found using $\underline{i}_1 = \underline{u}_1/\underline{Z}_1$, where $\underline{Z}_1$ represents the input impedance. Once the current is found, the phasor $\underline{\psi}_1$ can be calculated using $\underline{\psi}_1 = (\underline{u}_1 - \underline{i}_1 R_1)\frac{1}{\mathrm{j}\omega}$. The calculation of the secondary current is more elaborate as it requires access to the phasor $\underline{e}_{2\mathrm{D}}$. The process of calculating this phasor is initiated by determining the phasor $\underline{e}'_{2\mathrm{S}}$, which according to Fig. 6.13 can be found using $\underline{e}'_{2\mathrm{S}} = \underline{u}_1 - \underline{i}_1 R_1 - \mathrm{j}\omega L_\sigma \underline{i}_1$.

The phasor $\underline{e}_{2\mathrm{S}} = \mathrm{j}\omega\underline{\psi}_{2\mathrm{S}}$ is found using $\underline{e}_{2\mathrm{S}} = \underline{e}'_{2\mathrm{S}}\frac{1}{k}$ and this phasor must be transformed to the delta form (see Eq. (4.66a)) with $\underline{e}_{2\mathrm{D}} = \underline{e}_{2\mathrm{S}}\sqrt{3}\,\mathrm{e}^{-\mathrm{j}\gamma/4}$. Once this phasor is found, the flux-linkage phasor $\underline{\psi}_{2\mathrm{D}}$ and secondary current phasor $\underline{i}_{2\mathrm{D}}$ can be calculated according to $\underline{\psi}_{2\mathrm{D}} = \underline{e}_{2\mathrm{D}}\frac{1}{\mathrm{j}\omega}$ and $\underline{i}_{2\mathrm{D}} = \underline{e}_{2\mathrm{D}}\frac{1}{R_{\mathrm{L}}+R_2}$.

6.4 Tutorials

6.4.1 Tutorial 1: Three-Phase Transformer with Load

This tutorial is concerned with implementing the generic model given in Fig. 6.7. The load which is connected to the secondary winding is formed by an inductance $L_{\mathrm{L}} = 1\,\mathrm{mH}$ and resistance $R_{\mathrm{L}} = 0.4\,\Omega$. As input to the model, a flux-linkage space vector $\vec{\psi}_1$ is assumed of the form $\vec{\psi}_1 = \hat{\psi}_1\,\mathrm{e}^{\mathrm{j}\omega t}$, where $\hat{\psi}_1 = 1.0\,\mathrm{Wb}$ and $\omega = 2\pi\,50\,\mathrm{rad/s}$. The ITF winding ratio is taken to be $\frac{n_1}{n_2} = 5$. Furthermore, the magnetizing inductance is set to $L_{\mathrm{m}} = 1.5\,\mathrm{H}$. Leakage inductance and winding resistances are ignored.

Make use of a PLECS `Polar->rect` module to generate the flux vector input, with components $\psi_{1\alpha}$ and $\psi_{1\beta}$ for the transformer. Use a `Constant` module to generate the ω (electrical frequency in rad/s) value and add an integrator which will give as output the variable ωt. Check your work by using an "XY" scope module. Under "simulation parameters," set the simulation time to 1 s and observe the result, which should be a circle with a radius equal to 1.0. Maintain the solver settings and run time as given above for the rest of the tutorial.

Create a sub module which will generate the voltage space vector $\vec{u}_1 = \mathrm{j}\omega\,\vec{\psi}_1$ as shown in Fig. 6.8. You will need to use a multiplier and gain module with gain j. The latter is realized in PLECS by using a `Gain` module. Select "Matrix gain

K*u" within this module and set the gain to [0 -1;1 0]. Next, add an `ITF_flux` module, magnetizing inductance L_m, and a series of modules which will allow you to calculate the secondary voltage vector $\vec{u}_2 = j\omega\,\vec{\psi}_2$, where $\vec{\psi}_2$ is the output vector from the ITF module. Use four vector to RMS converters with display units to show the RMS primary/secondary voltage and RMS primary/secondary current values. Furthermore, add two additional sub-modules (as discussed in tutorial 5.5.4) which calculate the primary real and reactive power values P(W) and Q(Var), respectively.

The last step is concerned with testing our transformer under load and no-load conditions. Add a `manual switch` which enables you to connect/disconnect the load. An example of a PLECS model which corresponds to the generic model in question is given in Fig. 6.14. Clear observable are two sub-modules `jwX` which are used to generate the voltage vectors $\vec{u}_1$ and $\vec{u}_2$ based on the generic model shown in Fig. 6.8. Also present in Fig. 6.14 is a `Polar->rect` module which generates the primary flux vector $\vec{\psi}_1$. Run the simulation and record the six display values. Repeat the exercise for the no-load case.

The values which should appear on the display modules for the variables under discussion are given in Table 6.1.

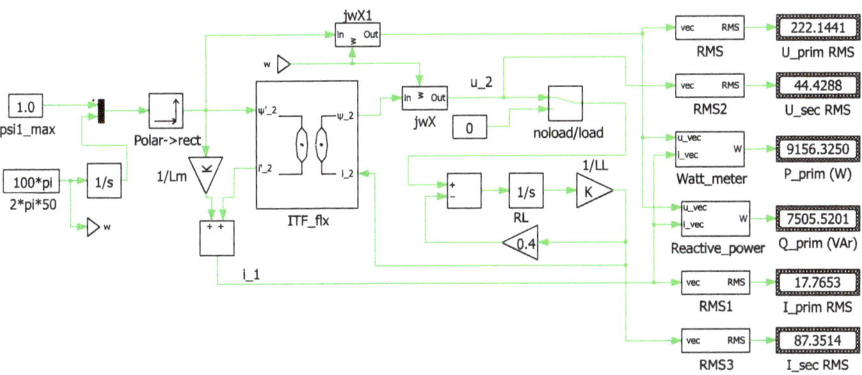

Fig. 6.14 PLECS model of transformer with load

Table 6.1 Simulation results transformer no-load/load

Parameters		No-load	Load
RMS primary voltage	U_{prim}	222.14 V	222.14 V
RMS secondary voltage	U_{sec}	44.42 V	44.42 V
RMS primary current	I_{prim}	0.47	17.76 A
RMS secondary current	I_{sec}	0.00 A	87.35 A
Real primary power	P_{prim}	0.00 W	9156.32 W
Reactive primary power	Q_{prim}	314.15 VA	7505.52 VAr

6.4.2 Tutorial 2: Phasor Analysis of a Three-Phase Transformer with Load

A phasor analysis should be carried out of the transformer configuration as discussed in the previous tutorial, to verify the simulation results obtained via the six display units shown in Fig. 6.14. The input for this analysis is taken to be the flux phasor $\underline{\psi}_1 = 1.0\,\mathrm{Wb}$. On the basis of this phasor we can calculate the remaining phasors $\underline{\psi}_2, \underline{u}_1, \underline{u}_2, \underline{i}_\mathrm{m}, \underline{i}_2, \underline{i}_1$. In addition, the real and reactive power value can be calculated. The exercise should be carried out for the no-load (means no *R-L* network connected to the secondary of the transformer) and load situation. A MATLAB file must be written for this exercise. An example of such an M-file is as follows:

M-file Code

```
%Tutorial 2, chapter 6
psi1_hat=1.0;                 % primary flux amplitude
psi1=psi1_hat;                % primary flux phasor
k=5;                          % n1/n2=5 winding ratio
Lm=1.5;                       % magnetizing inductance
w=2*pi*50;                    % frequency rad/s
%%%%%%%%%%%%%%%%%%%%%%%%%%%%
u1=j*w*psi1_hat;              % primary supply voltage phasor
U1=abs(u1)/sqrt(2);           % primary RMS voltage
psi2=1/k*psi1;                % secondary flux phasor
u2=j*w*psi2;                  % secondary  voltage phasor
U2=abs(u2)/sqrt(2);           % secondary RMS voltage
%
%%%%%%no-load case
im=psi1/Lm;                   % magnetizing current phasor
i1n=im;                       % no secondary current
I1n=abs(i1n)/sqrt(2);         % primary current no-load
Pn=3/2*real(u1*conj(i1n));    % primary real power
Qn=3/2*imag(u1*conj(i1n));    % primary reactive power
%%%%%%%%%%%%%%%%%%%%%%%%%%%%%%%%%%%%%%%%%
%%%%%%%% load case
LL=1e-3;                      % load inductance
RL=0.4;                       % load resistance
i2=u2/(RL+j*w*LL);            % secondary load current phasor
I2=abs(i2)/sqrt(2);           % secondary  RMS current
i2r=i2/k;                     % primary referred current phasor
i1=im+i2r;                    % primary current phasor load
I1=abs(i1)/sqrt(2);           % primary current load
P=3/2*real(u1*conj(i1));      % primary real power (load)
Q=3/2*imag(u1*conj(i1));      % primary reactive power (load)
```

The results obtained after running the M-file should match closely with those given in Table 6.1.

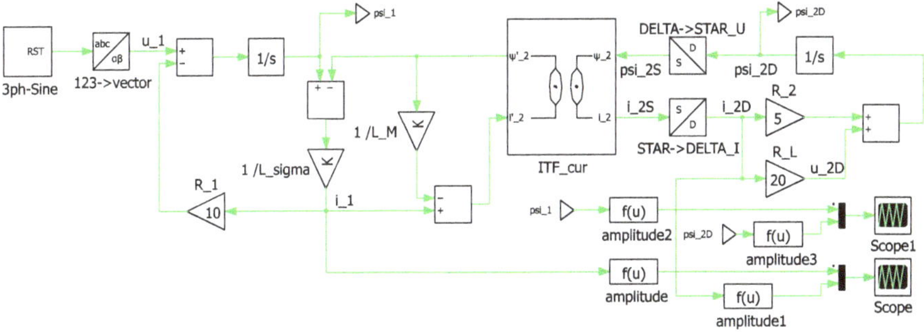

Fig. 6.15 PLECS model of transformer with resistive load

6.4.3 *Tutorial 3: Three-Phase Star/Delta Configured Transformer with Load*

A PLECS implementation of the example outlined in Sect. 6.3.1 is considered in this tutorial. A three-phase supply source as given in Fig. 4.47a is to be used, with an RMS phase voltage of $U_1 = 220$ V and angular supply frequency of $\omega = 2\pi f$ rad/s, where $f = 50$ Hz. The primary and secondary winding resistances are set to $R_1 = 10\,\Omega$ and $R_2 = 5\,\Omega$, respectively. The leakage and magnetizing inductance are equal to $L_\sigma = 10$ mH and $L_M = 300$ mH, respectively, while the load resistance of the delta connected load is set to $R_L = 20\,\Omega$. An ITF winding ratio of $k = 2$ is assumed.

The simulation is to be run for a period of 0.4 s and outputs of the simulation should be the amplitude of vectors $\vec{i}_1$, $\vec{i}_{2D}$, $\vec{\psi}_1$, and $\vec{\psi}_{2D}$. An example of a possible implementation of this problem is given in Fig. 6.15. The modules used in this example for the conversion of supply voltages to space vector format and those used for the star/delta conversions have been discussed in Chap. 4. Two `Scope` modules have been added to plot the results in the form of the absolute value versus time of the vectors $\vec{i}_1$, $\vec{i}_{2D}$, $\vec{\psi}_1$, and $\vec{\psi}_{2D}$. Figure 6.16 shows the current and flux plots which should appear after running your simulation.

The results show the presence of a transient in the current and flux waveforms after the transformer has been connected to the supply. A detailed observation of the results show that the steady-state currents and steady-state flux values are equal to $|\vec{i}_1| = 7.42$ A, $|\vec{i}_{2D}| = 8.08$ A and $|\vec{\psi}_1| = 0.77$ Wb, $|\vec{\psi}_{2D}| = 0.64$ Wb, respectively.

6.4.4 *Tutorial 4: Phasor Analysis of a Three-Phase Star/Delta Configured Transformer with Load*

The aim of this tutorial is to undertake a phasor analysis, as discussed in Sect. 6.3.2, to confirm the steady-state results obtained with the dynamic model discussed in the previous tutorial. The phasor analysis is to be given in the form of an M-file.

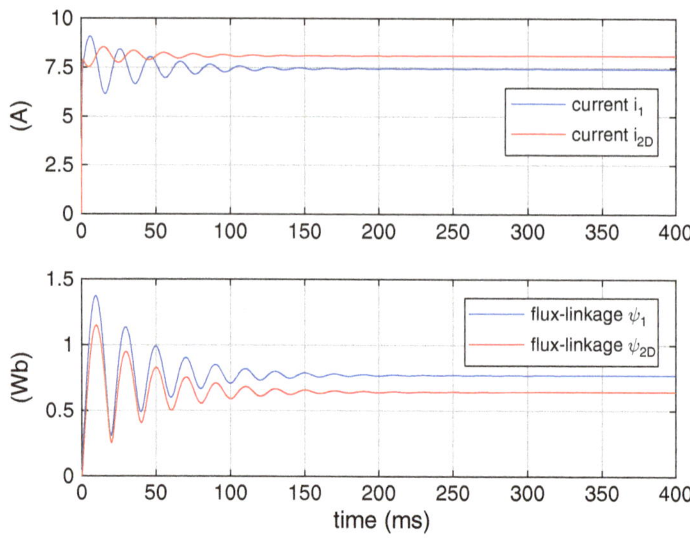

Fig. 6.16 Results: primary and secondary currents, $|\vec{i}_1|$, $|\vec{i}_{2D}|$ and flux-linkage, $|\vec{\psi}_1|$, $|\vec{\psi}_{2D}|$

An example of an M-file which shows this analysis is given at the end of this tutorial. The results, in the form of the variables $|\underline{i}_1|$, $|\underline{i}_{2D}|$, $|\underline{\psi}_1|$, and $|\underline{\psi}_{2D}|$, should match (within 1 %) the steady-state results obtained from the dynamic simulation.

M-file Code

```
%Tutorial 4, chapter 6
close all
%parameters
L_sig=10e-3;                        % leakage  inductance
L_M=300e-3;                         % magnetizing inductance
R_1=10;                             % primary resistance
R_2=5;                              % secondary resistance
R_l=20;                             % load resistance
k=2;                                % winding ratio ITF
%phasor analysis
U=220;                              % RMS phase voltage
gamma=2*pi/3;
w=100*pi;                           % angular freq
u_1=U*sqrt(2);                      % phasor amplitude
X_sig=j*w*L_sig;                    % magnetizing reactance
X_M=j*w*L_M;                        % leakage reactance
Rt=R_2+R_l;                         % sum load resistance
Rtref=k^2*Rt/3;                     % delta load/3,star,
                                      and referred
Zp=Rtref*X_M/(Rtref+X_M);           % XM/Rtref
Z1=R_1+X_sig+Zp;                    % total prim. impedance
```

```
i_1ph=u_1/Z1;                               % primary current
                                              phasor
i_1=abs(i_1ph);                             % abs value primary
                                              current
e_1ph=u_1-i_1ph*R_1;
psi_1ph=e_1ph/(j*w);                        % primary flux phasor
psi_1=abs(psi_1ph);                         % abs value primary
                                              flux
e2rSph=u_1-i_1ph*(R_1+X_sig);
e2Sph=e2rSph/k;                             % phasor secondary side
                                              (star)
e2Dph=e2Sph*sqrt(3)*(cos(-gamma/4)+j*sin(-gamma/4));
psi_2Dph=e2Dph/(j*w);                       % secondary flux phasor
                                              (delta)
psi_2D=abs(psi_2Dph);                       % abs value secondary
                                              flux
i_2Dph=e2Dph/Rt;                            % secondary current
                                              phasor
i_2D=abs(i_2Dph);                           % abs value secondary
                                              current
```

Chapter 7
Introduction to Electrical Machines

7.1 Introduction

This chapter considers the basic working principles of the so-called classical set of machines. This set of machines represents the asynchronous (induction), synchronous, DC machines, and variable reluctance machines. The latter will be discussed in detail, in the book "Advanced Electrical Drives" by the same authors. Of these classical machines, the asynchronous machine is most widely used in a large range of applications. Note that the term "machine" is used here, which means that the unit is able to operate as a motor (converting electrical power into mechanical power) or as a generator (converting mechanical power into electrical power). The machine can be fed via a power electronic converter or connected directly to an AC or DC supply.

Central to this chapter is the development of an *ideal rotating transformer* (IRTF), which is in fact a logical extension of the two-phase ITF module discussed in Chap. 6. We will then look to the conditions required for producing constant torque in an electrical machine. This in turn will allow us to derive the principle of operation for the three classical machine types. A universal two-phase model concept will be introduced at the end of this chapter which forms the backbone of the machine models presented in this book.

Electronic supplementary material The online version of this chapter (doi: 10.1007/978-3-319-29409-4_7) contains supplementary material, which is available to authorized users.

© Springer International Publishing Switzerland 2016

A. Veltman et al., *Fundamentals of Electrical Drives*, Power Systems,
DOI 10.1007/978-3-319-29409-4_7

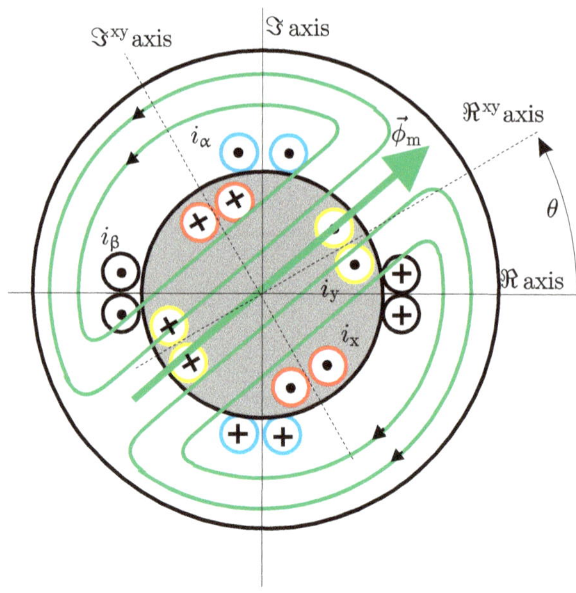

Fig. 7.1 Two-phase IRTF model

7.2 Ideal Rotating Transformer Concept

The fundamental building block for rotating machines used in this book is the IRTF module, which is directly derived from the two-phase space vector ITF concept given in Fig. 6.1 [13] . The new IRTF module, shown in Fig. 7.1, differs in two points. Firstly, the inner (secondary) part of the transformer is assumed to be able to rotate freely with respect to the outer (primary) side. The airgap between the two components of this model remains infinitely small. Secondly, the number of "effective" turns on the primary and secondary is assumed to be equal, i.e., $n_1 = n_2 = n$. Furthermore, the windings are (like the ITF) taken to be sinusoidally distributed (see Appendix A). This implies that the winding representation as shown in Fig. 7.1 is only symbolical as it shows where the majority of conductors for each phase are located. In the following, the primary and secondary will be referred to as the stator and rotor, respectively. A second complex plane (in addition to the stator based complex plane) with axis $\Re^{xy}$, $\Im^{xy}$ is introduced in Fig. 7.1 which is tied to the rotor. Note the use of the superscript "xy" which indicates that a vector is represented in *rotor coordinates*. For a stationary coordinate system, as used on the stator side, we sometimes use the superscript "αβ." However, in most cases this superscript is omitted to simplify the mathematical expressions. Hence, no superscript implies a stationary coordinate based vector.

The angle between the stationary and rotating complex plane is given as θ and this is in fact the machine shaft angle of rotation (relative to the stationary part of the motor). If the angle of rotation θ is set to zero, then the IRTF module is reduced

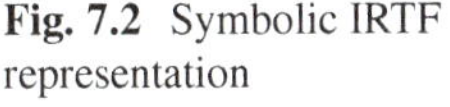

Fig. 7.2 Symbolic IRTF representation

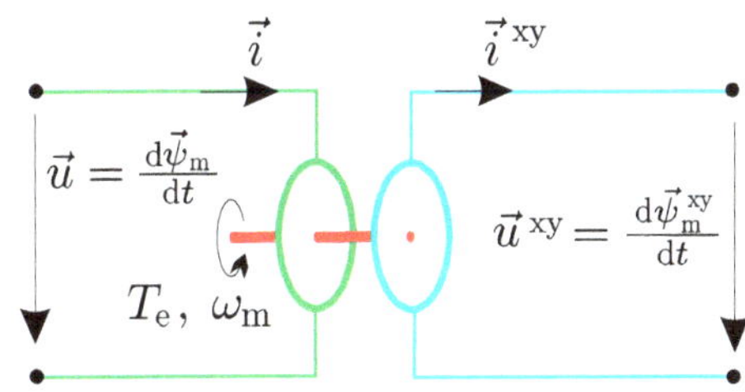

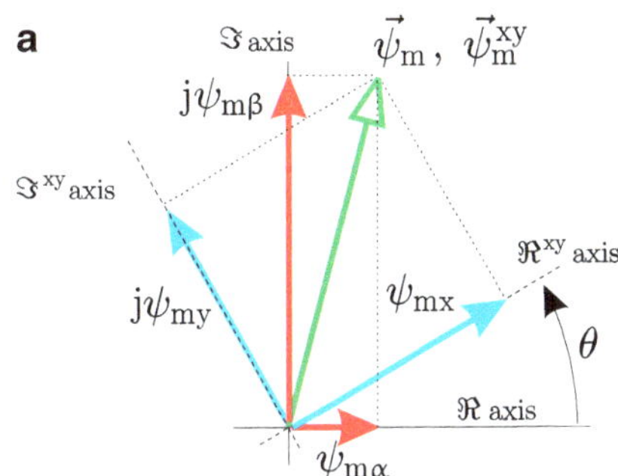

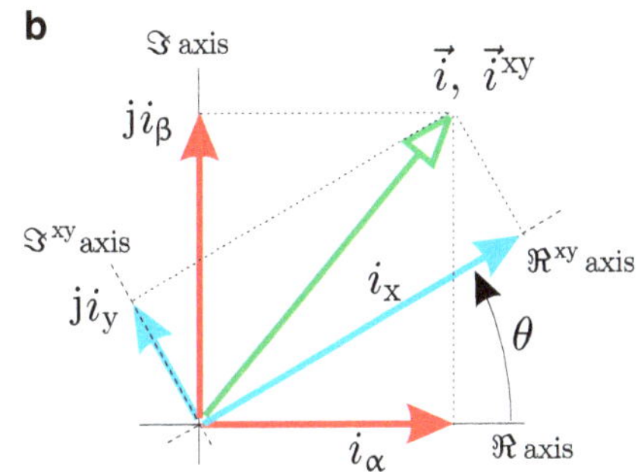

Fig. 7.3 Flux-linkage and current space vector diagrams. (**a**) Flux-linkage space vector. (**b**) Current space vector

to the two-phase ITF concept (with $n_1 = n_2$) as given by Fig. 6.1. The symbolic representation of the IRTF module as given in Fig. 7.2 shows similarity with the ITF module (see Fig. 6.2a).

The IRTF is a three-port unit (stator circuit, rotor circuit, and machine shaft). In Fig. 7.2, a symbolic shaft (shown in "red") is introduced, which links the rotor and stator circuits. The mechanical variables T_e and ω_m appear at the stator side of this symbolic representation as they can be observed by an observer linked to the stator coordinate system.

The flux linked with the stator and rotor is equal to $\vec{\psi}_m = n\vec{\phi}_m$ and can be expressed in terms of the components seen by each winding, namely

$$\vec{\psi}_m = \psi_{m\alpha} + j\psi_{m\beta} \tag{7.1a}$$

$$\vec{\psi}_m^{\,xy} = \psi_{mx} + j\psi_{my} \tag{7.1b}$$

An illustration of the flux-linkage seen by the rotor and stator winding is given in Fig. 7.3a. The relationship between the stator and rotor oriented flux-linkage space vectors can be expressed according to

$$\vec{\psi}_m^{\,xy} = \vec{\psi}_m\, e^{-j\theta} \tag{7.2}$$

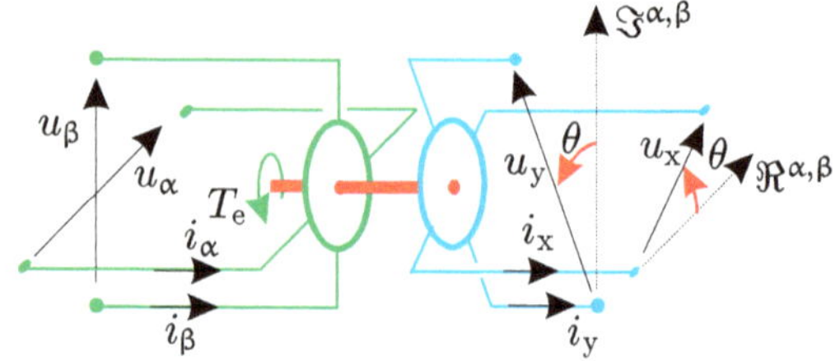

Fig. 7.4 Axonometric IRTF representation

The relationship between the rotor and stator oriented current space vectors as shown in Fig. 7.3b can be written as

$$\vec{i} = \vec{i}^{\text{xy}}\, e^{j\theta} \tag{7.3}$$

Figure 7.3 emphasizes the fact that there is only one flux-linkage and one current space vector present in the IRTF. The components of these vectors can be projected, in a rotating or stationary complex reference frame. The rotation relationship between the variables i_α, i_β and i_x, i_y, θ of Eq. (7.3) can also be written as

$$\begin{bmatrix} i_\alpha \\ i_\beta \end{bmatrix} = \begin{bmatrix} \cos\theta & -\sin\theta \\ \sin\theta & \cos\theta \end{bmatrix} \begin{bmatrix} i_x \\ i_y \end{bmatrix} \tag{7.4}$$

The symbolic diagram of Fig. 7.2 can also be shown in terms of its space vector components as was done for the ITF case (see Fig. 6.3). The result, given in Fig. 7.4, shows that the rotor side of the IRTF now rotates *with* the rotor shaft, given that it is physically attached to it. The energy balance for the IRTF module is found by using the power expressions (5.30) and (5.33) which are directly linked with the incremental energy $dW = p dt$. The input (stator) power p_{in} for the IRTF module is of the form

$$p_{\text{in}} = \frac{3}{2}\Re\left\{\vec{u}\left(\vec{i}\right)^*\right\} \tag{7.5}$$

Note that $\vec{u}$ and $\vec{i}$ are taken to be time dependent complex vectors as indicated by Eq. (5.8). Equation (7.5) can, with the aid of $\vec{u} = {}^{d\vec{\psi}_m}/_{dt}$, be expressed in terms of the stator (input) incremental energy

$$dW_{\text{in}} = \frac{3}{2}\Re\left\{d\vec{\psi}_m\left(\vec{i}\right)^*\right\} \tag{7.6}$$

Similar to the ITF, the IRTF is defined with a positive electrical and positive (rotor) electrical "power out" convention (see Fig. 3.4). Unlike the ITF, the IRTF has a second output power component formed by the product of the shaft torque T_{e} (Nm) and shaft speed ω_{m} (rad/s). If this product is positive, the machine is said to operate as motor. The total (mechanical plus electrical) output power is now of the form

$$p_{\mathrm{out}} = \frac{3}{2}\Re\left\{\vec{u}^{\,\mathrm{xy}}\left(\vec{i}^{\,\mathrm{xy}}\right)^{*}\right\} + T_{\mathrm{e}}\omega_{\mathrm{m}} \tag{7.7}$$

The incremental mechanical and electrical output energy linked with the rotor side of the IRTF can, with the aid of $\vec{u}^{\,\mathrm{xy}} = \mathrm{d}\vec{\psi}_{\mathrm{m}}^{\mathrm{xy}}/\mathrm{d}t$, be written as

$$\mathrm{d}W_{\mathrm{out}} = \frac{3}{2}\Re\left\{\mathrm{d}\vec{\psi}_{\mathrm{m}}^{\mathrm{xy}}\left(\vec{i}^{\,\mathrm{xy}}\right)^{*}\right\} + T_{\mathrm{e}}\mathrm{d}\theta \tag{7.8}$$

The overall IRTF incremental energy balance can, with the aid of Eqs. (7.6) and (7.8) and using the energy conservation law, be written as

$$\frac{3}{2}\Re\left\{\mathrm{d}\vec{\psi}_{\mathrm{m}}\left(\vec{i}\right)^{*}\right\} - \frac{3}{2}\Re\left\{\mathrm{d}\vec{\psi}_{\mathrm{m}}^{\mathrm{xy}}\left(\vec{i}^{\,\mathrm{xy}}\right)^{*}\right\} = T_{\mathrm{e}}\mathrm{d}\theta \tag{7.9}$$

The LHS of Eq. (7.9) shows the electrical incremental energy IRTF components. In Eq. (7.9), the term $\mathrm{d}\vec{\psi}_{\mathrm{m}}^{\mathrm{xy}}$ may be developed further using Eq. (7.2) and the differential "chain rule," which leads to

$$\mathrm{d}\vec{\psi}_{\mathrm{m}}^{\mathrm{xy}} = \mathrm{e}^{-\mathrm{j}\theta}\mathrm{d}\vec{\psi}_{\mathrm{m}} - \mathrm{j}\vec{\psi}_{\mathrm{m}}\mathrm{e}^{-\mathrm{j}\theta}\mathrm{d}\theta \tag{7.10}$$

Multiplication of Eq. (7.10) by the vector $\left(\vec{i}^{\,\mathrm{xy}}\right)^{*} = \vec{i}^{*}\mathrm{e}^{\mathrm{j}\theta}$ (see Eq. (7.3)) gives

$$\mathrm{d}\vec{\psi}_{\mathrm{m}}^{\mathrm{xy}}\left(\vec{i}^{\,\mathrm{xy}}\right)^{*} = \mathrm{d}\vec{\psi}_{\mathrm{m}}\vec{i}^{*} - \mathrm{j}\vec{\psi}_{\mathrm{m}}\vec{i}^{*}\mathrm{d}\theta \tag{7.11}$$

Substitution of Eq. (7.11) into Eq. (7.9) leads to the following expression for the electro-mechanical torque on the rotor.

$$T_{\mathrm{e}} = \frac{3}{2}\Re\left\{\mathrm{j}\vec{\psi}_{\mathrm{m}}\vec{i}^{*}\right\} \tag{7.12}$$

Equation (7.12) can, with the aid of expression $\Re\left\{\mathrm{j}\vec{a}\vec{b}^{*}\right\} = \Im\left\{\vec{a}^{*}\vec{b}\right\}$, be rewritten as

$$T_{\mathrm{e}} = \frac{3}{2}\Im\left\{\vec{\psi}_{\mathrm{m}}^{*}\vec{i}\right\} \tag{7.13}$$

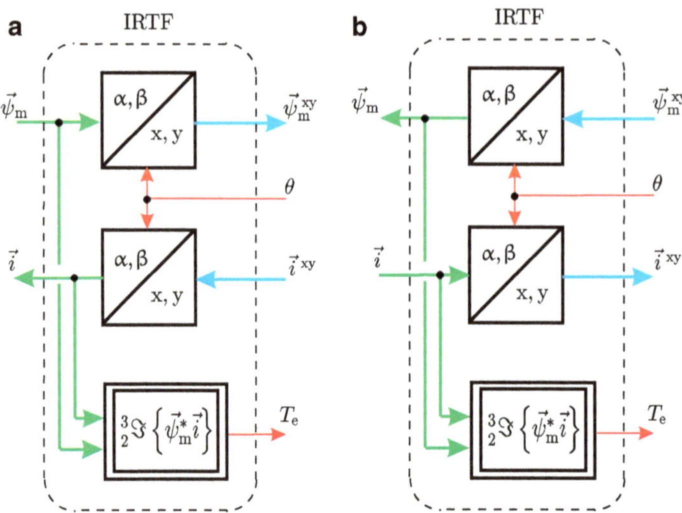

Fig. 7.5 Generic representations of IRTF module. (**a**) IRTF-flux. (**b**) IRTF-current

Hence, the torque acting on the rotor is at its maximum value when the two vectors $\vec{\psi}_{\mathrm{m}}$ and $\vec{i}$, shown in Fig. 7.3, are perpendicular with respect to each other. Under these circumstances the torque is directly related to the product of the rotor radius and Lorentz force. The latter is proportional to the magnitudes of the flux and current vectors. The generic diagram of the IRTF module that corresponds to the symbolic representation shown in Fig. 7.2 is based on Eqs. (7.2), (7.3), and (7.13). The IRTF generic module as given in Fig. 7.5a is shown with a stator to rotor coordinate flux conversion module and rotor to stator current conversion module. The two coordinate conversion modules can also be reversed as shown in Fig. 7.5b. The IRTF version used is application dependent as will become apparent at a later stage. The torque computation is not affected by the version used. Nor, for that matter, is the torque affected by the choice of coordinate system. The rotor angle θ required for the IRTF module must be derived from the mechanical equation set of the machine, which is of the form

$$T_{\mathrm{e}} - T_{\mathrm{l}} = J\frac{\mathrm{d}\omega_{\mathrm{m}}}{\mathrm{d}t} \tag{7.14a}$$

$$\omega_{\mathrm{m}} = \frac{\mathrm{d}\theta}{\mathrm{d}t} \tag{7.14b}$$

with T_{l} and J representing the load torque and inertia of the rotor/load combination, respectively (as discussed in Sect. 1.4.2). Finally, it is noted that the IRTF has a

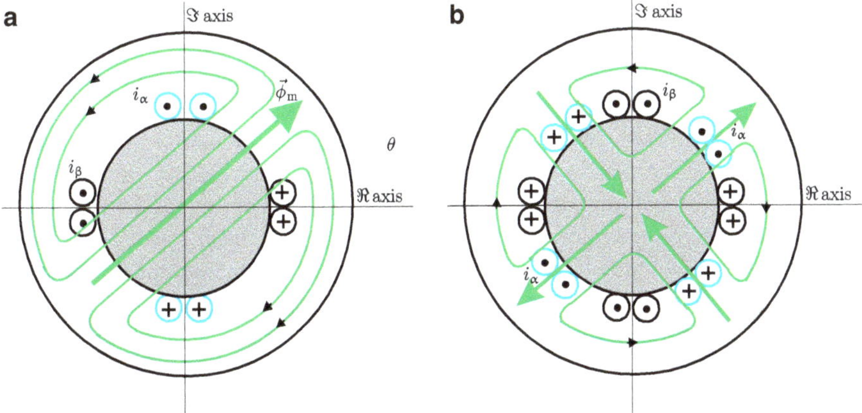

Fig. 7.6 Two- and four-pole flux distributions. (**a**) Two-pole flux distribution. (**b**) Four-pole flux distribution

unity winding ratio, which implies that an *inductance component (unlike a resistive component) may be moved from one side to the other without having to change its numerical value*.

7.2.1 IRTF Extension to "Multi-Pole" Representation

Up to now we have considered magnetic structures which have two magnetic poles, for example, the bar magnet given in Fig. 1.9a. The flux distribution (shown symbolically) of such a magnet is very similar to that shown in Fig. 7.6a, which is in fact a simplified version of Fig. 7.1, given that the rotor windings are not shown for didactic reasons. These rotor windings are however essential to the correct functioning of the IRTF. The model according to Fig. 7.6a has two magnetic poles or one pole pair ($p = 1$). Many electrical machines have more than one pole pair because often a more efficient winding configuration can be realized in, for example, a four-pole ($p = 2$) machine. The flux distribution which will occur in a four-pole machine is shown (symbolically) in Fig. 7.6b. The corresponding four-pole (sinusoidally distributed) phase windings, shown symbolically in Fig. 7.6b, are markedly different when compared to the two-pole case. In the latter case the majority of "α" windings are located on the imaginary axis. Furthermore, the angle between the phase winding halves is equal to π rad (see Fig. 7.6a). In addition, the two-phase windings are mechanically displaced by an angle of $\pi/2$ rad (90°). In the four-pole case the angle between phase winding halves is reduced to $\pi/4$ rad. In addition, each phase winding now has the majority of its sinusoidally distributed windings at four locations as shown in Fig. 7.6b. Furthermore, the two-phase windings are now mechanically displaced (with respect to each other) by an angle of $\pi/4$ rad.

This change in winding configuration has important implications for the so-called rotational speed (see Appendix A for more details on this concept). For example, in the two-pole machine we can rotate the resultant magnetic field by an angle of $\pi/2$ radians by way of a voltage excitation sequence of the two stator windings over a given time. When we apply the same excitation sequence to the four-pole machine, a magnetic field rotation of $\pi/4$ radians would occur. Vice versa if we move from a multi-pole to a two-pole model we need to double (when starting from a four-pole model) the rotor angle. The process of modeling the behavior of a multi-pole machine with a two-pole IRTF model can thus be initiated by introducing (as a first step) a "gain" module with gain p to the rotor angle input side of the IRTF model.

The torque per ampère produced by the multi-pole machine will increase when a larger number of pole pairs are used. A qualitative explanation of this statement is as follows: in the two-pole machine, as shown in Fig. 7.6a, a set of current carrying conductors is positioned along the circumference where the flux is at its highest level. This in turn causes a force on these windings and consequently a torque on the rotor. In the four-pole model as shown in Fig. 7.6b, there are four high flux concentration areas (magnetic poles), which implies that we can double the number of current carrying conductors (with half the cross-sectional area, given that total available area for the windings remains unchanged). Hence, the torque produced per ampère on the rotor will be *doubled* when compared to the two-pole case. This leads to the important conclusion that a (second) gain module, with gain p, must be added to the torque output of a two-pole generic IRTF model. For simulation purposes the two gain modules with gain p are combined as may be observed from Fig. 7.7. The relationship that exists between the multi-pole and two-pole torque/angle variables may be written as

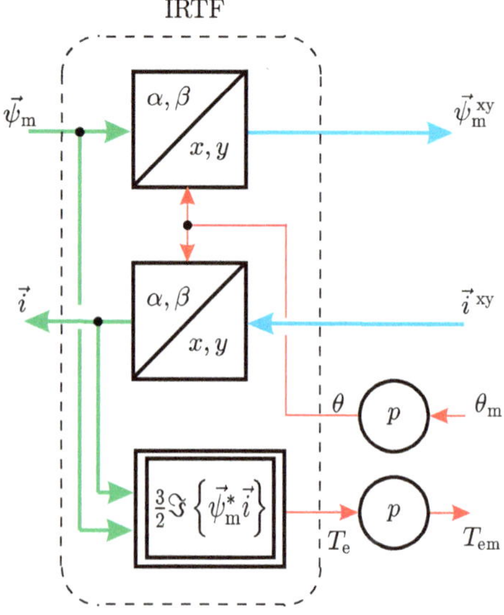

Fig. 7.7 Use of IRTF module for multi-pole pair models

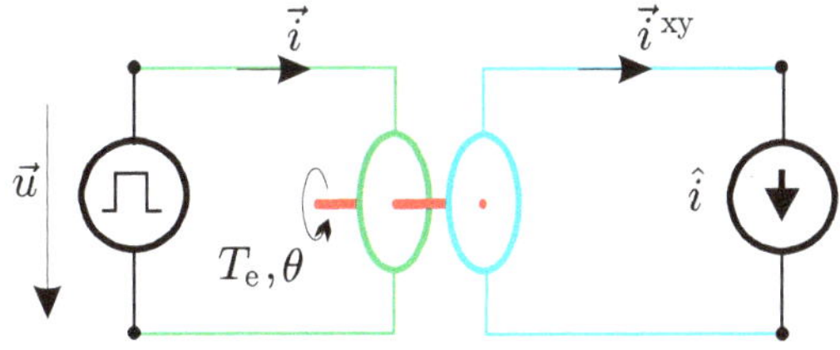

Fig. 7.8 IRTF module connected to voltage and current source

$$T_\mathrm{em} = p\,T_\mathrm{e} \tag{7.15a}$$

$$\theta_\mathrm{m} = \frac{1}{p}\,\theta \tag{7.15b}$$

where T_em and θ_m represent the shaft torque and shaft angle, respectively. Note that changing the number of poles does not affect the rated torque of the machine, given that the latter is constrained by the rotor volume, as was discussed in Sect. 1.7 on page 21.

7.2.2 *IRTF Example*

The IRTF module forms the backbone to the electrical machine concepts presented in this book. Consequently, it is particularly important to fully understand this concept. In the example given here we will discuss how stator currents and torque can be produced in the event that stator windings are connected to a voltage source and the rotor windings to a current source as shown in Fig. 7.8. In the following discussion, we will use Fig. 7.1 in a stylized form for didactic reasons. Furthermore, we will assume that we can hold the motor shaft at any desired position. The voltage source shown in Fig. 7.8 delivers a voltage pulse to the "α" winding at $t = t_0$ as shown in Fig. 7.9. The flux-linkage $\psi_{\mathrm{m}\alpha}$ versus time waveform which corresponds with the applied pulse is also shown in Fig. 7.9. The "β" stator winding is short circuited with initial condition $\psi_{\mathrm{m}\beta} = 0$. We will consider events after $t = t_0 + T$ in which case a flux distribution will be present in the IRTF where the majority of the flux is concentrated along the "$\Re^{\alpha\beta}$" axis, as shown in Fig. 7.10a. Furthermore, the flux-linkage value will be equal to $\psi_{\mathrm{m}\alpha} = \hat{\psi}_\mathrm{m}$. The corresponding space vector representation will be of the form $\vec{\psi}_\mathrm{m} = \hat{\psi}_\mathrm{m}$.

In addition, we would like to realize a rotor excitation of the form $\vec{i}^{\,\mathrm{xy}} = \mathrm{j}\hat{i}$, which implies that the "y" rotor winding must carry a current $i_\mathrm{y} = \hat{i}$. We have omitted for didactic reasons the "x" winding from Fig. 7.10 because this winding is not in use (open circuited). We will now examine the IRTF model and corresponding space vector diagrams for the excitation conditions indicated above and three rotor

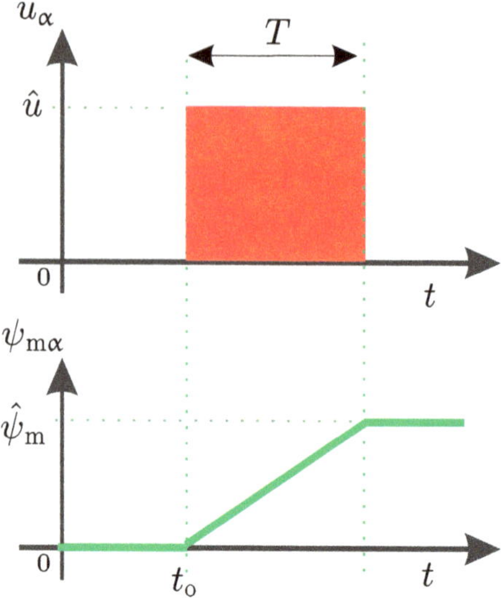

Fig. 7.9 Voltage excitation and flux-linkage for the "α" winding

positions. For each case (rotor position) we will assume that the rotor is initially set to the required rotor position after which the stator and rotor excitation is applied, as discussed above. The aim is to provide some understanding with respect to the currents which will occur on the stator side and the nature of torque production based on first principles. To assist us, two "contours," namely "x" and "y" are introduced in Fig. 7.10, which are linked to the rotating complex plane. These contours are helpful in determining the currents which must appear on the stator side. If we assume that such a contour represents a flux tube, then there would need to be a corresponding resultant MMF within the contour in case the latter would contain some form of magnetic reluctance. The magnetic reluctance of the IRTF model is assumed to be zero (infinite permeability material and infinitely small airgap). Hence the MMF "seen" inside either contour must always be zero in the IRTF.

- Rotor position $\theta = 0$: if we consider the "x" contour in Fig. 7.10a, then it coincides with the flux distribution that exists in the model. The MMF seen by this contour is equal to ni_α. The excited rotor winding cannot contribute (given both halves are in the contour) to this contour. Hence, the current i_α must be zero. By observing the "y" contour and the MMF "seen," we note the presence of the imposed rotor current $i_y = \hat{i}$. The number of winding turns on rotor and stator is equal, hence a stator current $i_\beta = \hat{i}$ (with the direction shown) must appear in the short-circuited β coil to ensure that the zero MMF condition with this contour is satisfied. The space vector representation of the current and flux, given in Fig. 7.10a, shows that they are $\pi/2$ radian apart. Note (again) that there is *only one* set of space vectors and their components may be projected onto either the rotor or stator complex plane. According to Eq. (7.13), the torque will under these circumstances (with the current vector leading the flux vector) be equal to $T_e = 3/2\,\hat{\psi}_m\hat{i}$. From first principles (see Fig. 1.8), we note that forces will be

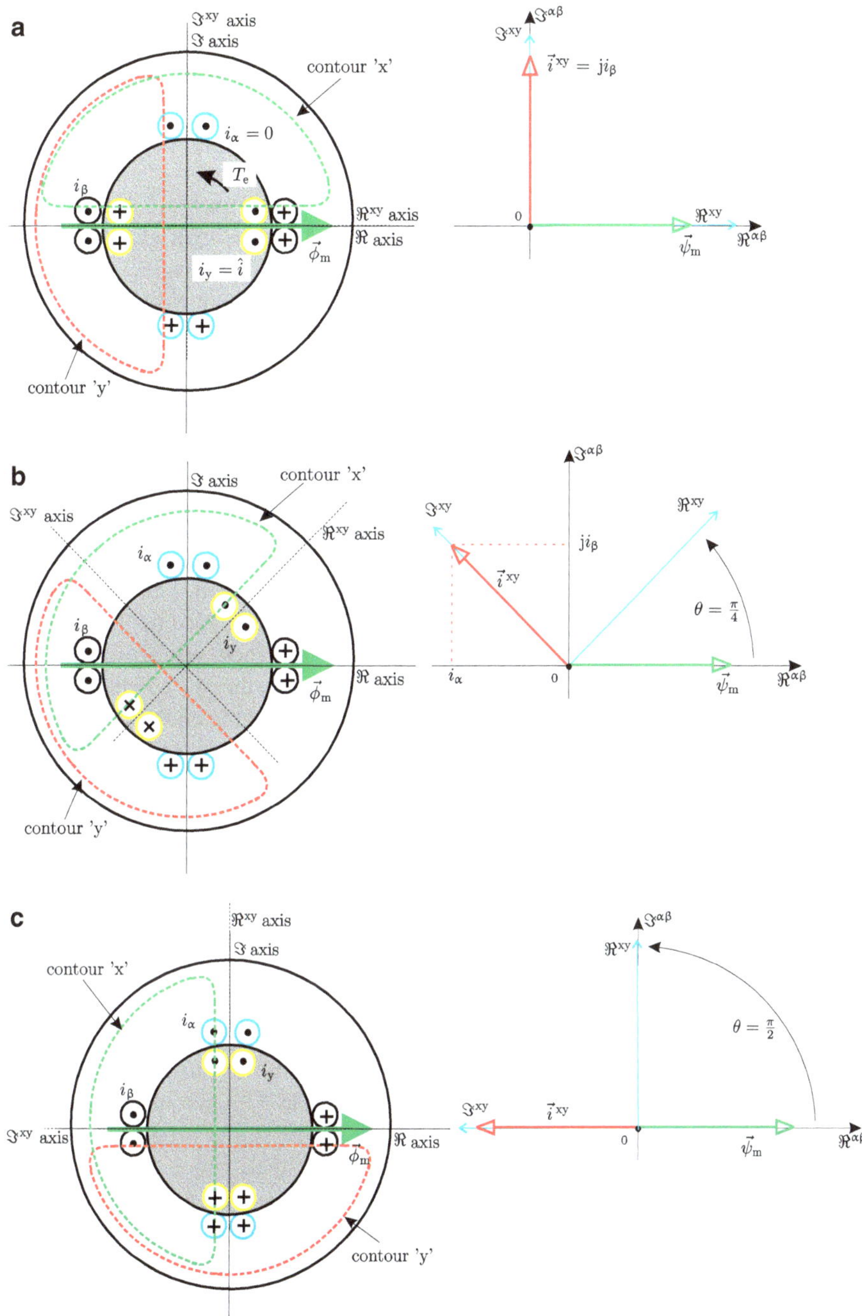

Fig. 7.10 IRTF symbolic model and space vector diagrams, for: $\theta = 0$, $\theta = \pi/4$, and $\theta = \pi/2$. (**a**) $\theta = 0$, $i_\alpha = 0$, $i_\beta = \hat{i}$, and $T_e = 3/2\,\hat{\psi}_m \hat{i}$. (**b**) $\theta = \pi/4$, $i_\alpha = -\hat{i}/\sqrt{2}$, $i_\beta = \hat{i}/\sqrt{2}$, and $T_e = 3/2\,\hat{\psi}_m \hat{i}/\sqrt{2}$. (**c**) $\theta = \pi/2$, $i_\alpha = -\hat{i}$, $i_\beta = 0$, and $T_e = 0$

exerted on the rotor winding in case the latter carries a current and is exposed to a magnetic field. In this case, the flux and flux density distributions are at their highest level along the "α" axis (see Appendix A). The "y" winding carries a current in the direction shown and this will cause a force on the individual conductors of the "y" rotor winding and thus a corresponding torque in the anti-clockwise (positive) direction.

- Rotor position $\theta = \pi/4$: if we consider the "x" contour in Fig. 7.10b, then we note that an MMF due to the "α" winding would be less than ni_{α}. The reason for this is that the contour also encloses part of "α" winding which gives a negative MMF contribution to the total MMF. The resultant MMF is found by integrating equation (A.9) over the angle range $\pi/4 \rightarrow \pi$ and $-\pi \rightarrow -3\pi/4$ and comparing the outcome of this integral with the integral over the range $0 \rightarrow \pi$. This analysis shows that the MMF is reduced by a factor $1/\sqrt{2}$ (when compared to the previous case). Hence the resultant MMF of the "α" winding is equal to $ni_{\alpha}/\sqrt{2}$. The same "x" contour also encloses part of the "β" winding and its MMF contribution is equal to $ni_{\beta}/\sqrt{2}$. There are no other contributions, which implies that the sum of these two MMFs is of the form $ni_{\alpha}/\sqrt{2} + ni_{\beta}/\sqrt{2} = 0$ because the MMF in the contour must be zero. From this analysis it follows that, for the given angle, the currents must be in opposition, i.e., $i_{\alpha} = -i_{\beta}$. If we now consider the MMF enclosed by the "y" contour, we note that the "α" winding will contribute an MMF component $-ni_{\alpha}/\sqrt{2}$ (now negative), while the "β" will contribute a component "$ni_{\beta}/\sqrt{2}$." Furthermore, the "y" winding will add a component $n\hat{i}$. The resultant MMF (the sum of these three contributions) gives together with the condition $i_{\alpha} = -i_{\beta}$, the required stator currents $i_{\alpha} = -\hat{i}/\sqrt{2}$ and $i_{\beta} = \hat{i}/\sqrt{2}$, respectively. The space vector representation shown in Fig. 7.10b confirms the presence of the two current components. Note also that the angle between the current and flux vectors is equal to $3\pi/4$ radian. The torque according to Eq. (7.13) and given the circumstances equals $T_{e} = 3/2\,\hat{\psi}_{m}\hat{i}/\sqrt{2}$. The torque must be less than the previous case, because a number of the conductors of the "y" winding now "see" a flux density value which is in opposition to that seen by the majority of the conductors.
- Rotor position $\theta = \pi/2$: if we consider the "x" contour in Fig. 7.10c, then we can conclude that the current i_{β} must be zero (the other winding cannot contribute, given that both halves are in the contour). The reason for this is that the MMF seen by this contour must be zero. By observing the "y" contour and the MMF "seen" by this contour we note the presence of the rotor current $i_{y} = \hat{i}$. The number of winding turns on rotor and stator is equal, hence a stator current $i_{\alpha} = -\hat{i}$ must appear to ensure that the zero MMF condition with this contour is satisfied. The space vector representation of the current and flux as given in Fig. 7.10c shows that they are π radian apart. The torque according to Eq. (7.13) will under these circumstances be equal to $T_{e} = 0$. From first principles we note that half the conductors of the "y" winding will experience a force in opposition to the other half, hence the net torque will be zero.

7.3 Conditions Required to Realize Constant Torque

The ability of a machine to produce a torque with a non-zero average component is of fundamental importance. In this section, we will consider the conditions under which electrical machines are able to produce a constant time independent torque, i.e., torque with zero torque ripple. For this analysis it is sufficient to reconsider the model according to Fig. 7.8. The rotor is again connected to a current source which in this case is assumed to be of the form $\hat{i}e^{j(\omega_r t+\rho_r)}$, where ω_r represents the angular velocity of this vector *relative* to the rotor of the IRTF. The current $\vec{i}^{xy}$ is therefore of the form

$$\vec{i}^{xy} = \hat{i}e^{j(\omega_r t+\rho_r)} \tag{7.16}$$

The stator is connected to a three-phase sinusoidal voltage source such that the voltage vector is taken to rotate at a constant angular velocity ω_s. The corresponding flux vector $\vec{\psi}_m$ can be found using $\vec{u} = {}^{d\vec{\psi}_m}/_{dt}$, which results in a rotating flux vector of the form

$$\vec{\psi}_m = \hat{\psi}_m e^{j\omega_s t} \tag{7.17}$$

The rotor angle θ is a function of rotor speed ω_m and load angle ρ_m and is (for constant speed operation) defined as

$$\theta = \omega_m t + \rho_m \tag{7.18}$$

The importance of the variable ρ_m will be discussed at a later stage. The induced voltage on the rotor side in Fig. 7.11 is in this case $\vec{e}_m^{xy} = {}^{d\vec{\psi}_m^{xy}}/_{dt}$, which, with the

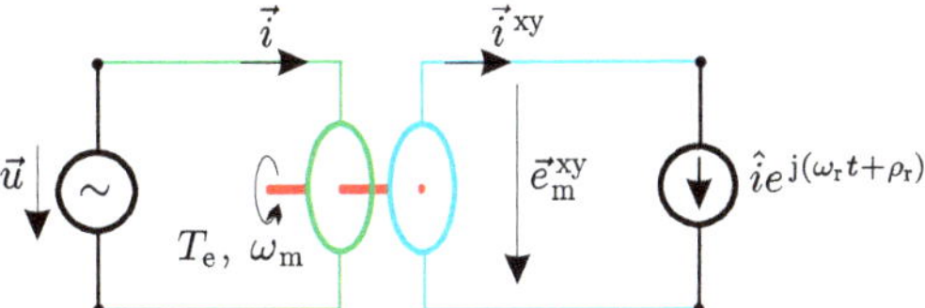

Fig. 7.11 Simplified two-phase machine model

aid of Eqs. (7.17), (7.2), and (7.18), may also be written as

$$\vec{e}_{\mathrm{m}}^{\,\mathrm{xy}} = \frac{\mathrm{d}\vec{\psi}_{\mathrm{m}}^{\,\mathrm{xy}}}{\mathrm{d}t} = \mathrm{e}^{\mathrm{j}((\omega_{\mathrm{s}}-\omega_{\mathrm{m}})t-\rho_{\mathrm{m}})}\frac{\mathrm{d}\hat{\psi}_{\mathrm{m}}}{\mathrm{d}t} + \mathrm{j}\,(\omega_{\mathrm{s}}-\omega_{\mathrm{m}})\,\vec{\psi}_{\mathrm{m}}^{\,\mathrm{xy}} \tag{7.19}$$

The term $\frac{\mathrm{d}\hat{\psi}_{\mathrm{m}}}{\mathrm{d}t}$ is taken to be zero given that a quasi-steady-state operation is assumed, i.e., we assume that the flux amplitude $\hat{\psi}_{\mathrm{m}}$ is constant. On the basis of this assumption, Eq. (7.19) reduces to

$$\vec{e}_{\mathrm{m}}^{\,\mathrm{xy}} = \mathrm{j}\,(\omega_{\mathrm{s}}-\omega_{\mathrm{m}})\,\vec{\psi}_{\mathrm{m}}^{\,\mathrm{xy}} \tag{7.20}$$

Note that the amplitudes of the vector $\vec{e}_{\mathrm{m}}^{\,\mathrm{xy}}$ and its stationary transformed counterpart $\vec{e}_{\mathrm{m}}$ can (and usually are) *different*. The torque produced by this machine is found using Eq. (7.13). Further mathematical handling of this torque equation and using Eqs. (7.17) (in rotor coordinates), (7.2), and (7.18) give

$$T_{\mathrm{e}} = \frac{3}{2}\,\Im\left\{\hat{\psi}_{\mathrm{m}}\hat{i}\,\mathrm{e}^{\mathrm{j}((\omega_{\mathrm{r}}+\omega_{\mathrm{m}}-\omega_{\mathrm{s}})t+\rho_{\mathrm{r}}+\rho_{\mathrm{m}})}\right\} \tag{7.21}$$

It is emphasized that the angle variables ρ_{r} and ρ_{m}, shown in Eq. (7.21), are not a function of time. Equation (7.21) is of prime importance as it shows that, with sinusoidal excitation, a time independent torque value can only be obtained in case the following condition is met

$$\omega_{\mathrm{s}} = \omega_{\mathrm{m}} + \omega_{\mathrm{r}} \tag{7.22}$$

If the speed condition, according to Eq. (7.22), is satisfied, then the torque expression is reduced to

$$T_{\mathrm{e}} = \frac{3}{2}\,\hat{\psi}_{\mathrm{m}}\hat{i}\sin\left(\rho_{\mathrm{r}}+\rho_{\mathrm{m}}\right) \tag{7.23}$$

Equations (7.22), (7.23), and (7.20) are useful in terms of explaining the basic classical machine concepts as will be done in the next three chapters. Prior to considering how these expressions are applied for the three types of machines it is instructive to look at the vector diagram given in Fig. 7.12 which corresponds to machine configurations discussed in this section. The diagram shows the stationary

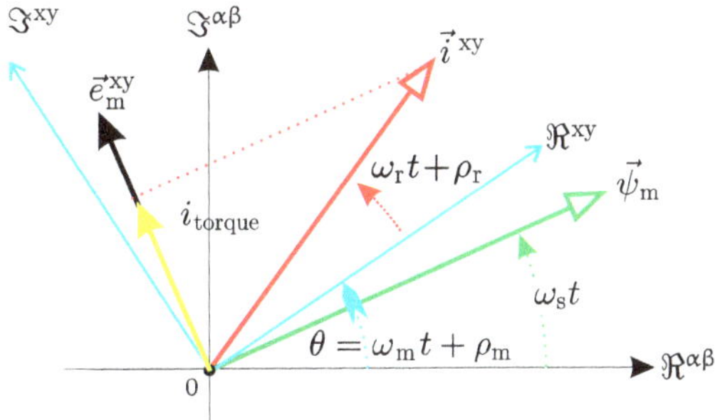

Fig. 7.12 Space vector diagram for machine model

complex plane ($\Re^{\alpha\beta}$, $\Im^{\alpha\beta}$) and the rotating (with constant rotor speed ω_m) rotor coordinate frame, which is displaced by an angle $\theta = \omega_m t + \rho_m$ with respect to the former. Also shown in the diagram is the rotating (with angular speed ω_s) flux vector $\vec{\psi}_m$, which is caused by the presence of the three-phase voltage supply connected to the stator. Finally, the current vector $\vec{i}^{xy}$ is shown which rotates at an angular speed ω_r *relative to the rotating reference frame*. This means that this vector rotates at a speed $\omega_r + \omega_m$ relative to the stator based (stationary) reference frame.

In the example shown, the current vector leads (we assume positive direction as anti-clockwise) the flux vector. This combination of the two vectors corresponds to a positive torque condition as will be shown below. A time independent torque will be realized if the angle between the current and flux vectors remains constant. This is the case when condition (7.22) is met. Furthermore, we also assume that the magnitude of the two vectors is constant. If condition (7.22) is met, the angle between the two vectors is equal to the sum of ρ_m and ρ_r. In the example shown, the sum of these two angles is taken to be positive, which according to Eq. (7.23) gives a positive torque, i.e., the machine acts as a motor. The induced voltage vector $\vec{e}_m^{xy}$ is orthogonal to the flux vector. The projection of the current vector on this voltage vector, shown as $i_{torque} = \hat{i} \sin(\rho_r + \rho_m)$ (see Fig. 7.12), is according to Eq. (7.23) proportional to the torque, provided the condition as given by Eq. (7.22) is met. The highest motor torque value, equal to $\hat{T}_e = {}^3/_2\, \hat{\psi}_m \hat{i}$, which can be delivered is realized when the rotor current vector leads the flux vector by $^{\pi}/_2$ rad, i.e., in phase with the voltage $^{d\psi_m}/_{dt}$. We will use this diagram again in the next three chapters in different forms consistent with the three main types of machines in use today. A tutorial at the end of this chapter is given to reinforce the concepts discussed in this section.

7.4 Universal IRTF Based Machine Model

The process of arriving at a universal two-phase machine model may be initiated by considering the model shown in Fig. 7.13. This model resembles the IRTF model as discussed in Sect. 7.2, with notable changes, namely the introduction of a finite airgap and an unequal stator to rotor turns ratio $^{n_s}/_{n_r}$. Furthermore, the currents in

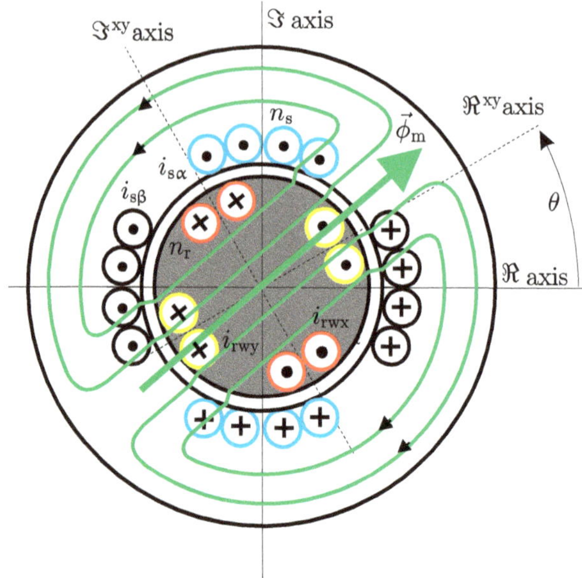

Fig. 7.13 Generalized two-phase machine model

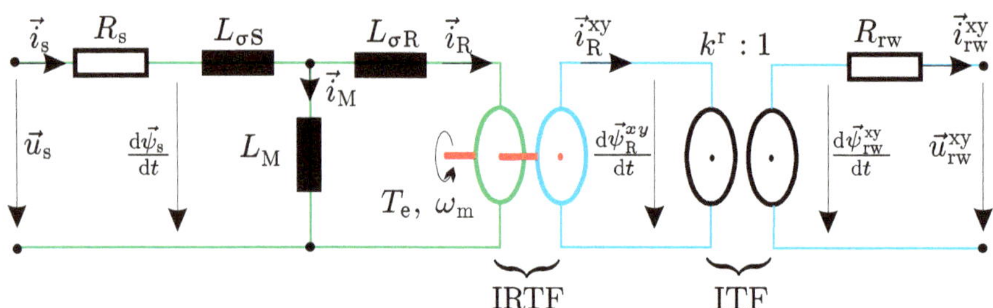

Fig. 7.14 Symbolic "universal" two-phase, IRTF/ITF machine model

the stator and rotor windings are defined as $i_{s\alpha}$, $i_{s\beta}$ and i_{rwx}, i_{rwy}, respectively. The task of determining a symbolic representation of the machine given in Fig. 7.13 can be undertaken by first considering the case where the rotor angle θ is locked to $\theta = 0$. Under such conditions, the machine is simply a two-phase transformer and the symbolic model topology given in Fig. 6.10 applies. The modeling process is completed by introducing the IRTF model, given in Fig. 7.2, to the two-phase ITF model, which leads to the symbolic model shown in Fig. 7.14. The location of the IRTF module has been chosen to accommodate the machine models to be discussed in subsequent chapters. However, the module can be relocated to suit a particular application, provided that only inductances are relocated from the stator to the rotor side and vice versa. Readily apparent in Fig. 7.14 are the stator and rotor leakage inductances $L_{\sigma S}$ and $L_{\sigma R}$ which, together with the magnetizing inductance L_M, form a universal inductance network, as was discussed in Sect. 3.7. The inductance parameters $L_{\sigma S}$, $L_{\sigma R}$, and L_M and the winding ratio k^r are a function of the transformation factor a as described for the single phase case (see Eq. (3.26) and equation set (3.27)), albeit the nomenclature needs to be adapted, which gives

$$L_{\sigma S} = L_m \left(\frac{L_s}{L_m} - a \right) \tag{7.24a}$$

$$L_{\sigma R} = a L_r \left(a - \frac{L_m}{L_r} \right) \tag{7.24b}$$

$$L_M = a L_m \tag{7.24c}$$

$$k^r = a \frac{n_s}{n_r} \tag{7.24d}$$

with $L_r = L_m + L_{\sigma r}$ and $L_s = L_m + L_{\sigma s}$, where $L_{\sigma s}$ and $L_{\sigma r}$ represent the *original* stator leakage inductance and stator *referred* rotor leakage inductance, respectively. Observation of Eq. (7.24) shows that the transformation variable a is bound by the condition

$$\frac{L_m}{L_r} \leq a \leq \frac{L_s}{L_m} \tag{7.25}$$

on the grounds that the inductance parameters must be greater or equal to zero. Note that a unity transformation value corresponds to the "original" three inductance network with

$$L_{\sigma S} = L_{\sigma s} \tag{7.26a}$$

$$L_{\sigma R} = L_{\sigma r} \tag{7.26b}$$

$$L_M = L_m \tag{7.26c}$$

Also shown in Fig. 7.14 is an ITF module with winding ratio k^r:1 as defined by Eq. (7.24d). The relationship between "universal" rotor parameters (with subscript R) and actual rotor parameters (with subscript rw) is given as

$$\vec{i}_R = \frac{\vec{i}_{rw}}{k^r} \tag{7.27a}$$

$$\vec{\psi}_R = k^r \vec{\psi}_{rw} \tag{7.27b}$$

In a number of machine modeling applications it is convenient to only use the universal model variables, in which case the ITF is simply ignored, provided that the original rotor resistance R_{rw} is referred to the "primary" side using

$$R_R = (k^r)^2 \; R_{rw} \tag{7.28}$$

The corresponding equation set for the model according to Fig. 7.14 is of the form

$$\vec{u}_s = \vec{i}_s R_s + \frac{d\vec{\psi}_s}{dt} \tag{7.29a}$$

$$\vec{\psi}_s = \vec{\psi}_M + \vec{i}_s L_{\sigma S} \tag{7.29b}$$

$$\vec{\psi}_R = \vec{\psi}_M - \vec{i}_R L_{\sigma R} \tag{7.29c}$$

$$\frac{\vec{\psi}_M}{L_M} = \vec{i}_s - \vec{i}_R \tag{7.29d}$$

$$\frac{d\vec{\psi}_{rw}^{xy}}{dt} = \vec{u}_{rw}^{xy} + \vec{i}_{rw}^{xy} R_{rw} \tag{7.29e}$$

$$\vec{i}_R = \vec{i}_R^{xy} \, e^{j\theta} \tag{7.29f}$$

$$\vec{\psi}_R^{xy} = \vec{\psi}_R \, e^{-j\theta} \tag{7.29g}$$

$$T_e = \frac{3}{2} \Im \left\{ \vec{\psi}_R^{*} \vec{i}_R \right\} \tag{7.29h}$$

The resultant model as indicated in Fig. 7.14 forms the basic concept which will be used in the following chapters to examine, with the aid of simulation models, a range of classic motor configurations as they exist today. Given the importance of this model it is instructive to summarize the underlying considerations, namely:

- Linear magnetic material stator/rotor,
- No magnetic losses,
- Sinusoidal distributed windings, and
- Two-pole machine.

7.4.1 Generic Model of a Universal IRTF Based Machine

It is instructive to consider a generic representation of the universal model given in Fig. 7.14. Such a generic model is of interest to demonstrate with the aid of a tutorial (see Sect. 7.5.3) that changing the transformation factor within the range given by Eq. (7.25) will not affect, for example, the torque of the machine. Note that in subsequent chapters more generic models will be introduced, with the ability to represent a three-inductance model with a two-inductance version.

The approach used to arrive at a generic model follows that of Sect. 3.6 for a single phase transformer. For this purpose of the analysis it is helpful to reconsider the relationship between the space vector based flux-linkages $\vec{\psi}_s$, $\vec{\psi}_R$ and currents $\vec{i}_s$, $\vec{i}_R$. These can, with aid of Eqs. (7.29b)–(7.29d), be written in matrix format as

$$\begin{bmatrix} \vec{\psi}_s \\ \vec{\psi}_R \end{bmatrix} = \begin{bmatrix} L_s & -L_M \\ L_M & -L_R \end{bmatrix} \begin{bmatrix} \vec{i}_s \\ \vec{i}_R \end{bmatrix} \tag{7.30}$$

with $L_R = L_M + L_{\sigma R}$ and $L_s = L_M + L_{\sigma S}$. For the development of an IRTF/ITF based model it is helpful to invert Eq. (7.30) which gives

$$\begin{bmatrix} \vec{i}_s \\ \vec{i}_R \end{bmatrix} = \underbrace{\frac{1}{L_s L_R - (L_M)^2} \begin{bmatrix} L_R & -L_M \\ L_M & -L_s \end{bmatrix}}_{[L^{-1}]} \begin{bmatrix} \vec{\psi}_s \\ \vec{\psi}_R \end{bmatrix} \tag{7.31}$$

which, together with Eq. (3.17), leads to the generic model given in Fig. 3.15 that corresponds to the symbolic model given in Fig. 7.15. Note that this generic model is not suitable for modeling an open circuited (with a passive load) rotor winding or for that matter a stator winding since the rotor current $\vec{i}_{rw}^{xy}$ and stator current $\vec{i}_s$ are outputs of this generic model. In applications where, for example, the rotor current is an input vector a different type of generic model should be used, as will become apparent in the next chapter.

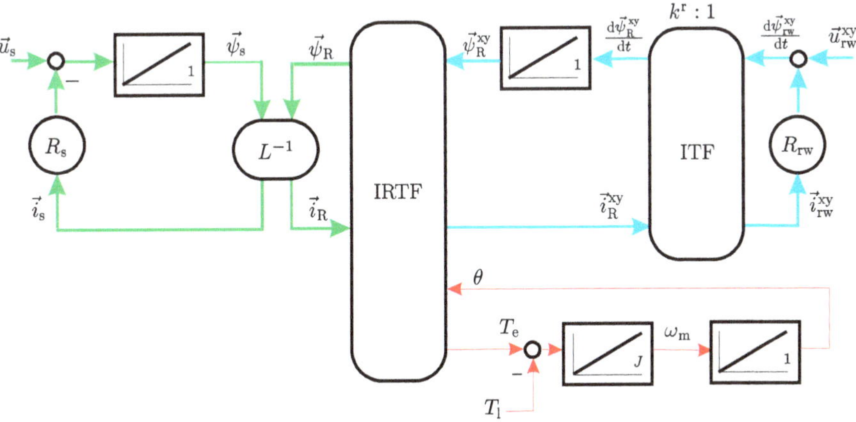

Fig. 7.15 Generic universal two-phase, two-pole, IRTF/ITF machine model

7.5 Tutorials

7.5.1 Tutorial 1: PLECS Based Model of an IRTF Module with Stator Flux Excitation

This tutorial is aimed at providing a better understanding of the IRTF module. As a first step, a PLECS model of the IRTF module must be built. The implementation follows the generic version given in Fig. 7.5a. An example of possible implementation is given in Fig. 7.16, where the IRTF sub-module `RRF->SRF` (which is a standard PLECS module) performs the conversion $\vec{i} = \vec{i}^{\mathrm{xy}} \mathrm{e}^{\mathrm{j}\theta}$. A similar (standard PLECS module) `SRF->RRF` is used to implement the conversion $\vec{\psi}_{\mathrm{m}}^{\mathrm{xy}} = \vec{\psi}_{\mathrm{m}} \mathrm{e}^{-\mathrm{j}\theta}$. Finally, the torque produced by the machine is calculated according to $T_{\mathrm{e}} = {}^{3}\!/\!_{2}\,\Im\left\{\vec{\psi}_{\mathrm{m}}^{\,*}\vec{i}\right\}$, which requires the vector components $\psi_{\mathrm{m}\alpha}$, $\psi_{\mathrm{m}\beta}$, i_{α}, and i_{β}. Further development of the torque equation leads to an expression which can be used in a `Fcn` module to generate the torque. In the second part of this tutorial a numerical implementation of the example as discussed in Sect. 7.2.2 is considered. The simulation diagram, as given in Fig. 7.17, shows the IRTF module developed in the first part of this tutorial. The flux vector, which is the input on the stator side, is of the form $\vec{\psi}_{\mathrm{m}} = 1.0\,\mathrm{Wb}$. This is a scalar, i.e., no imaginary component. The rotor side is connected to a current source (as shown in Fig. 7.8) and we assume that the vector is equal to $\vec{i}^{\mathrm{xy}} = \mathrm{j}2\,\mathrm{A}$, i.e., only an imaginary component is present. The rotor angle is provided via an integrator which in turn has as input the angular frequency $2\pi\,50\,\mathrm{rad/s}$. Hence if we run the simulation for 20 ms, the rotor will have moved

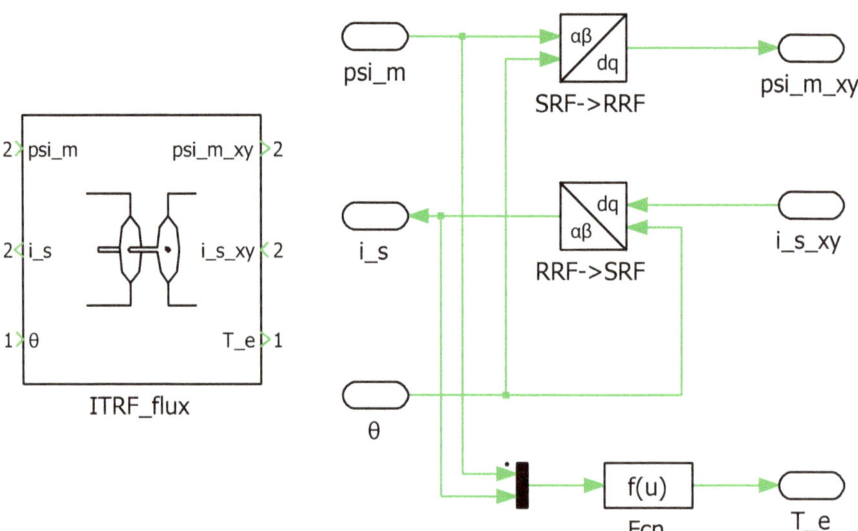

Fig. 7.16 PLECS model of IRTF module

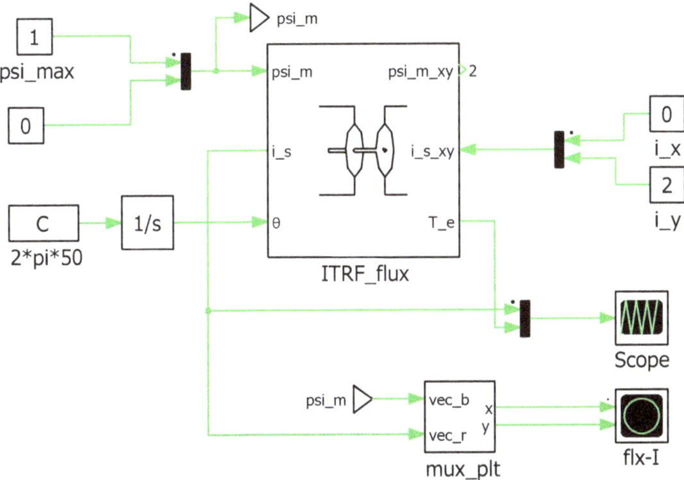

Fig. 7.17 PLECS model of IRTF simulation

one full rotation. For this reason, we will set the "run time" for our simulation to 20 ms. Furthermore, in the dialog box "Simulation Parameters," set the "Max step size" to `1e-8`, which will slow down the simulation process. This will make it easier to observe the vectors $\vec{i}_{\mathrm{s}}$, $\vec{\psi}_{\mathrm{m}}$ during said simulation. To view the results a `Scope` and "XY Plot" module `flx-I` should be added. The latter module is used to observe the current and flux vectors over the course of the simulation. Examine the results produced carefully to determine if the IRTF module is working correctly. To assist you with this task it is helpful to look carefully at the example given in Sect. 7.2.2. The output produced after running your simulation should look similar to those given in Figs. 7.18 and 7.19. Shown in Fig. 7.19 are the stator flux vector ("green") $\vec{\psi}_{\mathrm{m}}$ and the stator current vector ("red") $\vec{i}_{\mathrm{s}}$ at the end of the simulation. In this case the flux remains stationary, whilst the current vector rotates, as is evident by the presence of the locus trajectory. Note that the simulation time was purposely chosen to ensure that the current vector would rotate 360°, hence the plot shown is identical for the beginning and end of the simulation interval. A module `mux_plt` (see Fig. 7.17) has been added to show two vectors in one "XY Plot."

7.5.2 *Tutorial 2: PLECS Based Model to Examine Constant Torque Operation Using an IRTF Module*

The IRTF example discussed in tutorial 1 is modified to explore how a constant (time independent) torque value can be obtained (see Sect. 7.3). The revised model, as given in Fig. 7.20, shows the IRTF-flux module with inputs $\vec{\psi}_{\mathrm{m}} = \hat{\psi}_{\mathrm{m}} \mathrm{e}^{j\omega_{\mathrm{s}} t}$ and

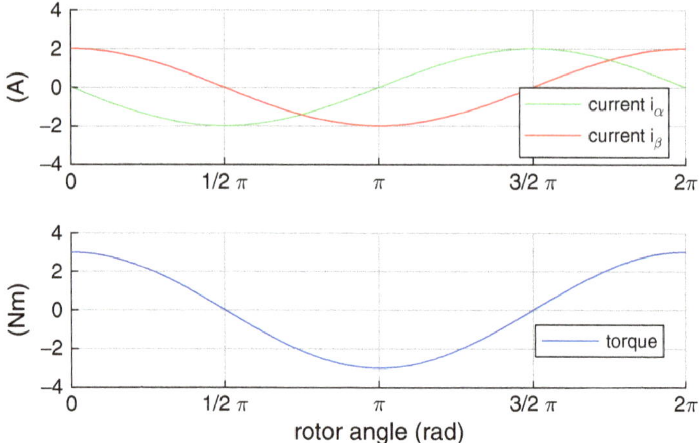

Fig. 7.18 Scope results of IRTF simulation

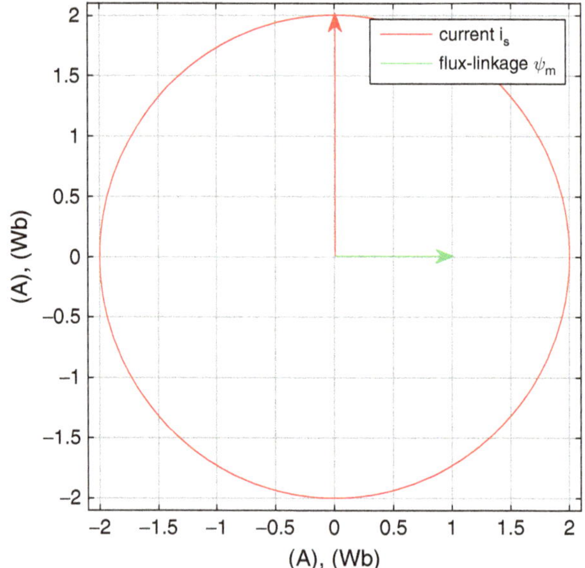

Fig. 7.19 "XY Plot" results of IRTF simulation

$\vec{i}^{xy} = \hat{i}e^{j(\omega_r+\rho_r)}$. The simulation model is directly based on the symbolic model given in Fig. 7.20 with the important change that the voltage source is replaced with a "flux" source $\vec{\psi}_m$.

The amplitudes for the flux and current vectors are arbitrarily set to $\hat{\psi}_m = 1.0$ Wb and $\hat{i} = 2$ A, respectively, which results in a peak torque value of $\hat{T}_e = 3.0$ Nm. The generation of the flux vector for the IRTF module is realized with the aid of a polar

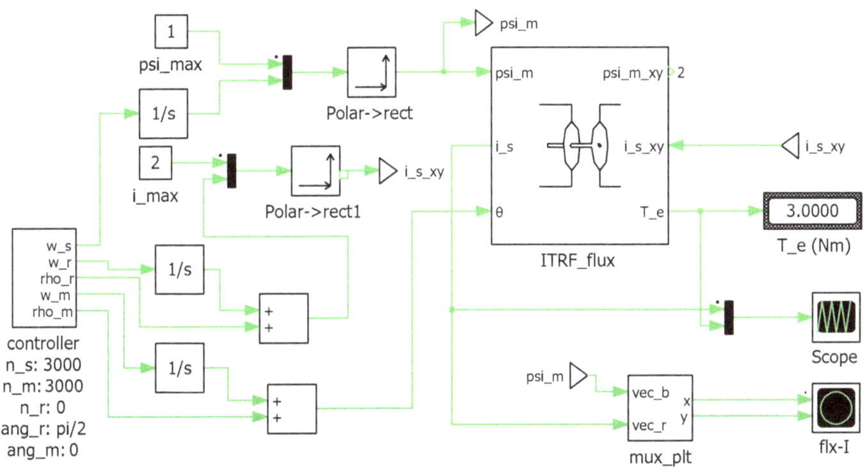

Fig. 7.20 Speed condition model: "DC brushless configuration" selected

to cartesian conversion module, as shown in Fig. 6.6. A similar approach is also used for the generation of the current vector $\vec{i}^{xy}$ with the minor change that a phase angle input ρ_r must be added as shown in Fig. 7.20.

The "controller" module sets the variables ω_s, ω_m, and ω_r (rad/s) and angles ρ_r and ρ_m (rad) for the simulation. The controller inputs are taken to the rotational speed (rpm) variables n_s, n_m, and n_r (rpm), which correspond to the angular frequencies given above. The controller multiplies these variables with a factor $2\pi/60$. In addition, the user must set values for the angles ρ_r and ρ_m (rad). As with the previous tutorial the shaft angle, speed, and load angle are defined by the user and taken to be of the form $\theta = \omega_m t + \rho_m$. The torque T_e produced by the machine (IRTF module) is examined using a "scope" and "display" module.

The "scope" module is used to verify that the torque is indeed time independent, while the display module is added to show directly the torque as a numerical value. In addition an "XY Plot" module has been added to observe the current $\vec{i}_s$ and stator flux vector $\vec{\psi}_m$ over the course of the simulation interval. The aim is to choose the controller variables in such a manner as to ensure that the machine delivers a constant torque level equal to the maximum value of 3.0 Nm. Four cases are considered which reflect the basic operation of electrical machines. Run your simulation for a "run" time of 20 ms and "Max step size" to `1e-8` (to be able to observe the vector during the simulation). Note that controller variables must be chosen in accordance with Eqs. (7.22) and (7.23). In the latter case the sum of the two angles ρ_r and ρ_m (rad) must be equal to $\pi/2$ rad because the aim is to set the output torque to its maximum value, i.e., $T_e = \hat{T}_e$. Observe the output of your simulation for the following cases: An example of the results which will

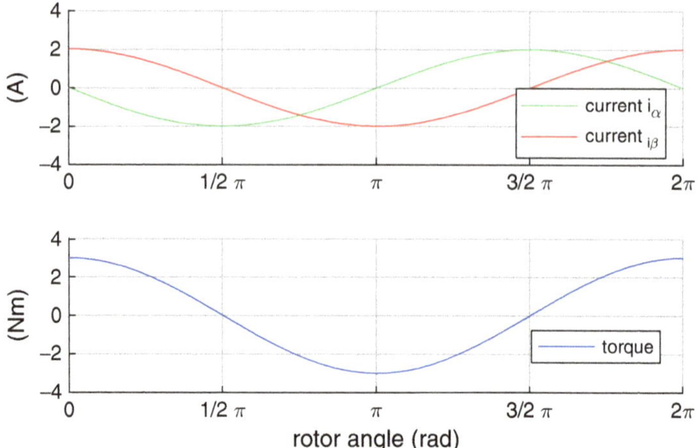

Fig. 7.21 Scope result of the simulation: "DC brushless" configuration

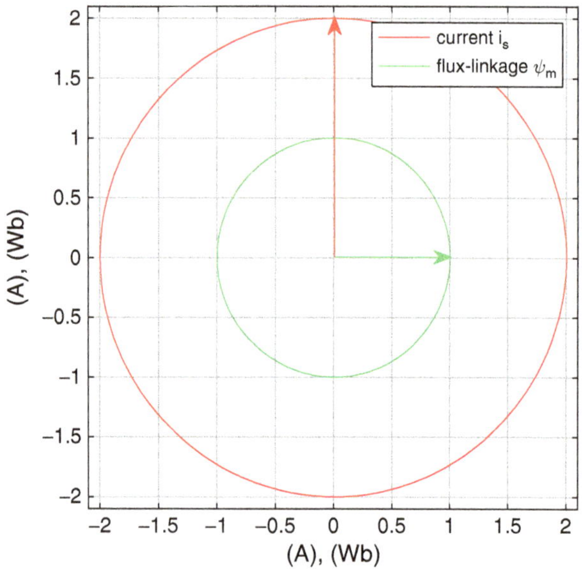

Fig. 7.22 "XY Plot" results of simulation: "DC brushless" configuration

appear on the `Scope` and "XY Plot" `flx-I` modules (shown in Fig. 7.20) is given in Figs. 7.21 and 7.22. These results reflect operation according to configuration "DC brushless" (see Table 7.1). Observation of Fig. 7.21 shows that the torque T_e is indeed constant and equal to 3.0 Nm. Furthermore, the stator current vector components i_α, i_β are also shown which point to a rotating vector as may also be confirmed by observation of the "XY Plot" during operation. Figure 7.22 shows the current and stator flux vectors at the end of the simulation. Clearly observable are

Table 7.1 Case studies

Machine type	n_s (rpm)	n_m (rpm)	n_r (rpm)	ρ_r (rad)	ρ_m (rad)
Synchronous	3000	3000	0	π	$-\pi/2$
Asynchronous	3000	2000	1000	$\pi/2$	0
DC brush	0	3000	−3000	$\pi/2$	0
DC brushless	3000	3000	0	$\pi/2$	0

Table 7.2 Parameters of a two-phase machine

Parameters		Value
Stator inductance	L_s	346.9 mH
Rotor inductance	L_r	346.9 mH
Magnetizing inductance	L_m	340.9 mH
Stator resistance	R_s	6.9 Ω
Rotor resistance	R_{rw}	0.03 Ω
Effective stator turns	n_s	100 t
Effective rotor turns	n_r	10 t
Inertia	J	0.001 kg m^2
Pole pairs	p	1
Initial rotor speed	ω_m^o	0 rad/s

the locus traces of both vectors which are circular as expected. Furthermore, the angle between said vectors is 90° with the current vector "leading" the flux vector which results in a positive and constant torque output.

7.5.3 *Tutorial 3: PLECS Based Model to Examine the Universal Machine Concept*

The purpose of this tutorial is to demonstrate the universal model transformation concept. For this purpose a PLECS based simulation model is to be developed which is based on the generic model shown in Fig. 7.15. For the purpose of the tutorial the secondary windings are to be short circuited, which implies that the condition $\vec{u}_{rw}^{xy} = 0$ holds. Note that the assumed machine configuration with a parameter set defined by Table 7.2 is in fact an asynchronous machine as will become apparent in Chap. 8. A constant load torque of $T_l = 10\,\text{Nm}$ is to be used for this simulation. To simplify the ensuing analysis, stator flux excitation will be assumed, in which case it is helpful to define the stator flux space as $\vec{\psi}_s = 1.0\,e^{j\omega_s t}$, with $\omega_s = 100\,\pi$ rad/s.

The simulation model shown in Fig. 7.23 satisfies the requirement for this tutorial. Clearly identifiable are the IRTF module and ITF module. The latter module `IFR_cur` has in this case been provided with an additional input `a_` which is connected to the `L-1 matrix` module that sets (among others) the transformation factor a. This implies that the ITF winding ratio is now defined as $k^r = a\,(n_s/n_r)$ which is in accordance with Eq. (7.24d). The `L-1 matrix` module

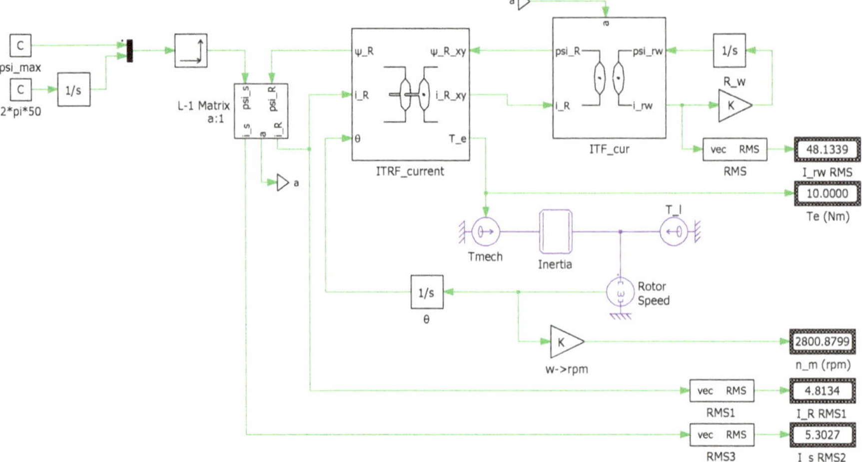

Fig. 7.23 PLECS simulation: two-phase IRTF/ITF machine, universal model concept verification

Table 7.3 Simulation results for different transformation factor values

Transformation factor	I_s (A)	I_R (A)	I_{rw} (A)	T_e (Nm)	n_m (rpm)
$a = \frac{L_m}{L_r} = 0.9827$	5.3027	4.891	48.133	10.0	2800.8
$a = 1$	5.3027	4.813	48.133	10.0	2800.8
$a = \frac{L_s}{L_m} = 1.018$	5.3027	4.728	48.133	10.0	2800.8

content is in accordance with the L^{-1} matrix expression given in Eq. (7.31). A set of dialog boxes is present in said `L-1 matrix` module, which are used to enter the machine inductance parameters and notably the transformation factor a. A set of space vector to RMS conversion modules `RMS` have been introduced, the outputs of which will be used to monitor the quasi-steady-state behavior of the machine when the transformation factor is changed. Also show in Fig. 7.23 are a set of PLECS mechanical control modules which are used to implement the load equation $T_e - T_l = J d\omega_m/dt$. The speed output is integrated, which yields the desired rotor angle of the IRTF module `IRTF_current`. The steady-state results obtained from the simulation model for three key transformation values are shown in Table 7.3. The first and third row entries shown in Table 7.3 correspond to a universal model configuration with $L_{\sigma R} = 0$ and $L_{\sigma S} = 0$, respectively. Also shown in the second row is the case $a = 1$ which corresponds to the original machine, hence $L_{\sigma S} = L_{\sigma s}$ and $L_{\sigma R} = L_{\sigma r}$. An important observation that can be made from the table is that the torque, shaft speed, RMS stator current I_s, and RMS rotor winding current I_{rw} are NOT affected by a change in transformation factor. Only the variable I_R changes, which is in line with theory presented as this variable is indirectly defined by Eq. (7.27a).

Chapter 8
Voltage Source Connected Synchronous Machines

8.1 Introduction

The synchronous machine has traditionally been used for power generation purposes. For motor applications (when connected to the power grid), a synchronous machine is ideal when the operating speed must remain constant, i.e., independent of load changes. Starting up, however, needs special measures.

Modern drives use a converter, which gives us more flexibility in terms of controlling the machine and enable the machine to self-start. Synchronous machines fed by power converters have become important players in the field of high performance drives.

In this chapter, we will look at the basic operation of the synchronous machine.

8.2 Synchronous Machine Configuration

The synchronous machine has a non-rotating component known as the stator, which is shown in Fig. 8.1. The stator consists of a "frame" within which a laminated stator core stack is positioned. This core has a series of slots that house the three-phase windings of the machine. Of these windings the so-called active sides (named thus because these are responsible for the energy conversion) are distributed appropriately in the core slots. The stator-coil end-winding-parts are at each end of the core stack, as shown in Fig. 8.1. The three-phase windings will, when connected to a three-phase supply source, produce a rotating magnetic field which is, as was discussed in the previous chapter, an essential requirement for

Electronic supplementary material The online version of this chapter (doi: 10.1007/978-3-319-29409-4_8) contains supplementary material, which is available to authorized users.

© Springer International Publishing Switzerland 2016

A. Veltman et al., *Fundamentals of Electrical Drives*, Power Systems,
DOI 10.1007/978-3-319-29409-4_8

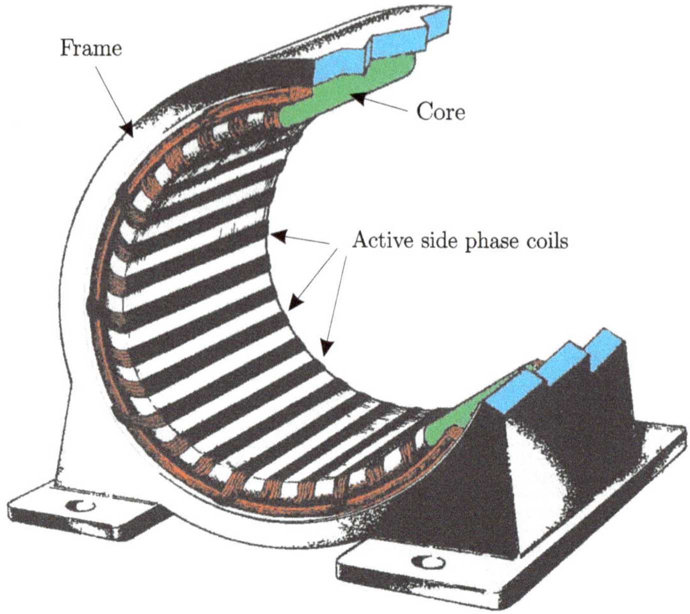

Fig. 8.1 Stator of three-phase synchronous machine

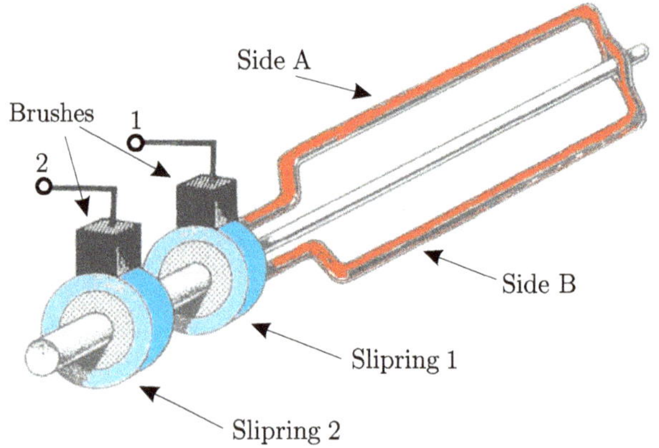

Fig. 8.2 Simple rotor for synchronous machine

producing constant torque. More details on machines and rotating fields are given in Bödefeld [1] or Hughes [5]. Note that the same stator is also used when discussing the asynchronous, or the so-called induction machine. The rotor configuration for the synchronous machines may take on several forms. A very simple configuration as shown in Fig. 8.2 conveys the basic structure. A more extensive discussion on the machine structure can be found in electrical machine design books. Shown in Fig. 8.2 is a rotor in the form of a single coil (referred to as the field winding)

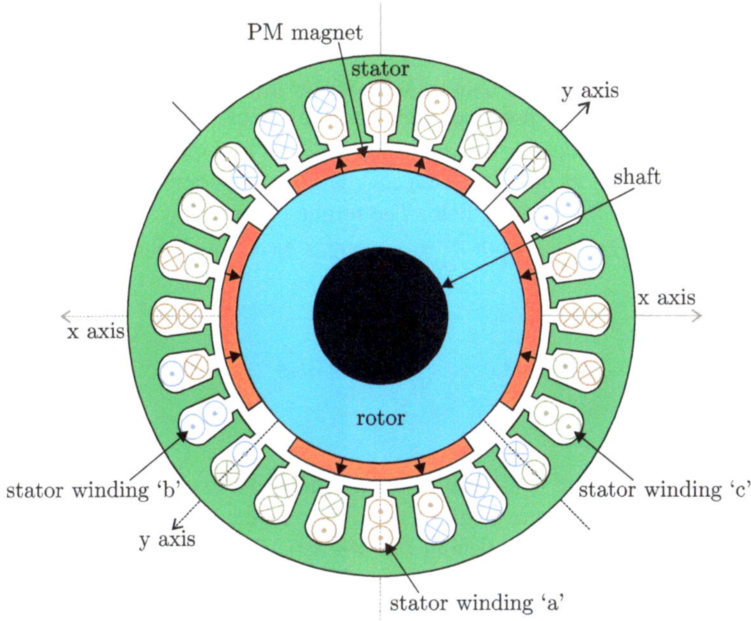

Fig. 8.3 PM synchronous machine with rotor magnets [2]

with sides "A" and "B" which are connected to slip rings 1 and 2, respectively. These copper slip rings are linked to a set of brushes which in turn are connected to a stationary DC power supply. The use of the slip ring/brush combination allows us to excite the rotor coil with a DC "field" current via a stationary source. Note that this rotor coil can be replaced by a permanent magnet which means that the slip ring/brush and the DC source can be avoided. However, the excitation under these circumstances cannot be varied. Such machines belong to the class of the so-called brushless DC machines. An example of such a machine is shown in Fig. 8.3, where the presence of magnets on the rotor is readily visible. Also present are the three-phase windings that are located in the slots of the stator.

8.3 Operating Principles

The synchronous machine with a slip ring/brush combination is, as was discussed in the previous section, fed on the rotor side with a DC field current. This implies that the current $\vec{i}^{\mathrm{xy}}$ (as introduced in Sect. 7.3) is given by Eq. (7.16), with $\omega_{\mathrm{r}} = 0$, hence $\vec{i}^{\mathrm{xy}} = \hat{i}\,\mathrm{e}^{\mathrm{j}\rho_{\mathrm{r}}}$. A choice remains with respect to the angle ρ_{r}; its value can be taken to be zero or π rad. The latter choice amounts to reversing the polarity of the DC current source shown in Fig. 7.11. This option is chosen here for reasons which will become apparent shortly, hence $\vec{i}^{\mathrm{xy}} = -\hat{i}$. The magnitude of the current $\hat{i}$ is in context of synchronous machines more commonly known as the field current i_{f}, i.e., $\vec{i}^{\mathrm{xy}} = -i_{\mathrm{f}}$. On the basis of the speed condition (see Eq. (7.22), with $\omega_{\mathrm{r}} = 0$), constant torque operation is only possible when the shaft speed is equal to the rotational speed ω_{s} of the rotating flux vector produced by the three-phase stator winding. The term "synchronous machine" reflects this type of operation, i.e., the rotor speed is synchronized to the rotating field. The torque produced by this machine can be calculated using Eq. (7.23) with $\vec{i}^{\mathrm{xy}} = -i_{\mathrm{f}}$ and $\rho_{\mathrm{r}} = \pi$ which yields

$$T_{\mathrm{e}} = -\frac{3}{2}\,\hat{\psi}_{\mathrm{m}} i_{\mathrm{f}} \sin\left(\rho_{\mathrm{m}}\right) \tag{8.1}$$

If a mechanical load is applied to the machine, the load angle ρ_{m} will be non-zero. This explains why this angle is referred to as the "load angle." When a load is applied, the rotor will lag behind the magnetic field ($\rho_{\mathrm{m}} < 0$), which leads to a positive torque value that matches the applied load. The load angle can for a given load torque be modified by varying the amplitude of the field or rotor current. The maximum torque $\hat{T}_{\mathrm{e}}$ that can be delivered by this machine is reached when the load angle reaches $\pi/2$ rad. Speed changes are implemented by changing the stator frequency ω_{s}, which nowadays requires the use of a power electronic converter.

The general space vector diagram, according to Fig. 7.12, changes to the form shown in Fig. 8.4 given the present choice of rotor excitation. Several interesting observations can be made with respect to Fig. 8.4. Firstly, the current vector $\vec{i}^{\mathrm{xy}}$ is tied to the negative real axis of the rotor given that the rotor is fed with a DC current i_{f}, which is in the opposite direction, as shown in Fig. 7.11. Secondly, the vectors

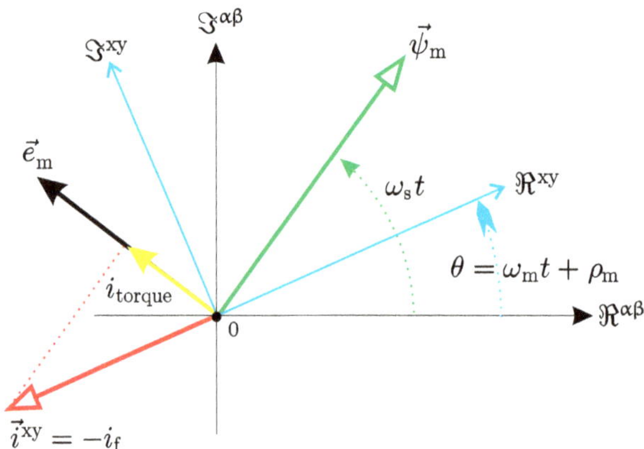

Fig. 8.4 Space vector diagram for synchronous machine, motoring operation shown with $\rho_{\mathrm{m}} < 0$

$\vec{\psi}_{\mathrm{m}}$ and $\vec{i}$ are stationary with respect to each other. The angle between the two can and will vary depending on the load torque. For example, an increase in the load torque will see the rotor slip back momentarily (in the clockwise direction) so that the machine adjusts its torque T_{e} to match the load torque T_{l}. As indicated above, the highest torque (known as the "pull out" torque) that can be delivered by the machine is equal to $\hat{T}_{\mathrm{e}} = {}^{3}/_{2}\,\hat{\psi}_{\mathrm{m}}\, i_{\mathrm{f}}$. The machine will stall (or rotate uncontrolled in the opposite direction) when a load torque above this value is applied. In this situation, the two vectors will no longer be stationary with respect to each other, i.e., the machine will deliver a pulsating torque with zero average value. Also shown in Fig. 8.4 is the voltage vector $\vec{e}_{\mathrm{m}}$, shown in stator coordinates (which is equal to the supply voltage vector $\vec{u}$), which leads the flux vector by ${}^{\pi}/_{2}$ rad. The projection of the current onto this vector, shown as i_{torque} in Fig. 8.4, is proportional to the torque.

If the option $\rho_{\mathrm{r}} = 0$ would have been selected, then the field current would be aligned with the positive real axis. Under the circumstances shown in Fig. 8.4 the torque would be negative, hence the rotor would rotate until the current vector would be diametrically opposite to its present position.

8.4 Zero Leakage Inductance and Zero Resistance Model

The universal model as shown in Fig. 7.14 is a convenient starting point for this type of model. Elimination of the leakage inductances $L_{\sigma\mathrm{S}}$, $L_{\sigma\mathrm{R}}$ from this model implies that transformation ratio will be equal to $a = 1$, given that $L_{\mathrm{s}} = L_{\mathrm{r}} = L_{\mathrm{M}} = L_{\mathrm{m}}$. A synchronous machine with a rotor based field winding is assumed, which is connected to a current source i_{fw} via a set of slip rings. Note that the ITF module shown in Fig. 7.14 has effectively been omitted in Fig. 8.5 by making use of a current source with current i_{f}. The latter source is located on the 'primary' side of the ITF module. The relationship between the actual field winding current i_{fw} and the ITF primary referred current i_{f} is therefore of the form $i_{\mathrm{fw}} = ({}^{n_{\mathrm{s}}}/_{n_{\mathrm{f}}})\, i_{\mathrm{f}}$, where n_{f} represents the number of field winding turns, which replaces the variable n_{r}. This implies that the ITF transformation ratio, as shown in the universal model, is defined as $k^{\mathrm{r}} = {}^{n_{\mathrm{s}}}/_{n_{\mathrm{f}}}$. For the synchronous machine the magnetizing inductance is relocated to the "rotor" side of the IRTF as may be observed from Fig. 8.5. The reason for this choice is that modeling of machines with saliency becomes a lot easier when the magnetizing inductance is located on the rotor side [2]. Hence, for consistency reasons the magnetizing inductance is also located on the rotor side of the IRTF.

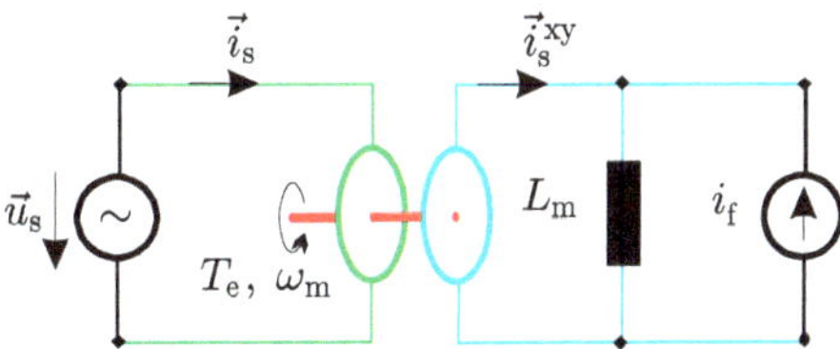

Fig. 8.5 Synchronous machine, zero resistance and zero leakage inductance

The equation set which corresponds to Fig. 8.5 is of the form

$$\vec{u}_s = \frac{d\vec{\psi}_m}{dt} \tag{8.2a}$$

$$\vec{\psi}_m = L_m \left(\vec{i}_s + i_f e^{j\theta}\right) \tag{8.2b}$$

$$T_e - T_l = J\frac{d\omega_m}{dt} \tag{8.2c}$$

$$\omega_m = \frac{d\theta}{dt} \tag{8.2d}$$

Note that in this equation set, the flux vector $\vec{\psi}_m$ [see Eq. (8.2b)] is given in its stator based form. This expression is in fact found by using Kirchhoff's current law on the rotor side which gives $\vec{\psi}_m^{xy} = L_m \left(\vec{i}_s^{xy} + i_f\right)$.

A rotating stator flux vector $\vec{\psi}_s = \hat{\psi}_s e^{j\omega_s t}$ is established as a result of the machine being connected to a three-phase grid with angular frequency ω_s. Note that in the present case (zero leakage inductance), the magnetizing flux space vector is equal to the stator flux vector hence $\vec{\psi}_m = \vec{\psi}_s$.

8.4.1 Generic Model

A generic representation of the (two-pole) synchronous machine in its present form is given in Fig. 8.6. The model follows directly from Eq. (8.2). Central to this model is the IRTF sub-module, which is represented by Fig. 7.5a. The load torque is provided by a sub-module $T_l(\omega_m)$, which assumes, for simplicity reasons, a relationship between load torque and speed (not position).

The model in question provides the user with a simple machine representation, that can be used to examine various operating modes (including steady-state operation) as will become apparent in the tutorial section at the end of the chapter.

8.5 Generalized Machine Model

For the development of a complete model which accommodates the leakage inductance it is helpful to reconsider the universal model shown in Fig. 7.14, in a more convenient (to reflect the synchronous machine topology) configuration. The model as given in Fig. 8.7 shows the magnetizing inductance L_m, stator leakage inductance $L_{\sigma s}$, and field winding leakage inductance $L_{\sigma fw}$. Also shown are the stator resistance R_s and field winding resistance R_{fw}. Note that the variables u_{fw}, i_{fw} which represent the field winding terminal voltage and field winding current, respectively,

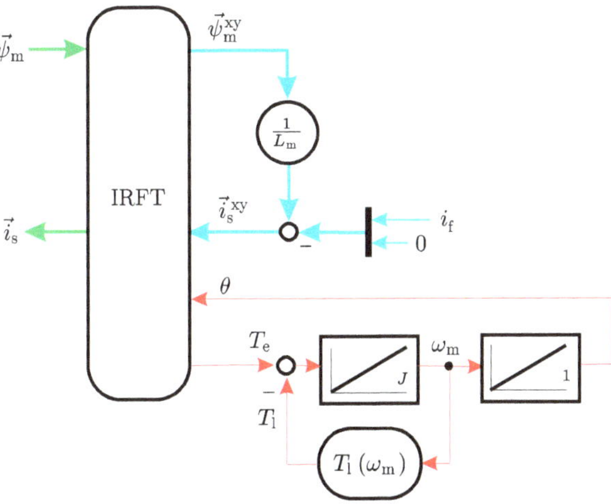

Fig. 8.6 Generic representation of a voltage source connected synchronous machine, which corresponds to Fig. 8.5

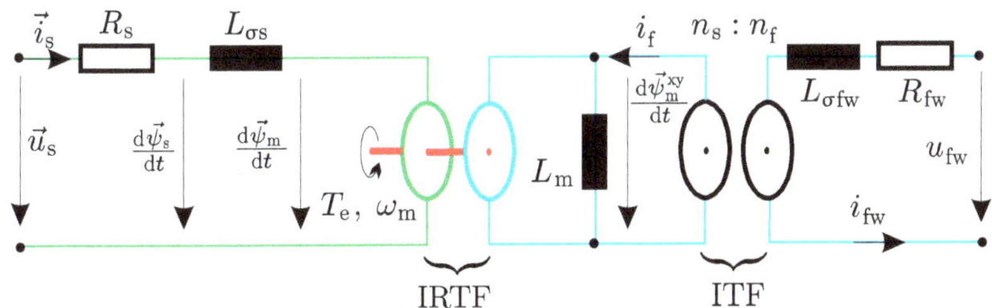

Fig. 8.7 Generalized synchronous machine model

are scalar quantities because they are linked to the "x" winding of the IRTF module. Note that the "y" rotor winding is not used and is therefore ignored in the subsequent analysis. The current $i_f = {}^{n_f}/_{n_s}\, i_{fw}$ shown in Fig. 8.7 represents the stator referred field current. A comparison between the synchronous machine model and the universal model with unity transformation variable $a = 1$ shown in Fig. 7.14 confirms the presence of the three inductance structure in both models. For the synchronous model the rotor leakage inductance is located on the field winding side, but could equally have been referred to the primary side of the ITF. For the purpose of the analysis in this chapter we will assume that a DC current source i_{fw} is connected to the "x" rotor (field) winding. This implies that the field winding leakage inductance and resistance do not need to be accommodated in the generic model to be developed. A further simplification of the present model can be undertaken by introducing a DC current source i_f on the "primary" side of this transformer to replace the current source i_{fw} connected to the "x" rotor (field) winding. The complete model shown in Fig. 8.8 is applicable to the so-called

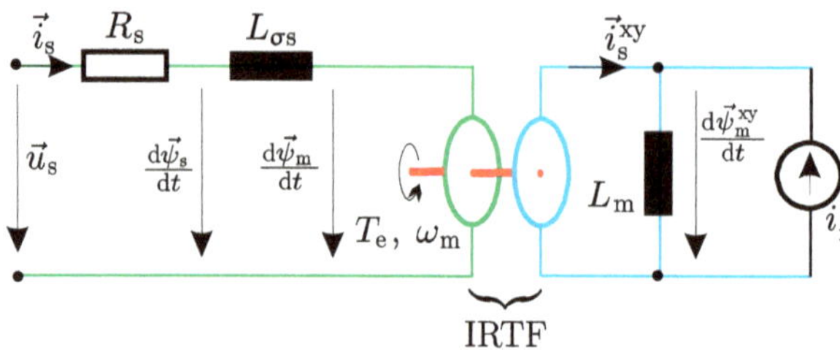

Fig. 8.8 Generalized synchronous machine model, with referred field winding current source

non-salient machines, which generally do not carry any damper windings (short-circuited windings on the rotor). Damper windings can be accommodated in this IRTF model but this is considered to be outside the scope of this book. Salient machines show different L_m values for the x and y direction as is discussed in our text "Advanced Electrical Drives" [2]. The equation set which corresponds to this machine model in its present form is given by Eq. (8.3).

$$\vec{u}_s = \vec{i}_s R_s + \frac{d\vec{\psi}_s}{dt} \tag{8.3a}$$

$$\vec{\psi}_s = \vec{i}_s L_{\sigma s} + \vec{\psi}_m \tag{8.3b}$$

$$\vec{\psi}_m^{xy} = L_m \left(\vec{i}_s^{xy} + i_f\right) \tag{8.3c}$$

$$T_e - T_l = J \frac{d\omega_m}{dt} \tag{8.3d}$$

$$\omega_m = \frac{d\theta}{dt} \tag{8.3e}$$

In order to move towards a generic model where we are able to group the two inductances $L_{\sigma s}$ and L_m it is helpful to introduce the flux $\psi_{mf} = L_m i_f$. This allows us to rewrite Eq. (8.3c) in the form given by Eq. (8.4).

$$\vec{\psi}_m^{xy} = L_m \vec{i}_s^{xy} + \psi_{mf} \tag{8.4}$$

Substitution of Eq. (8.4) into Eq. (8.3b) (written in rotor coordinate format) leads to Eq. (8.5), which for completeness contains the complete set needed to derive a generic model of this machine.

$$\vec{u}_s = \vec{i}_s R_s + \frac{d\vec{\psi}_s}{dt} \tag{8.5a}$$

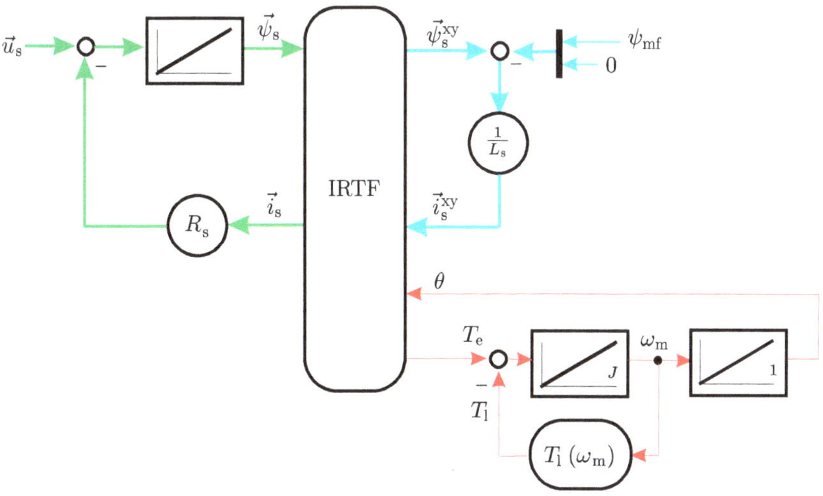

Fig. 8.9 Full generic synchronous machine model, which corresponds to the model given in Fig. 8.8, where the current source i_f has been replaced by a "field" flux source $\psi_{mf} = L_m\, i_f$

$$\vec{\psi}_s^{xy} = \vec{i}_s^{xy} \underbrace{(L_{\sigma s} + L_m)}_{L_s} + \psi_{mf} \tag{8.5b}$$

$$T_e - T_l = J \frac{d\omega_m}{dt} \tag{8.5c}$$

$$\omega_m = \frac{d\theta}{dt} \tag{8.5d}$$

In Eq. (8.5b) the sum of the two inductances known as the stator inductance L_s appears as intended.

8.5.1 *Generic Model*

The generic model of the two-pole non-salient synchronous machine without damper winding is directly found using Eq. (8.5). An example of implementation as given in Fig. 8.9 shows the presence of a gain module which represents the inverse stator inductance $1/L_s$. Observation of Fig. 8.9 shows that the IRTF module has as "inputs" the flux vector $\vec{\psi}_s$ and stator current vector $\vec{i}_s^{xy}$. The variable ψ_{mf} represents the stator referred magnetizing flux which for a machine with a field winding will be different to the stator referred field flux linkage ψ_f. The difference between the two is attributed to the field winding leakage inductance. For machines where the excitation is provided by a permanent magnet the field winding leakage inductance will be zero, which implies that $\psi_{mf} = \psi_f$.

The mechanism through which a stator current will occur is readily shown using this generic model. Note that the reasoning presented here differs from that given in Sect. 8.3, given the way in which the IRTF module is now used. Both approaches to describing machine operation must of course give the same result. For simplicity we will ignore the stator resistance in this discussion. If we assume that the stator is connected to a three-phase sinusoidal supply, then this leads to a rotating stator flux vector $\vec{\psi}_{s}$. The rotor winding carries a current i_{f} (which is a referred value, the actual field winding current is i_{fw}). If we assume that the rotor rotates at the same speed as the stator flux field, then the rotor and stator flux vectors will move at the same speed. If we initially assume that the magnitude of both vectors is equal and aligned (which implies $\vec{\psi}_{s}(0) = \psi_{mf}$), then no stator current will be present. Under these circumstances the vector into the gain module $1/L_{s}$ (see Fig. 8.9) is zero, hence the current $\vec{i}_{s}$ will be zero. If, for example, we increase the field current, then the field flux ψ_{mf} will increase and a stator current component will occur which will lead the voltage vector by $\pi/2$ radians. No torque will be realized under these circumstances.

If we return to our initial conditions (flux vectors equal magnitude and aligned), then the application of a mechanical load will momentarily cause the rotor to slow down, until an angle between the two flux vectors occurs. The difference vector between the two vectors is proportional to the stator current. Hence, a current vector will occur which means that the machine will produce a torque to counteract the new load torque (after transient effects have died down following the load torque change).

8.6 Steady-State Characteristics

In this section we look at the steady-state performance of the synchronous machine in case we connect the stator windings to a three-phase sinusoidal source. This implies that the stator phase voltage equals the grid voltage with its fixed amplitude and frequency. We also assume the shaft speed to run at synchronous speed, although different shaft angles are possible with respect to the rotating field in the stator. Consequently the torque and field current are the only independent variables left at this stage. Steady-state analysis provides insight with regard to the trajectory of the stator current vector and load angle when either of these independent variables is varied.

The approach taken is to consider the simplified model first and develop the so-called Blondel diagram and torque angle curves on the basis of the phasor equation set applicable to this model. This model is then extended to a full machine model. Synchronous operation is assumed which means that the condition $\omega_{s} = \omega_{m}$ is met. The stator is connected to a three- phase sinusoidal voltage supply which is represented by the space vector $\vec{u}_{s}$. Up to now we have chosen the flux vector of the form $\vec{\psi}_{m} = \hat{\psi}_{m}e^{j\omega_{s}t}$ which corresponds to a supply vector $\vec{u}_{s} = j\omega_{s}\hat{\psi}_{m}e^{j\omega_{s}t}$. The corresponding phasor representations are according to Eq. (4.60) of the form $\underline{\psi}_{m} = \hat{\psi}_{m}$ and $\underline{u}_{s} = j\omega_{s}\hat{\psi}_{m}$.

The phasor diagrams which are linked to AC machines will be discussed in line with the general convention where the supply voltage phasor is chosen along the real axis, i.e., $\underline{u}_s = \hat{u}_s$. With our present choice of flux vector the voltage phasor is along the imaginary axis. If we re-define (for the purpose of examining the steady-state performance only) the relationship between space vectors and phasors as

$$\vec{x} = \underline{x}\,e^{j(\omega_s t + \frac{\pi}{2})} \tag{8.6}$$

then the flux and voltage phasor will be of the form $\underline{\psi}_m = \hat{\psi}_m e^{-j\frac{\pi}{2}}$ and $\underline{u}_s = \omega_s \hat{\psi}_m$, respectively, i.e., rotated clockwise so that the voltage phasor is real, as preferred for steady-state analysis.

8.6.1 Steady-State Characteristics, Simplified Model

The steady-state characteristics of the non-salient synchronous machine are studied with the aid of Fig. 8.6. The magnetizing flux vector $\vec{\psi}_m$ will also rotate at the same speed but will lag the voltage vector $\vec{u}_s$ by $\pi/2$ radians. This vector is derived from the voltage vector using $\vec{u}_s = d\vec{\psi}_m/dt$, or $\underline{u}_s = j\omega_s \underline{\psi}_m$ in phasor form.

The basic characteristics of the machine relate to the stator current of the two-pole machine in phasor form and the torque load angle curve. The flux phasor $\underline{\psi}_m$ according to Eq. (8.2b) can, with the aid of Eqs. (8.6) and (7.18) (with $\omega_m = \omega_s$), be written as

$$\underline{\psi}_m = L_m \left(\underline{i}_s + i_f e^{j(\rho_m - \frac{\pi}{2})} \right) \tag{8.7}$$

The current $\underline{i}_s$ can, with the aid of Eq. (8.7) and expression $\underline{\psi}_m = \hat{u}_s/j\omega_s$, be written as

$$\underline{i}_s = \frac{\hat{u}_s}{j\omega_s L_m} - i_f\,e^{j(\rho_m - \frac{\pi}{2})} \tag{8.8}$$

Equation (8.8) can also be represented in terms of an equivalent circuit as given by Fig. 8.10. Equation (8.8) may also be rewritten in the following form

$$\underline{i}_s = \underbrace{\frac{\hat{u}_s}{j\omega_s L_m}}_{\underline{i}_{s1}} - \underbrace{\frac{\hat{u}_s\, k_F\, e^{j\rho_m}}{j\omega_s L_m}}_{\underline{i}_{s2}} \tag{8.9}$$

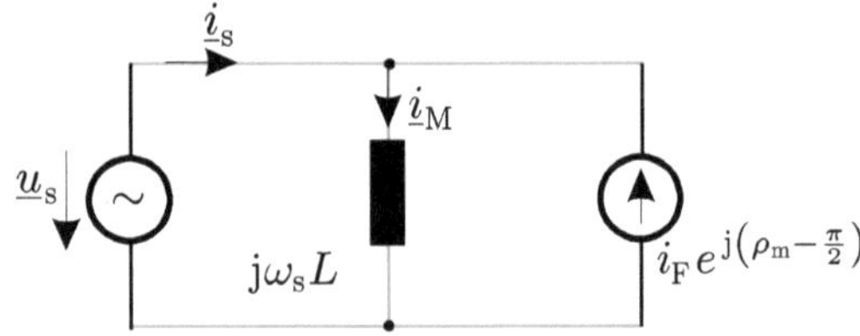

Fig. 8.10 Simplified synchronous machine, phasor based model

in which the factor k_F is defined as

$$k_\mathrm{F} = \frac{\omega_\mathrm{s} L_\mathrm{m}\, i_\mathrm{f}}{\hat{u}_\mathrm{s}} \tag{8.10}$$

which can also be written in the form given by Eq. (8.11).

$$k_\mathrm{F} = \frac{L_\mathrm{m}\, i_\mathrm{f}}{\hat{\psi}_\mathrm{m}} \tag{8.11}$$

If the value of k_F is greater than 1, the machine is said to be operating under "over-excited" conditions. For k_F value less than 1 a so-called under-excited machine operating condition is present.

The current phasor $\underline{i}_\mathrm{s}$ can be plotted as a function of the load angle ρ_m with k_F as parameter. This type of diagram as given in Fig. 8.11 shows the two current components according to Eq. (8.9) together with lines of constant output power p_out. These output power curves are found by making use of the energy balance equation, which for the current machine with zero stator resistance is of the form

$$\frac{3}{2}\,\Re\left\{\underline{u}_\mathrm{s}\underline{i}_\mathrm{s}^*\right\} = p_\mathrm{out} \tag{8.12}$$

where $p_\mathrm{out} = T_\mathrm{e}\omega_\mathrm{m}$. The power balance states that the output power is in this case equal to the input power. This expression may be reduced to

$$\frac{3}{2}\,\hat{u}_\mathrm{s}\Re\left\{\underline{i}_\mathrm{s}^*\right\} = p_\mathrm{out} \tag{8.13}$$

where use is made of $\underline{u}_\mathrm{s} = \hat{u}_\mathrm{s}$. Equation (8.13) shows that lines of constant output power are represented by vertical lines in Fig. 8.11. It is noted that for a given shaft speed the output power is proportional to the output torque.

The diagram according to Fig. 8.11 is known as a "Blondel" diagram and is particularly useful for identifying a range of operating situations. In particular the user is able to gain insight as to how, for example, changes to the load torque or field current will affect the stator current and power factor. A number of these are itemized as follows.

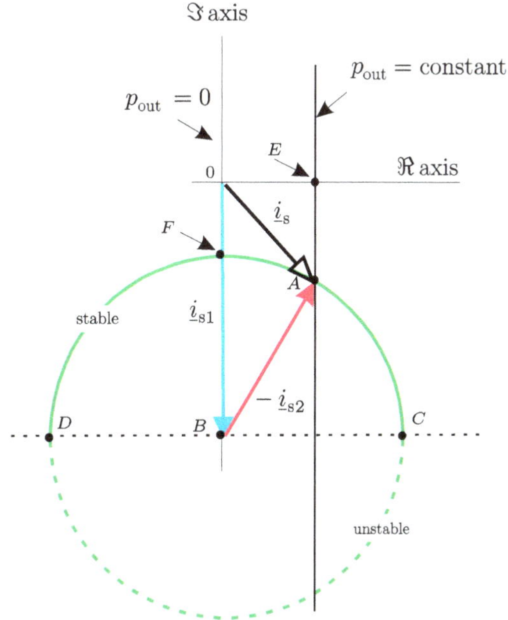

Fig. 8.11 Blondel diagram of synchronous machine with $R_s = 0$, $\rho_m < 0$, and $k_F < 1$

- Motor operation is realized in the first and fourth quadrant of the complex plane.
- Generator operation is achieved if the current vector end point "A" is located in the second or third quadrant of the complex plane.
- When the load angle is changed a circle will appear which represents the stator current end point trajectory. The circle radius is proportional to the current i_f, where point "B" represents the case for $i_f = 0$. In case the load torque is zero and $i_f > 0$, the operating point will be located at, for example, point "F". As the load torque is increased ($\rho_m < 0$) the motoring region is entered and the output power will increase. The horizontal distance from the operating point to the zero power line determines the output power level. The larger this distance, the higher the output power. This means that motor operation with a current phasor end point positioned at "C" gives the highest output power level achievable for a given field current. This operating point also represents the limit for stable motor operation. If the load torque is increased beyond this value, then the load angle is increased (in absolute terms) further, which leads to a smaller rather than larger output torque. The motor will now loose synchronization with the grid. A similar reasoning is also applicable for generator operation in which case the limit of stable operation is identified by point "D". The entire operating trajectory for unstable operation is also shown in Fig. 8.11.
- For a given output power level at, for example, point "A" it is possible to change the excitation current as to minimize the stator current amplitude. This is achieved at point "E" which corresponds to unity power factor. Under these conditions the machine will be over-excited, i.e., $k_F > 1$.

- The Blondel diagram according to Fig. 8.11 is shown with $k_F = 0.7$. The current end point trajectory from point $A \rightarrow E$ is realized in case k_F is increased (by changing i_f) to the value 1.06 while maintaining a constant output power level.

In addition to using the Blondel diagram it is desirable to have access to a cartesian type diagram in the form of the output power/load angle diagram. This type of diagram is readily obtained by making use of, for example, Eqs. (8.1) and (8.11), with $p_{out} = T_e \omega_s$ and $\hat{\psi}_m = \hat{u}_s/\omega_s$, which leads to the following output power expression

$$p_{out} = -\frac{3\hat{u}_s^2 k_F}{2\omega_s L_m} \sin \rho_m \tag{8.14}$$

This expression can be further developed by introducing a normalization factor $3\hat{u}_s^2/2\omega_s L_m$ which gives

$$p_{out}^n = -k_F \sin \rho_m \tag{8.15}$$

A graphical representation of Eq. (8.15) is shown in Fig. 8.12 with $k_F = 0.7$. The parameters used here are identical to those used for Fig. 8.11. This means that a number of operating points given in the Blondel diagram can also be shown here for comparison purposes. A second torque/load angle curve is included in Fig. 8.12 which represents the case $k_F = 1.06$ which as was discussed earlier allows operation with unity power factor. Observation of Fig. 8.12 shows that the stable operating range of load angles corresponds to region $C - D$. The concept of the so-called stable operation is readily illustrated by considering operation at, for example, point "A". If the mechanical load power is increased, the machine can respond by producing more output power which in turn leads to a more negative load angle. If this example is carried out for operation at, for example, point "C", then an increase in load power cannot be met by increase in machine output power, in which case de-synchronization will occur.

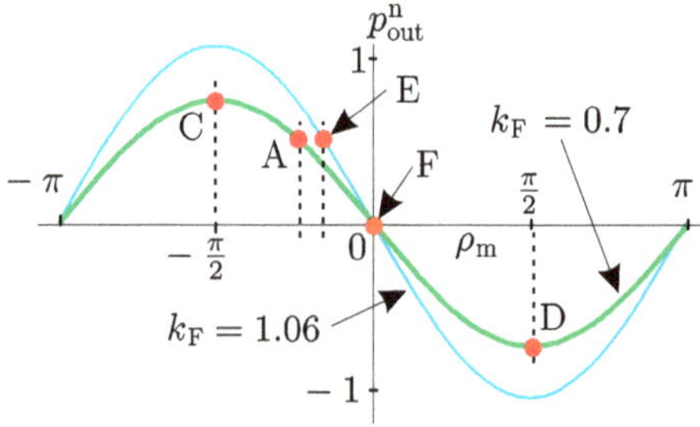

Fig. 8.12 Normalized output power versus load angle curves, with $R_s = 0$

8.6.2 Steady-State Characteristics, Full Model

In reality machines have a finite stator resistance and stator inductance. It is therefore necessary to consider the Blondel diagram and load torque/angle curves for the more general case.

The Blondel diagram is found by making use of equation set (8.5), which may be converted to phasor form using Eqs. (8.6) and (7.18) (with $\omega_m = \omega_s$), which leads to Eq. (8.16).

$$\underline{u}_s = \underline{i}_s R_s + j\omega_s \underline{\psi}_s \tag{8.16a}$$

$$\underline{\psi}_s = L_s \underline{i}_s + \psi_{mf}\, e^{j\left(\rho_m - \frac{\pi}{2}\right)} \tag{8.16b}$$

Elimination of the flux phasor $\underline{\psi}_s$ from Eq. (8.16) leads to the following current phasor expression.

$$\underline{i}_s = \frac{\hat{u}_s\left(1 - k_F\, e^{j\rho_m}\right)}{R_s + j\omega_s L_s} \tag{8.17}$$

Note that this expression contains the variable k_F as introduced in expression (8.10). This variable in its general format is defined by Eq. (8.19), as will be discussed shortly. Expression (8.17) may also be presented in terms of a circuit representation as given in Fig. 8.13. Observation of this figure shows that the current $\underline{i}_s$ can also be found by application of a Superposition Theorem strategy which considers the current component for each voltage source separately (in this case the supply source $\underline{u}_s$ and back-emf $k_F\hat{u}_s$) and then adding (in complex format) the two terms, which gives

$$\underline{i}_s = \underbrace{\frac{\hat{u}_s}{R_s + j\omega_s L_s}}_{\underline{i}_{s1}} - \underbrace{\frac{k_F \hat{u}_s\, e^{j\rho_m}}{R_s + j\omega_s L_s}}_{\underline{i}_{s2}} \tag{8.18}$$

The excitation factor k_F as defined by Eq. (8.10) is expressed in its more general form (in terms of $\psi_{mf} = L_m i_f$) as given by Eq. (8.19).

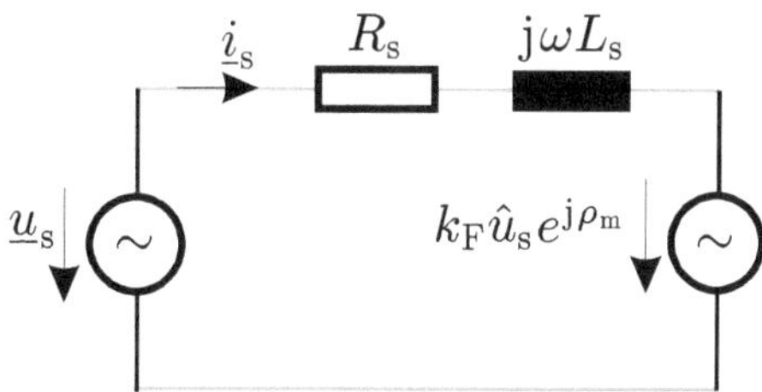

Fig. 8.13 Synchronous phasor based machine model

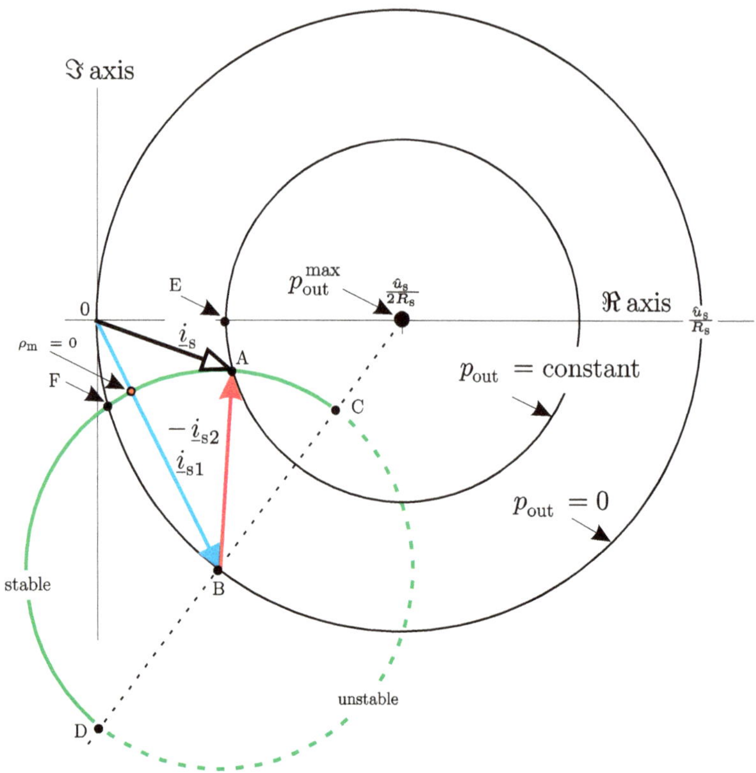

Fig. 8.14 Blondel diagram of synchronous machine with $k_F = 0.7$ and ${}^{\omega_s L_s}/_{R_s} = 1.96$

$$k_F = \frac{\omega_s \psi_{mf}}{\hat{u}_s} \tag{8.19}$$

A graphical representation of Eq. (8.18) as given in Fig. 8.14 on page 216 shows the two current components together with curves of constant output power p_{out}. The speed is kept quasi-constant while the load angle is able to vary. The curves of constant torque, i.e., constant output power, are found using the energy balance equation

$$\frac{3}{2}\Re\left\{\underline{u}_s \underline{i}_s^*\right\} - \frac{3}{2}\Re\left\{\underline{i}_s \underline{i}_s^*\right\} R_s = p_{out} \tag{8.20}$$

In this case, a dissipative term is introduced which was not present in the previous energy equation (8.12). Expression (8.20) can be written in a more simplified normalized form which after some mathematical handling gives

$$\left(x - \frac{\hat{u}_s}{2R_s}\right)^2 + y^2 = \frac{1}{R_s}\left(\frac{\hat{u}_s^2}{4R_s} - \frac{2}{3}p_{out}\right) \tag{8.21}$$

with

$$x = \Re\left\{\underline{i}_s\right\}$$
$$y = \Im\left\{\underline{i}_s\right\}$$

Equation (8.21) states that we are able to represent circles of constant output power in the complex plane as well as the stator current phasor. These circles are centered on the $\Re$ axis with coordinates $(\hat{u}_s/2R_s, 0)$ and radius

$$r_{out} = \sqrt{\frac{1}{R_s}\left(\frac{\hat{u}_s^2}{4R_s} - \frac{2}{3}p_{out}\right)} \tag{8.22}$$

Equation 8.22 implies that the zero output power circle has a radius of $r_{out} = \hat{u}_s/2R_s$. Furthermore, the machine has a maximum output power level

$$p_{out}^{max} = \frac{3\hat{u}_s^2}{8R_s} \tag{8.23}$$

which in Fig. 8.14 is found at coordinates $(\hat{u}_s/2R_s, 0)$ of the complex plane with $r_{out} = 0$. It is noted that for a given shaft speed the output power is proportional to the output torque. The general Blondel diagram with $R_s > 0$ as given by Fig. 8.14 has a number of interesting operating points which are itemized below.

- If the field current i_f is set to zero (point "B"), then the stator current will be positioned on the zero output power circle.
- Motor operation is present within the circle constrained by $p_{out} = 0$.
- For a given field current (and corresponding k_F value) and variable load angle a circle will appear which represents the stator current end point trajectory. The radius of this circle is proportional to the current i_F. When the load torque is zero the operating point will be located at point "F". As load torque is increased, the motoring region is entered and the output power will increase. The distance from the operating point to the maximum output power point p_{out}^{max} determines the output power level. The shorter this distance, the higher the output power. This means that operation with a current phasor end point positioned at "C" gives the highest output power level achievable for the given field current. This operating point also represents the limit for stable operation.

- For a given output power level at, for example, point "A", it is possible to change the excitation to minimize the stator current. This is achieved at point "E" which corresponds to unity power factor.
- The Blondel diagram according to Fig. 8.14 is shown with $k_F = 0.7$ and ${}^{\omega_s L_s}/_{R_s} = 1.96$. The current end point trajectory can be made to intersect with the maximum power point in case k_F is increased to the value 1.1.
- Lines of constant power are represented as circles in this case which have their center at coordinates $({}^{\hat{u}_s}/_{2R_s}, 0)$.

The output power/load angle diagram is found using Eqs. (7.29h) and (8.18). An observation of the generic diagram (Fig. 8.9) shows that the IRTF module now calculates the torque using the flux vector $\vec{\psi}_s$ and current vector $\vec{i}_s^{xy} = \vec{i}_s \, e^{-j\theta}$. Both vectors can be converted to phasor form using Eqs. (8.6) and (7.18) (with $\omega_m = \omega_s$), which leads, with the aid of Eq. (8.16b), to expression (8.24).

$$T_e = \frac{3}{2} \Im \left\{ j\psi_{mf} \, \underline{i}_s \, e^{-j\rho_m} \right\} \tag{8.24}$$

The output power, defined as $p_{out} = T_e \omega_m$, is then found using Eqs. (8.24), (8.18), and (8.19). The output power expression in a normalized form is given by Eq. (8.25)

$$p_{out}^{n} = -\frac{4k_F}{\sqrt{1 + \left(\frac{\omega_s L_s}{R_s}\right)^2}} \left\{ \sin\left(\rho_m + \gamma - \frac{\pi}{2}\right) - k_F \sin\left(\gamma - \frac{\pi}{2}\right) \right\} \tag{8.25}$$

where $\gamma = \arctan\left({}^{\omega_s L_s}/_{R_s}\right)$. The normalization used in Eq. (8.25) is of the form

$$p_{out}^{n} = \frac{p_{out}}{p_{out}^{max}} \tag{8.26}$$

A graphical representation of Eq. (8.25) as function of the load angle ρ_m is shown in Fig. 8.15 with $k_F = 0.7$ and ${}^{\omega_s L_s}/_{R_s} = 1.96$.

The parameters used here are identical to those used for Fig. 8.14. This means that a number of operating points given in the Blondel diagram can also be shown here for comparison purposes. A second torque/load angle curve is included in Fig. 8.15 that represents the case $k_F = 1.1$, which as was discussed earlier, allows operation at the maximum power point.

Several interesting observations can be made with respect to the output power versus load angle diagram.

- The peak values for motor and generator operation are generally not equal. Equality of the peak values is achieved in case the term ${}^{\omega_s L_s}/_{R_s} \to \infty$. This is readily apparent for the zero resistance case (Fig. 8.12).

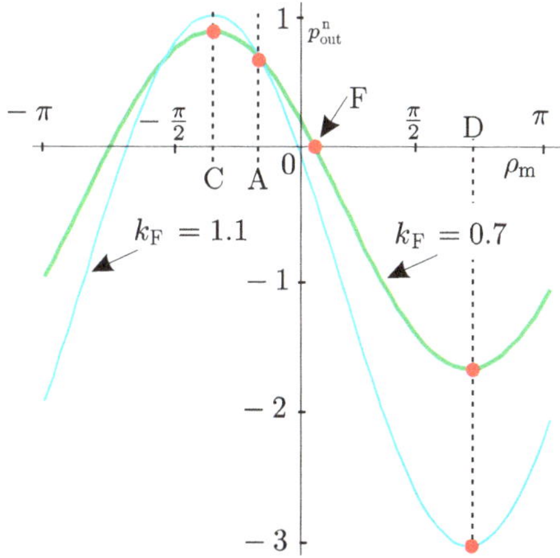

Fig. 8.15 Normalized output power versus load angle curves

- Zero load angle does generally not correspond to zero output power. Only for $k_{\mathrm{F}} = 1$ is this the case.
- The stable operating range of load angle corresponds to region $C - D$.

8.7 Tutorials

8.7.1 Tutorial 1: Simplified Grid Connected Synchronous Machine

This tutorial considers a three-phase IRTF based synchronous machine in its simplified form, i.e., no stator resistance, rotor resistance, or leakage inductance. The aim is to build a PLECS model of this machine in accordance with the generic model given in Fig. 8.6. This model can be used to examine the steady-state characteristics of this simplified machine. An example of such a model is given in Fig. 8.16. The machine in question has a magnetizing inductance of $L_{\mathrm{m}} = 1\,\mathrm{H}$, and an inertia of $J = 10\,\mu\mathrm{kg\,m^2}$. The rotor is connected to a DC current source which provides a (referred) current i_{f}. Furthermore, a rotating magnetizing flux vector $\vec{\psi}_{\mathrm{m}}$ is assumed which is the result of connecting the machine on the stator side to a three-phase sinusoidal voltage source. The use of a rotating flux vector is in line with the approach taken in the previous tutorial on space vector transformers (see Sect. 6.4.1 on page 168). The machine is connected to a mechanical load via which the load torque can be altered under motor operation. For this part of the model use is made of modules for the PLECS "Mechanical" library. We will now consider some of the modules shown in Fig. 8.16 in more detail.

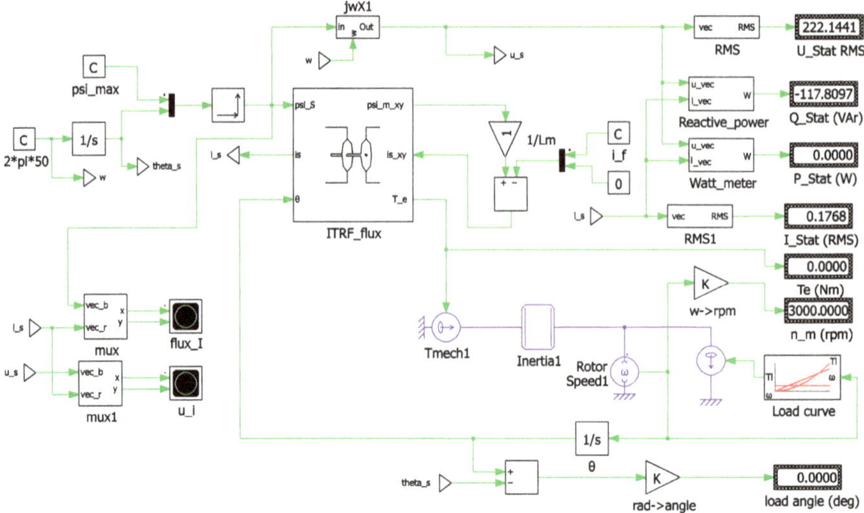

Fig. 8.16 PLECS model (simplified) of synchronous machine

A PLECS standard "Polar to Rectangular" module shown in Fig. 8.16 is used to generate the input flux vector

$$\vec{\psi}_{\mathrm{m}} = \hat{\psi}_{\mathrm{m}}\, \mathrm{e}^{\mathrm{j}\rho} \tag{8.27}$$

where $\hat{\psi}_{\mathrm{m}} = 1.0\,\mathrm{Wb}$. This flux vector may also be written as $\vec{\psi}_{\mathrm{m}} = \psi_{\mathrm{m}\alpha} + \mathrm{j}\psi_{\mathrm{m}\beta}$. The angle $\rho = \omega_{\mathrm{s}} t$ is produced in exactly the same way as discussed in tutorial (Sect. 6.4.1). Check your work by using an "XY" scope module.

Run the simulation for 1s and observe the result, which should be a circle with a radius of 1.0, given that the flux vector is of the form $\vec{\psi}_{\mathrm{m}} = 1.0\,\mathrm{e}^{\mathrm{j}100\pi t}$.

Add a module which will generate the voltage space vector $\vec{u}_{\mathrm{s}} = \mathrm{j}\,\omega_{\mathrm{s}}\,\vec{\psi}_{\mathrm{m}}$ which is based on the model given in Fig. 6.8. Use the PLECS sub-module `jwX` developed in an earlier tutorial example (see Sect. 6.4.1).

Add a sub-module which allows you to calculate the RMS value of the three-phase waveforms which corresponds to a vector $\vec{x}$. Connect the output of the vector $\vec{u}_{\mathrm{s}}$ via an RMS converter to a "Display module," which will give the RMS stator voltage value.

The value on this display follows from the flux amplitude and frequency, namely $\hat{u}_{\mathrm{s}} = \hat{\psi}_{\mathrm{m}}\omega_{\mathrm{s}}$, where $\hat{u}_{\mathrm{s}}$ represents the amplitude of the voltage vector. The readout value is therefore equal to $U_{\mathrm{s}} = \omega_{\mathrm{s}}\hat{\psi}_{\mathrm{m}}/\sqrt{2}$, which in numerical terms is equal to $U_{\mathrm{s}} = 1.0 \cdot 100\pi/\sqrt{2} = 222.1\,\mathrm{V}$.

The IRTF module as developed in Sect. 7.5.1 on page 194 is directly applicable to this tutorial. The magnetizing current component $\vec{i}_{\mathrm{m}}^{\,\mathrm{xy}} = \vec{\psi}_{\mathrm{m}}^{\,\mathrm{xy}}/L_{\mathrm{m}}$ must also be added in order to calculate the primary current vector $\vec{i}_{\mathrm{s}}^{\,\mathrm{xy}}$. This vector is given as $\vec{i}_{\mathrm{s}}^{\,\mathrm{xy}} = \vec{i}_{\mathrm{m}}^{\,\mathrm{xy}} - i_{\mathrm{f}}$.

It is helpful to also add a series of modules which will allow you to show the torque, rotational speed (rpm), and load angle. The latter is the angle between the shaft and flux vector, its value is shown in degrees, i.e., you need to convert from radians to degrees (factor $^{180}/_{\pi}$). If a mechanical load is applied, then a NEGATIVE load angle will occur. In addition to the above, use a vector to RMS converter with display unit to show the RMS stator current value.

We will also need to add two additional sub-modules which will give us the stator real and reactive power values known as P (W) and Q (VAr), respectively. The inputs to these modules will be the space vectors $\vec{u}_s$ and $\vec{i}_s$. The last part of this tutorial is concerned with the load side of the machine. The generic diagram (Fig. 8.6) shows the implementation of the mechanical equation set which links load torque T_l, shaft torque T_e, and inertia J with the shaft speed ω_m and rotor angle θ. Figure 8.16 shows how the mechanical equation set is implemented in PLECS, where use is made of mechanical control blocks. Note that the inertia module must have its initial condition set to $\omega_s = 100\,\pi$ rad/s. This means that the simulation starts with the machine rotating at synchronous speed. The load torque control module `load curve` controls the mechanical `Torque` block. With this module the user can select a "constant," "linear," or "quadratic" load torque versus speed characteristic. The relationship between load torque and speed is therefore of the form

$$T_l = T^* \tag{8.28a}$$

or

$$T_l = k_{L1}\omega_m \tag{8.28b}$$

or

$$T_l = \begin{cases} \text{if } \omega_m \geq 0 & k_{L2}\omega_m^2 \\ \text{if } \omega_m < 0 & -k_{L2}\omega_m^2 \end{cases} \tag{8.28c}$$

where $k_{L1} = {}^{T^*}/_{\omega^*}$ and $k_{L2} = {}^{T^*}/_{(\omega^*)^2}$. In this example we can use the quadratic load curve as given by Eq. (8.28c). The value for ω^* must be set to 100π, while T^* is a variable which must be adjusted in this tutorial. This variable is in fact the steady-state load torque value. An example of an implementation of Eq. (8.28) in PLECS is given in Fig. 8.17. Figure 8.17 shows the use of a `Selector` module, which is controlled by the variable "type," which has a value linked to the selected load/speed curve. The modules `Tn` and `wn` are constants which represent the variables T^* and ω^*, respectively.

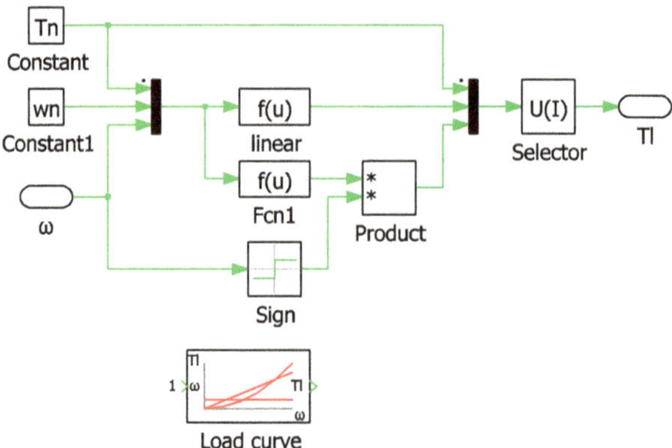

Fig. 8.17 PLECS model: load torque module

Table 8.1 Simulation results synchronous machine: no-load

Parameters	$i_f = 1.0$ A	$i_f = 0.833$ A
RMS stator voltage U_s	222.14 V	222.14 V
RMS stator current I_s	0.00 A	0.11 A
Real stator power P_s	0.00 W	0.00 W
Reactive stator power Q_s	0.00 VAr	78.69 VAr
Load angle ρ_m	0.00°	0.00°
Shaft speed ω_m	3000.00 rpm	3000.00 rpm

8.7.2 *Tutorial 2: Steady-State Analysis of a Simplified Synchronous Machine Operating Under No-Load Conditions*

This tutorial is concerned with using the PLECS model as developed in the previous tutorial. The model will be used to examine the steady-state behavior of the machine operating as a motor under no-load conditions. Set the load torque to zero, i.e., $T^* = 0$ Nm, in your load torque module. Set the field current to $i_f = 1.0$ A and run your simulation with a `Stop` time of `1s` and `Max step size` of `1e-5s` (to reduce the simulation speed in order to observe the space vectors on a scope).

Change the field current value to 0.833 A and rerun the simulation. Observe the result on the display modules and confirm these results via a steady-state phasor analysis in the form of a MATLAB file. The results which should appear on the numerical display modules for the two simulation runs are given in Table 8.1.

It is instructive to consider the vector plot `u-i` gives in this model which shows the scaled voltage vector $\vec{u}_s/300$ and current $\vec{i}_s$ for the two selected field current values at the end of the simulation interval. Observation of Fig. 8.18, subplot (a) confirms that the current vector is zero (as shown by a "dot") when the field current is equal to $i_f = 1.0$ A. When the field current is reduced, as shown in subplot (b), a "lagging" stator current vector appears which is consistent with the results shown in Table 8.1.

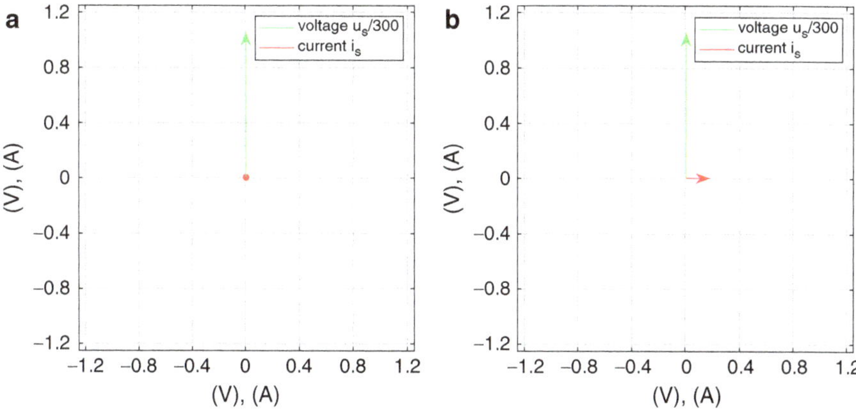

Fig. 8.18 Vector plots showing $\vec{u}_s/300$ ("*green*") and current $\vec{i}_s$ for different field current values under no-load conditions. (**a**) $i_f = 1$ A. (**b**) $i_f = 0.833$ A

Note also that the phase angle between the two vectors is 90° hence the real power P_s is zero. The M-file as given below shows the phasor analysis for this problem. The file must be run for the two i_f current values in order to obtain the results shown in Table 8.1.

M-Code

```
%Tutorial 2, chapter 8
%steady state analysis
Lm=1;                                  % Magnetizing inductance
ws=2*pi*50;                            % frequency rad/s
psim_hat=1.0;                          % flux vector amplitude
psim_ph=-j*psim_hat;                   % input flux phasor
im_ph=psim_ph/Lm;                      % magnetizing current
us_ph=j*ws*psim_ph;                    % stator voltage phasor
U_s=abs(us_ph)/sqrt(2);                % RMS value phase voltage
T_e=0;                                 % no-loadcase,zero torque
i_f=1.0;                               % field current
rho_m=-asin(T_e/(psim_hat*i_f));       % load angle rad
rho_mD=rho_m*180/pi;                   % load angle in degrees
rh=rho_m-pi/2;
is_ph=im_ph-i_f*(cos(rh)+j*sin(rh)); % stator current phasor
                                         calculation
I_s=abs(is_ph)/sqrt(2);                % RMS value phase current
P=3/2*real(us_ph*conj(is_ph));         % real stator power
Q=3/2*imag(us_ph*conj(is_ph));         % reactive stator power
wm=ws;                                 % shaft speed rad/s
nm=wm*60/(2*pi);                       % shaft speed rad/s
```

8.7.3 Tutorial 3: Steady-State Analysis of a Simplified Synchronous Machine Operating Under Variable Load Conditions

This tutorial considers the steady-state behavior of the model as discussed in tutorials 1 and 2 under varying load conditions. For this example, maintain a field current value of 1 A and vary the load torque. The maximum load torque which our machine can handle is 1.5 Nm. Explain where this value comes from.

Answer: the peak flux level is $\hat{\psi}_{\mathrm{m}} = 1.0\,\mathrm{V\,s/rad}$, the field current is $i_{\mathrm{f}} = 1.0\,\mathrm{A}$, which according to Eq. (8.1) gives a maximum torque of 1.5 Nm. Change the load torque in ten steps in the range 0–1.5 Nm and record (after doing a simulation run for each STEP) the following data from the display modules: load angle ρ_{m}, shaft torque T_{e}, RMS stator current, real and reactive stator power.

The data as given in Table 8.2 should appear from your simulation.

Build an M-file which will display the data from your simulation in the form of four subplots: $T_{\mathrm{e}}(\rho_{\mathrm{m}})$, $I_{\mathrm{s}}(\rho_{\mathrm{m}})$, $P_{\mathrm{s}}(\rho_{\mathrm{m}})$, $Q_{\mathrm{s}}(\rho_{\mathrm{m}})$. In addition, plot the current phasor $\underline{i}_{\mathrm{s}}$ in the form of a "Blondel" diagram. Note that the angle between the current and voltage phasor can be calculated using your real and reactive power readings. Add to these plots the results as calculated via a steady analysis. Show these calculations in the same M-file.

An example of the results which should appear from this M-file is given in Figs. 8.19 and 8.20.

Also shown (not to scale) in Fig. 8.20 by way of reference is the stator voltage phasor $\underline{u}_{\mathrm{s}}$. Clearly noticeable from Fig. 8.20 is that the locus of the stator current phasor $\underline{i}_{\mathrm{s}}$ is part of a circle which has its center at 0, -1.0 A. An example of an M-file implementation is given below.

Table 8.2 Simulation results Synchronous machine: no-load→load

T_{e} (Nm)	ρ_{m} (°)	P_{s} (W)	Q_{s} (VAr)	I_{s} (A)
0.0	0	0	0	0
0.15	−5.73	47.12	2.36	0.07
0.30	−11.53	94.24	9.52	0.14
0.45	−17.45	141.37	21.70	0.21
0.60	−23.57	188.40	39.34	0.28
0.75	−30.00	235.61	63.13	0.36
0.90	−36.87	282.74	94.24	0.44
1.05	−44.42	329.86	134.70	0.53
1.20	−53.13	376.99	188.49	0.63
1.35	−64.16	424.11	265.83	0.75
1.50	−89.28	471.20	465.35	0.99

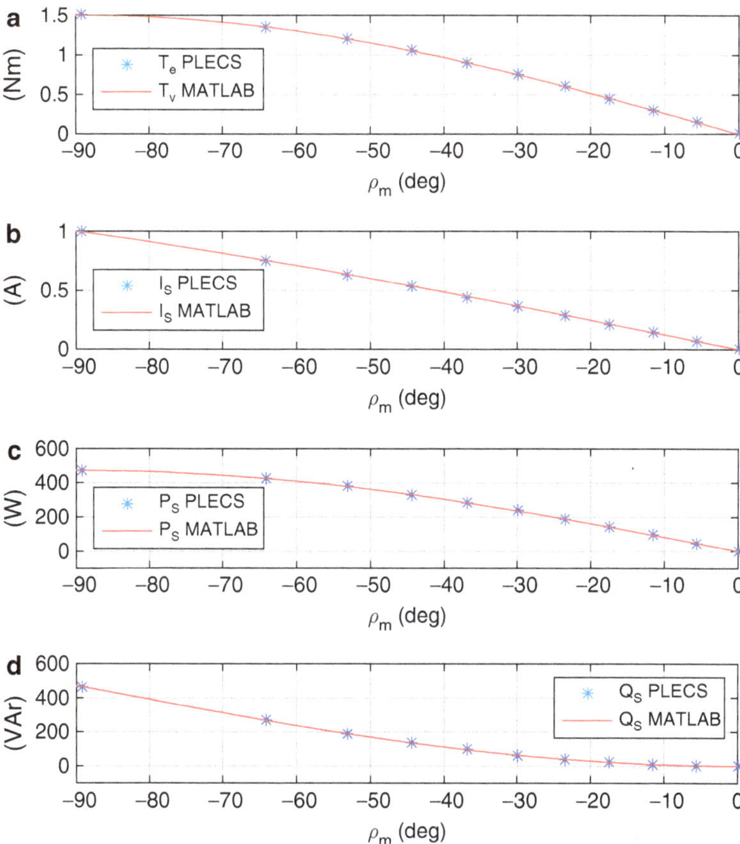

Fig. 8.19 PLECS/MATLAB result: $T_e(\rho_m), I_s(\rho_m), P_s(\rho_m), Q_s(\rho_m)$

M-Code

```
%Tutorial 3, Chapter 8
%Tutorial synchronous machine-simplified model
%steady state analysis
clear all
close all
Lm=1;                                   % Magnetizing
                                           inductance
ws=2*pi*50;                             % frequency rad/s
psim_hat=1.0;                           % flux vector
                                           amplitude
psim_ph=-j*psim_hat;                    % input flux phasor
im_ph=psim_ph/Lm;                       % magnetizing current
us_ph=j*ws*psim_ph;                     % stator voltage
                                           phasor
```

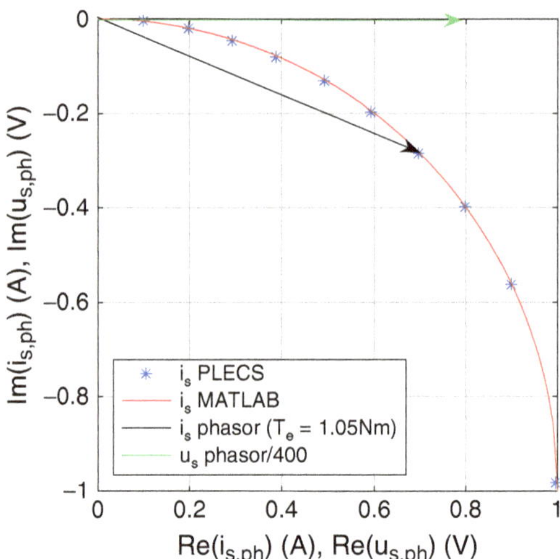

Fig. 8.20 PLECS/MATLAB result: Blondel diagram

```
U_s=abs(us_ph)/sqrt(2);                    % RMS value phase
                                             voltage
%%%%%%%%%%%%%%%%%%%%%%%%%%%%%%%%%%%%%%%%%
%%%%%Data PLECS
Te=[0:1.5/10:1.50];                        % selected  load
                                             torque values
%measured, from display, load angle (deg)
rhoM=[0 -5.73 -11.53 -17.45 -23.57 -30.00 -36.87 -44.42 -53.13
                                             -64.16 -89.28];
%measured, from display, real power (W)
PM=[0 47.12 94.24  141.37 188.4 235.61 282.74 329.86 376.99
                                             424.11 471.2];
%measured, from display, reactive power (VA)
QM=[0 2.36 9.52 21.70 39.34 63.13 94.24 134.70 188.49 265.83
                                             465.35];
%measured, from display, RMS stator current (A)
IsM=[0 0.07 0.14 0.21 0.28 0.36 0.44 0.53 0.63 0.75 0.99];
%%%%%%%%%%%%%%%%%%%%%%%%%%%%%%%%%%%%%%%%%%%%%%%%%%%%%%%%%%%
%display measured results
figure (1)
subplot(4,1,1)
plot(rhoM,Te,'*')
grid
xlabel('(a) \rho_m (deg)')
ylabel('(Nm)')
subplot(4,1,2)
plot(rhoM,IsM,'*')
grid
xlabel('(b) \rho_m (deg)')
ylabel('(A)')
xlabel('(b) \rho_m (deg)')
subplot(4,1,3)
plot(rhoM,PM,'*')
```

```
grid
ylabel('(W)')
xlabel('(c) \rho_m (deg)')
subplot(4,1,4)
plot(rhoM,QM,'*')
grid
ylabel('(VAr)')
xlabel('(d) \rho_m(deg)')
%%%%%%%%%%%%%%plot Blondel diagram
rhos=atan(-QM./PM); Is_phM=IsM*sqrt(2).*(cos(rhos)+j*sin(rhos));
ISRE=real(Is_phM); ISIM=imag(Is_phM);
figure(2)
plot(ISRE,ISIM,'*')
grid
%%%%%%%%%%%%%%%%%%Add theoretical results phasor analysis to
   plots
T_e=[0:1.5/100:1.5];                             % Te values for
                                                    analysis
i_f=1.0;                                         % field current
rho_m=-asin(T_e/(1.5*psim_hat*i_f));             % load angle rad
rho_mD=rho_m*180/pi;                             % load angle in
                                                    degrees
%%%plot results
figure(1)
subplot(4,1,1)
hold on
plot(rho_mD, T_e,'r')
legend('T_e PLECS',' T_e MATLAB')
%%%%%%%%%%%%%%%%%
rh=rho_m-pi/2;
is_ph=im_ph-i_f*(cos(rh)+j*sin(rh));             % stator current
% phasor calculation
I_s=abs(is_ph)/sqrt(2);                          % RMS value phase
                                                    current
subplot(4,1,2)
hold on
plot(rho_mD, I_s,'r')
legend('I_s PLECS',' I_s MATLAB')
%%%%%%%%%%%%%%%%%%%%%%%%%%%
P=3/2*real(us_ph*conj(is_ph));                  % real stator
                                                    power
subplot(4,1,3)
hold on
plot(rho_mD, P,'r')
legend('P_s PLECS',' P_s MATLAB')
Q=3/2*imag(us_ph*conj(is_ph));                  % reactive stator
                                                    power
subplot(4,1,4)
hold on
plot(rho_mD, Q,'r')
legend('Q_s PLECS',' Q_s MATLAB')
%%%%%%%%%%Blondel diagram
figure(2)
hold on
```

Table 8.3 Simulation results synchronous machine: $i_f \subset [0.5, \ldots, 1.75]$ A

i_f (A)	I_s (A)	Q_s (VA)	ρ_m (°)
0.5	0.57	329.86	−53.13
0.75	0.38	172.27	−32.23
1.0	0.28	39.34	−23.58
1.25	0.31	−86.83	−18.66
1.50	0.42	−210.0	−15.47
1.75	0.57	−331.59	−13.21

```
isRE=real(is_ph); isIM=imag(is_ph);
figure(2)
plot(isRE,isIM,'r')
axis([0 1 -1 0])
axis equal
legend('PLECS','MATLAB')
xlabel('Re(is_{ph}) (A)')
ylabel('Im(is_{ph}) (A)')
```

8.7.4 *Tutorial 4: Steady-State Analysis of a Simplified Synchronous Machine Operating with Constant Load and Variable Field Current Conditions*

This tutorial is an extension of the previous tutorial. However, in this case constant torque operation is assumed while varying the field current. Set the load torque to 0.6 Nm and vary the field current in the range of 0.5–1.75 A with an incremental step of 0.25 A. At each incremental step, run your simulation and record the RMS stator current, stator reactive power, and load angle. Plot these variables versus the field current.

The data as shown in Table 8.3 should appear from your simulation.

Build an M-file which will display the data from your simulation in the form of three subplots: $I_s(i_f)$, $Q_s(i_f)$, $\rho_m(i_f)$. In addition, plot the stator current phasor $\underline{i}_s$ in the form of a "Blondel" diagram.

Add to these plots the results as calculated via a steady-state (phasor) analysis. Show these calculations in the same M-file. An example of the results which should appear from this M-file is given in Figs. 8.21 and 8.22.

The stator voltage phasor $\underline{u}_s$ is again added to the Blondel diagram shown in Fig. 8.22. An important observation which can be made from Fig. 8.22 is that the reactive power can be controlled by varying the field current, *without* affecting the output power of the machine. Indeed, the reactive power can be changed from inductive $Q > 0$ (stator current phasor lags the voltage phasor) to capacitive $Q < 0$ (stator current leads the voltage phasor). An example of an M-file implementation for this tutorial is given below:

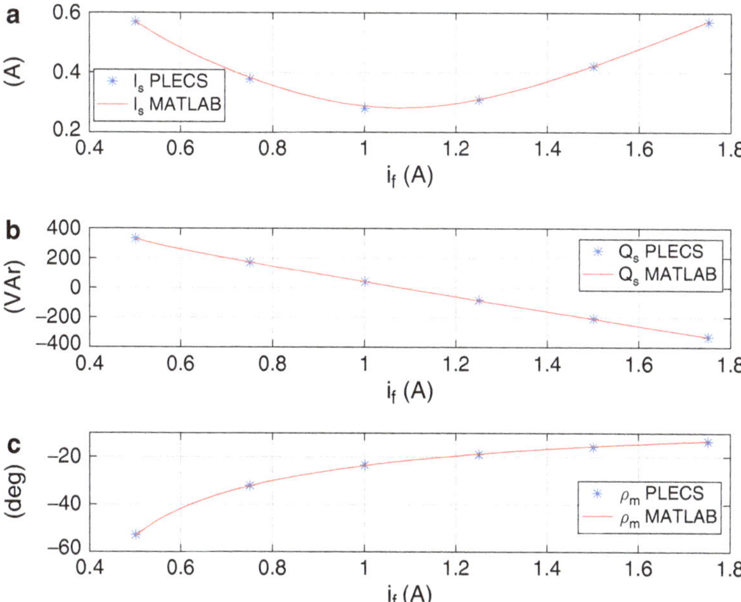

Fig. 8.21 PLECS/MATLAB result: $I_s(i_f)$, $Q_s(i_f)$, $\rho_m(i_f)$

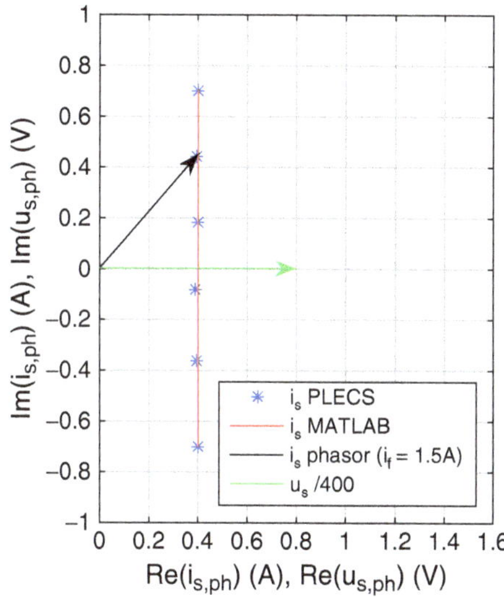

Fig. 8.22 PLECS/MATLAB result: Blondel diagram, constant torque $T_e = 0.6\,\mathrm{Nm}$

M-Code

```
%Tutorial 4, Chapter 8
%Tutorial synchronous machine-simplified model
%steady state analysis: constant torque
```

```
close all
Lm=1;                                  % Magnetizing inductance
ws=2*pi*50;                            % frequency rad/s
psim_hat=1.0;                          % flux vector amplitude
psim_ph=-j*psim_hat;                   % input flux phasor
im_ph=psim_ph/Lm;                      % magnetizing current
us_ph=j*ws*psim_ph;                    % stator voltage phasor
U_s=abs(us_ph)/sqrt(2);                % RMS value phase voltage
%%%%%%%%%%%%%%%%%%%%%%%%%%%%%%%%%%%%%%%%%
%%%%%%Data PLECS
TeM=0.6;                               %torque
ifM=[0.5:0.25:1.75];                   % selected  field current
                                          values
%measured, from display, load angle (deg)
rhoM=[-53.13 -32.23 -23.58 -18.66 -15.47 -13.21];
%measured, from display, real power (W)
PM=188.49;
%measured, from display, reactive power (VA)
QM=[329.86 172.27 39.34 -86.83 -210.0 -331.59];
%measured, from display, RMS stator current (A)
IsM=[0.57 0.38 0.28 0.31 0.42 0.57];
%%%%%%%%%%%%%%%%%%%%%%%%%%%%%%%%%%%%%%%%%%%%%%%%%%%%%%%%%%
%display measured results
figure (1)
subplot(3,1,1)
plot(ifM,IsM,'*')
grid
xlabel('(a) i_f (A)')
ylabel('(A)')
subplot(3,1,2)
plot(ifM,QM,'*')
grid
ylabel('(VAr)')
xlabel('(b) i_f (A)')
subplot(3,1,3)
plot(ifM,rhoM,'*')
grid
ylabel('(deg)')
xlabel('(c) i_f (A)')
%%%%%%%%%%%%%%%plot Blondel diagram
rhos=atan(-QM./PM); Is_phM=IsM*sqrt(2).*(cos(rhos)+j*sin
  (rhos));
ISRE=real(Is_phM); ISIM=imag(Is_phM); figure(2)
plot(ISRE,ISIM,'*')
grid
%%%%%%%%%%%%%%%%%%Add theoretical results phasor analysis to
   plots
%Constant torque T=0.5 Nm
i_f=[0.5:0.025:1.75];                   % field current values
   for analysis
T_e=0.6;% Torque
rho_m=-asin(T_e./(1.5*psim_hat.*i_f));   % load angle rad
rho_mD=rho_m*180/pi;                    % load angle in degrees
rh=rho_m-pi/2;
```

```
is_ph=im_ph-i_f.*(cos(rh)+j*sin(rh));% stator current phasor
   calculation
I_s=abs(is_ph)/sqrt(2);                  % RMS value phase current
P=3/2*real(us_ph*conj(is_ph));           % real stator power
Q=3/2*imag(us_ph*conj(is_ph));           % reactive stator power
%%%plot results
figure(1)
subplot(3,1,1)
hold on
plot(i_f, I_s,'r')
legend('I_s PLECS',' I_s MATLAB')
%%%%%%%%%%%%%%%%%
subplot(3,1,2)
hold on
plot(i_f, Q,'r')
legend('Q_s PLECS','Q_s MATLAB')
%%%%%%%%%%%%%%%%%%%%%%%%%%%
subplot(3,1,3)
hold on
plot(i_f, rho_mD,'r')
legend('rho_m PLECS','rho_m MATLAB')
%%%%%%%%%%Blondel diagram
figure(2)
hold on
isRE=real(is_ph); isIM=imag(is_ph);
figure(2)
plot(isRE,isIM,'r')
axis equal
axis([0 1  -1.0 1.0])
legend('PLECS','MATLAB')
xlabel('Re(is_{ph}) (A)')
ylabel('Im(is_{ph}) (A)')
```

8.7.5 *Tutorial 5: PLECS Based Model of a Permanent Magnetic Machine Connected to a Three-Phase Voltage Source*

The aim of this tutorial is to examine the operation of a general purpose synchronous machine model under load at a given shaft speed. A three-phase sinusoidal variable voltage/frequency voltage source is required because it is assumed that the machine is at standstill at the start of the simulation. Typically this type of excitation is referred to as voltage/frequency (V/f) control as provided by a power electronic drive shown in our book "Applied Control of Electrical Drives" [10]. Hence to start drive operation from standstill, a frequency and supply voltage ramp function will need to be implemented in order to achieve steady-state synchronous operation at a given speed. The practical alternative is to accelerate the machine mechanically (as a generator) to its required speed and then connect the stator windings to the three-phase supply. This process, referred to as 'synchronization', must be done

Table 8.4 Parameters for PM synchronous motor for Teknic motor [10]

Parameters		Value
Stator inductance	L_s	180 μH
Stator resistance	R_s	0.43 Ω
PM flux amplitude	ψ_f	6.42e−3 Wb
Inertia	J	0.02e−3 kg m^2
Pole pairs	p	4
Initial rotor speed	ω_m^0	0 rad/s

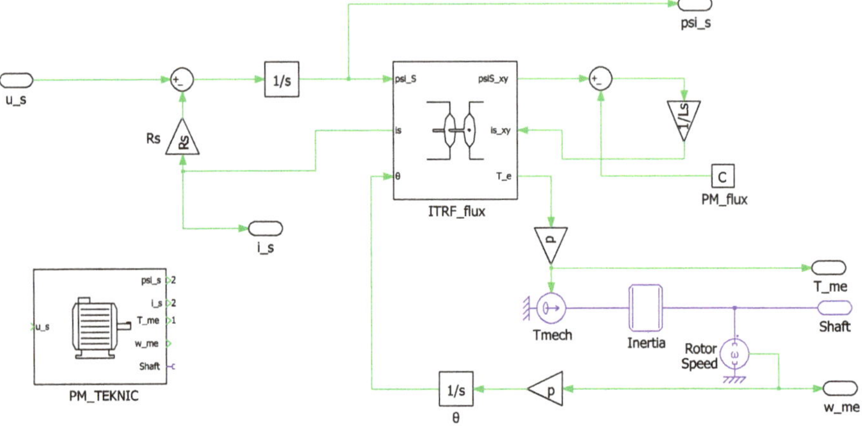

Fig. 8.23 PLECS model: synchronous machine

carefully to ensure that the phase voltages from machine and supply are in phase and equal in magnitude before the connection is made. Synchronization is typically used for mains connected synchronous machines, given that the frequency of the supply is constant. For this example a three-phase eight-pole permanent magnet machine is considered [10] with a topology shown in Fig. 8.3 (which has eight magnetic poles) and a set of parameters as given in Table 8.4. Note that the field flux ψ_{mf} [as introduced in Eq. (8.4)] is for permanent magnet machines also equal to ψ_f given that there is no leakage flux component due to a rotor winding, hence $L_{\sigma rfw} = 0$. The generic model according to Fig. 8.9 can be directly implemented in terms of a PLECS model as shown in Fig. 8.23. The IRTF sub-module for this example needs to be version: `IRTF_flux` (see Fig. 7.5a). The reason for this is that a stator based flux vector and rotor based current vector are to be used as IRTF inputs. For this machine model the stator flux vector $\vec{\psi}_s$ is also an output variable. As mentioned above, an eight-pole machine model is assumed, which means that two additional gain modules with gain p must be added to the model as discussed in Sect. 7.2.1 on page 181. The machine in question was designed for operation at 6000 rpm, which correspond to a stator frequency of 400 Hz. However such operation typically takes place under so-called field oriented control [2, 10]. In this example a simplified control technique is used which warrants (for reasons of stability) a lower steady-

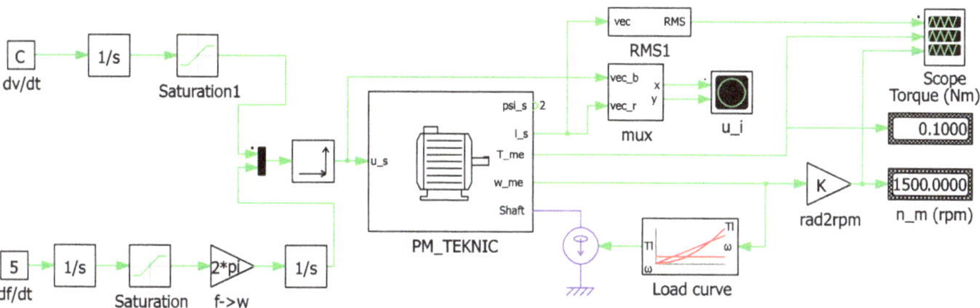

Fig. 8.24 PLECS model: synchronous machine connected to V/f source

state operation frequency of 100 Hz. A steady-state peak phase voltage of 6.0 V is assumed, in order to limit the peak currents to values less than 7.1 A rated. Both frequency and voltage must be ramped up to the steady-state value over a time interval of 20 s (this is to ensure that the machine remains in synchronous operation during the ramp up). A quadratic load module is to be used which generates a load torque of 100 mNm at 1500 rpm (which is the designated steady-state operating speed). The simulation model given in Fig. 8.24 shows the "star" configured machine, which is connected to a variable frequency/voltage source that meets the requirements set out above. Input to the `PM_Teknic` machine model is a voltage vector $\vec{u}_s$, which is generated by a PLECS `Polar to Rectangular` module. This module has as input the voltage amplitude $\hat{u}_s$ and vector angle $\rho_s = \omega_s t$, where the latter represent the stator frequency in rad/s. Both frequency and amplitude are ramped up over a period of 20 s to steady-state values of 100 Hz and 6 V, respectively. The `Saturation` modules are used to limit the output of the ramp-up integrators to the required values. The load module shown in Fig. 8.24 as discussed in Sect. 8.7.1 is used with a quadratic load torque/speed curve. The torque and speed reference values for this module must be set to $T^* = 0.1$ Nm and $\omega^* = {}^{2\pi n_s}/_{1500}$ rad/s, respectively. The output of this unit is connected to a "mechanical" torque module which acts as the load for the machine. A `Scope` module is used to observe the RMS phase current, shaft torque, and speed over the 25 s run time interval. Furthermore, an "XY" Plot module `u_i` is provided to show the stator voltage and current space vectors. An example of the results which appear after running this simulation is given in Figs. 8.25 and 8.26.

An observation of these simulation results reveals some interesting details, namely

- The machine accelerates to the synchronous speed of 1500 rpm from standstill as required. A linear ramp-up is apparent, which is conform the ramp rate set by the reference frequency.
- Clearly observable is the quadratic load curve which leads to the steady-state load torque of 0.1 Nm at $t = 20$ s and shaft speed of 1500 rpm.
- The RMS current increases until a steady-state value of 4 A is reached.
- The vector plot (b) given in Fig. 8.26 shows the voltage and current space vectors under load conditions at the end of the simulation period. Clearly observable

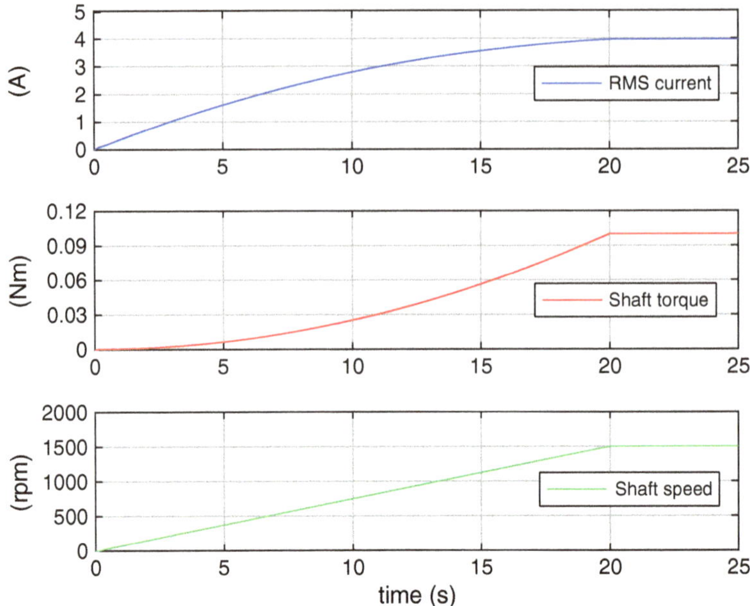

Fig. 8.25 Scope results: RMS phase current I_s, Shaft torque T_m, and Shaft speed n_m, under load conditions

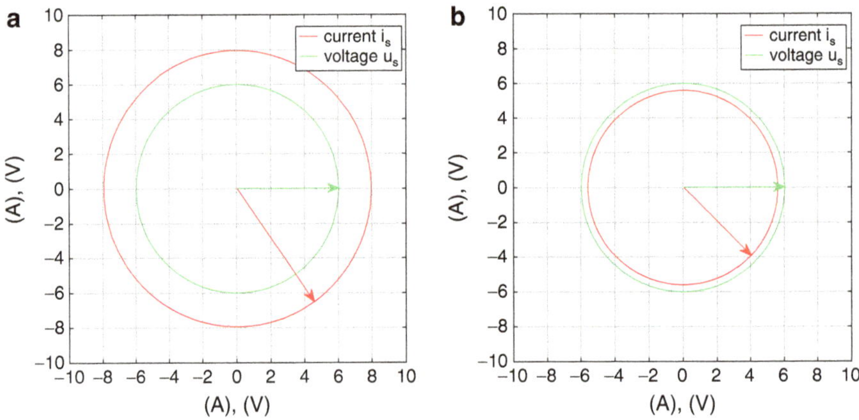

Fig. 8.26 Vector plots showing stator voltage $\vec{u}_s$ ("*green*") and stator current $\vec{i}_s$ during steady-state operation under no-load/load conditions. (**a**) No-load. (**b**) Load

is the current which has an amplitude equal to RMS current time $\sqrt(2)$. For comparison purposes the simulation was repeated without a load, in which case the vector plot (b) as given in Fig. 8.26 was found. A comparison between the two vector plots shows that the current amplitude and the angle between the voltage and current vectors is *larger* under *no-load* conditions.

8.7.6 Tutorial 6: Steady-State Analysis of a Permanent Magnetic Machine Connected to a Three-Phase Voltage Source

It is instructive to examine the steady operation of the PM machine by way of a phasor analysis according to the theory presented in Sect. 8.6.2. More specifically it is considered important to understand the relationship between current and voltage phasors under steady-state operation as the machine moves from no-load to load conditions. Construct a Blondel diagram for the Teknic machine as presented in the previous tutorial using the steady conditions discussed and present in this diagram:

- the operating trajectory for motor operation
- a circle which represents the power output of the motor
- the maximum available power out and its value of the motor
- the operating points which correspond to no-load operation, operation with the specified 0.1 Nm load torque and the maximum load torque that can be realized by the motor, given the steady-state excitation conditions in use.

Show your calculations in the form of an M-file. In addition, use this M-file to plot the output power versus speed curve [see Eq. (8.25)] as shown in Fig. 8.15, for the Teknic machine in use.

An example of the results obtained from the M-file is shown in Fig. 8.27 in the form of a Blondel diagram. Central to this figure is the zero output power circle with diameter $\hat{u}_s/R_s = 13.95$, within which motor operation takes place. Aligned with the $\Re$ axis is the voltage phasor $\underline{u}_s = 6$. When the machine is not synchronized machine operation at point B takes place, which is on the zero output power curve with $\underline{i}_s = \underline{i}_{s1}$. Under synchronous machine operation the operation curve is formed by the circle with radius $|\underline{i}_{s2}|$. Motor operation from no-load, steady-state torque $T_l = 0.1$ and maximum possible torque is given by operating points F, A, and C, respectively. The operating point A of the drive is found by calculating the phasor $\underline{i}_{s1}$ and radius of the circle $|\underline{i}_{s2}|$ using Eq. (8.18). In addition the radius of the output power curve r_{out} using Eq. (8.22), the output power under steady conditions, and the parameters/excitation of the machine. The intersection of the output power circle $p_{out} = 15.7$ with the operating circle $F \rightarrow C$ leads to the operating point of the drive A, which matches with the vector plot shown in Fig. 8.26, subplot (b). Under no-load the machine operating point moves to F, which confirms the earlier observation that the phase current amplitude and angle between the voltage/current phasors increase under no-load conditions (see Fig. 8.26).

A further observation is that the maximum power possible for this machine is less than the theoretical maximum power level of the machine, which for the present excitation is equal to $p_{out}^{max} = 31.4\,W$ [see Eq. (8.23)]. This may be deduced by the fact that the maximum power operating point C does not coincide with the p_{out}^{max} point. A similar observation may also be deduced by calculating the normalized output power versus load characteristic for the machine using Eq. (8.25), which for

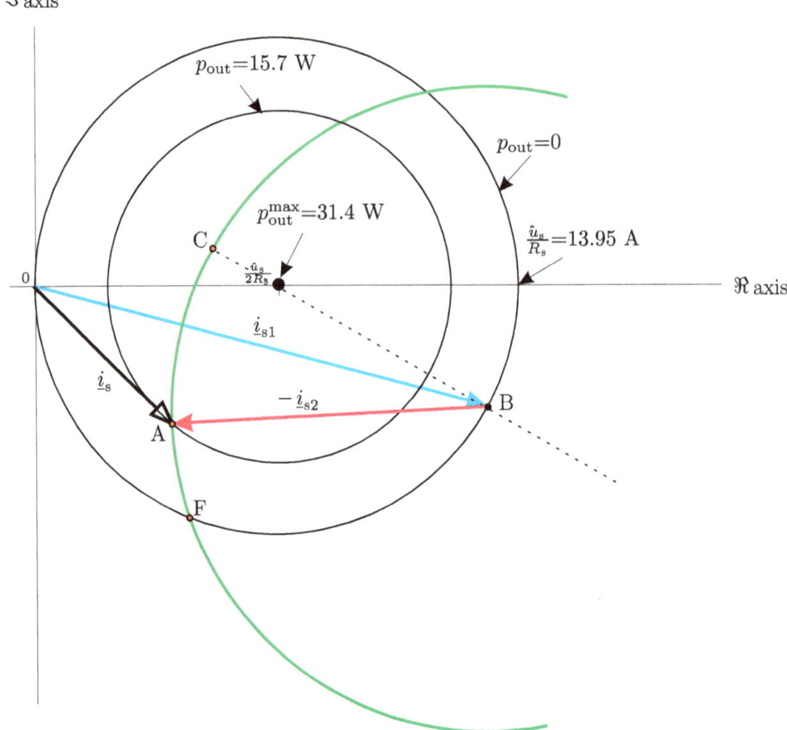

Fig. 8.27 Blondel diagram Teknic motor with $f_s = 100$ Hz and $\hat{u}_s = 6$ V

this machine is given by Fig. 8.28. In this figure the operating points F, A, and C shown earlier in Fig. 8.27 are also introduced, where it is noted that operating point C in Fig. 8.28 is at $p^{\rm n}_{\rm out} = 0.89$. Hence the maximum achievable power of the machine at point C is equal to 27.9 W.

The M-file given below shows the steady-state analysis linked with this problem.

M-Code

```
%Tutorial 8, chapter 8
%steady-state analysis
clear all
close all
p=4;                                    % pole pair
T_m=100e-3                              % shaft torque (Nm)
T_e=T_m/p                               % shaft torque
   electrical
Ls=180e-6;                              % stator inductance (H)
Rs=0.43;                                % stator resistance
psiF=6.42e-3;                           % amplitude PM flux
```

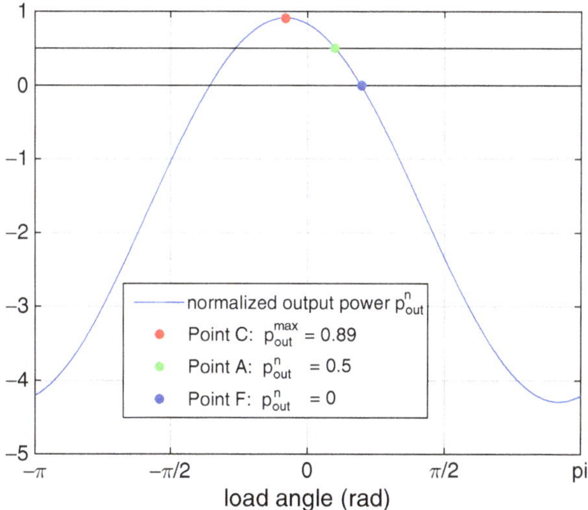

Fig. 8.28 Normalized output power versus load angle: $p^{\mathrm{n}}_{\mathrm{out}}\ (\rho_{\mathrm{m}})$

```
   phasor
freq=100;                                    % frequency (Hz)
ws=2*pi*freq;
us_ph=6;                                     %phase/peak voltage
   amplitude
us_hat=abs(us_ph);
%calculate  Blondel diagram data
is1_ph=us_ph/(Rs+j*ws*Ls)                    %eqn (8.12) part 1
is1=abs(is1_ph);
rho_isi=180/pi*angle(is1_ph);
kF=ws*psiF/us_hat                           %KF factor eqn
                                               (8.19)
is2=abs(kF*us_hat/(Rs+j*ws*Ls))             %eqn (8.12) part 2
   (circle)
pout=T_e*ws;                                %output power (w)
r_pout=sqrt(1/Rs*(us_hat*us_hat/(4*Rs)-(2/3)*pout))  % via
                                                    eqn (8.21)
pout_MAX=3*us_hat*us_hat/(8*Rs);            %max power equation
                                                (8.23)
%%% calculate power/load angle curve
rhom=[-pi:pi/20:pi];
%%%%%%%calculate load angle using Te(rho_m) eqn  in theory
pN_out=pout/pout_MAX;                     % normalized output
                                              power
sig=ws*Ls/Rs;                             % electrical time
                                              constant machine
gamma=atan(sig);
term1=kF*sin(gamma-pi/2)-pN_out/(4*kF)*sqrt(1+sig^2);
rho_m=-gamma+pi/2+asin(term1);            % load angle in use.
%calculate current phasor check
```

```
is_ph=us_hat*(1-kF*(cos(rho_m)+j*sin(rho_m)))/(Rs+j*ws*Ls);
is_hat=abs(is_ph);
%%%%%check powfig
pn=-4*kF/sqrt(1+sig^2)*(sin(rhom+gamma-pi/2)-kF*sin
   (gamma-pi/2));
%plot result
plot(rhom,pn,'b')
grid
legend('normalized output power')
xlabel('load angle (rad)')
```

8.7.7 Demo Lab 1: Voltage/Frequency PM Drive

Experimental verification of theoretical models, such as those discussed in this chapter, is important. It is for this reason that a number of demonstration (demo) laboratory examples have been included in this book. The demonstration setup as shown in Fig. 8.29 consists of two mechanically connected machines and a Texas Instruments LAUNCHXL-F28069M kit as used in our book "Applied Control" [10]. A Texas Instruments LVSERVOMTR (Teknic) PM machine (top black machine in Fig. 8.29) which is part of the 2MTR-DYNO Dual PM motor kit is shown together with the LVACIMTR (EMsynergy) induction machine which will be used in a forthcoming chapter. In this laboratory it is not used, hence the PM motor is operating with a friction load (due to the bearings of both machines) only. Attached to the Texas Instruments LAUNCHXL-F28069M module are two (6 cm × 6 cm) BOOSTXL-DRV8301 modules which provide the excitation for both machines if required. Each BOOSTXL-DRV8301 houses a DRV 8301 module that contains the MOSFET power electronics devices, voltage/current sensing, and protection circuitry. Observation of Fig. 8.29 shows that the LAUNCHXL-F28069M setup has a small footprint which (in this case) translates to a relative inexpensive drive setup. Note that a 24 V DC power supply must also be acquired for this laboratory. Furthermore, Fig. 8.29 also shows an oscilloscope and DC current probe, both of these are "optional." However, the presence of this type of equipment certainly facilitates the learning experience.

For this demonstration the LVSERVOMTR (Teknic) PM machine which is the motor used in the previous two tutorials is to be connected to the aft (furthest away from the USB connector) BOOSTXL-DRV8301 module. A voltage/frequency controller is used [10] which can generate a three-phase sinusoidal supply with the required voltage amplitude and frequency. This controller, implemented in embedded software VisSim[14], is shown in Fig. 8.30. Central to this figure is a target interface module which controls the microcontroller (MCU) present on the LAUNCHXL-F28069M module. A set of sliders are used to set the required 100 Hz frequency and 6 V amplitude of the three-phase supply voltage for the PM motor. Furthermore, a `Plot` module is present which shows the scaled (by a factor 48) voltage $u_{n\alpha}$ ("blue") and scaled (by a factor 20) current $i_{n\alpha}$ over a period of 60 ms.

Fig. 8.29 Demonstration drive setup [10]

The following critical observations of the results shown on the scope module (see Fig. 8.30) can be made in conjunction with the results shown earlier in tutorials Sects. 8.7.5 and 8.7.6, namely:

- Synchronous operating speed of the drive is 1500 rpm which corresponds to a frequency of 100 Hz and a period time of 100 ms, which is indeed the case.
- The drive was set to operate with voltage space vector $\vec{u}_s = u_{s\alpha} + j\,u_{s\beta}$ with amplitude $|\vec{u}_s| = 6.0$ V. Shown on the plot module is the scaled real component $u_{n\alpha} = {}^{u_{s\alpha}}/_{48}$, which is indeed 0.125.
- The scale current component $i_{n\alpha} = {}^{i_{s\alpha}}/_{20}$ has an amplitude of 0.35 which implies that the amplitude of the current space vector is equal to $|\vec{i}_s| = 7.0$ A. Inspection of the no-load XY scope simulation result (see Fig. 8.26) and corresponding Blondel diagram (see Fig. 8.27, operating point F) suggest that the current amplitude should be 8.0 A instead of the 7.0 A found experimentally. However, as mentioned above, the drive is operating with a friction load, which accounts for the fact that lower current is found. The operating point on the Blondel diagram (see Fig. 8.27) will be on the "green" circle between operating points

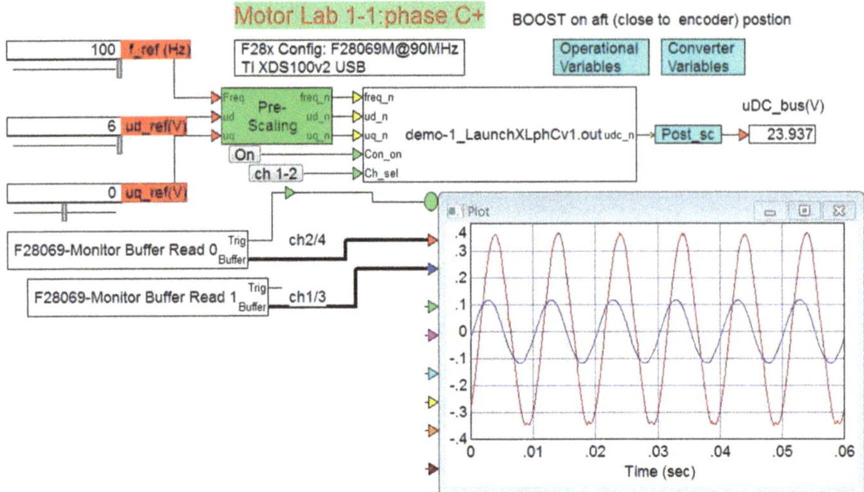

Fig. 8.30 VisSim[14] based V/f drive demonstration controller [10]

F and A, where the exact position corresponds to a phase amplitude of 7.0 A. Some trigonometric analysis shows that the angle between the voltage (located on the real axis) and current phasors is in that case equal to 53°.

- The expected phase angle between the scales voltage and current waveforms should as mentioned in the previous bullet be 53°, which with the current operating conditions corresponds to a time delay of 14.7 ms. Observation of the result shows that this phase lag is indeed present.

An overall conclusion is that the results derived with the DEMO lab confirm the theoretical and simulated results shown. Further experimental results could be obtained by, for example, using the second BOOST converter, which could be used to implement a field oriented induction drive [2, 10] that acts as a dynamometer for the PM machine. In which case behavior of the machine under load could also be verified. Note however that drive operation under voltage/frequency control is relatively inefficient hence the reader needs to be aware of thermal constraints, i.e., rapid heating up of the PM machine under test.

Chapter 9
Voltage Source Connected Asynchronous (Induction) Machines

9.1 Introduction

The induction machine is by far the most commonly used machine around the globe. Induction machines consume approximately one-third of the energy used in industrialized countries. Consequently this type of machine has received considerable attention in terms of its design and application.

The induction machine is one of the older electric machines with its invention being attributed to Tesla, then working for Westinghouse, in 1888. However, as with most great inventions there were many contributors to the development of this machine. The fundamental operation principle of this machine is based on the magnetic induction principle discovered by Faraday in 1831.

In this chapter we will look at this type of machine in some detail. As with the synchronous machine a simple symbolic and generic diagram will be discussed, which in turn is followed by a more extensive dynamic model of this type of machine. Finally, a steady-state analysis will be discussed where the role of the machine parameters will become apparent.

9.2 Machine Configuration

The stator with its three-phase winding as given in Fig. 8.1 on page 202 is used to develop a rotating magnetic field in exactly the same way as realized with the synchronous machine.

The rotor of an induction machine usually consists of a laminated steel rotor stack as shown in Fig. 9.1. The metal shaft is through the center of this stack. The

---

Electronic supplementary material The online version of this chapter (doi: 10.1007/978-3-319-29409-4_9) contains supplementary material, which is available to authorized users.

© Springer International Publishing Switzerland 2016

A. Veltman et al., *Fundamentals of Electrical Drives*, Power Systems,
DOI 10.1007/978-3-319-29409-4_9

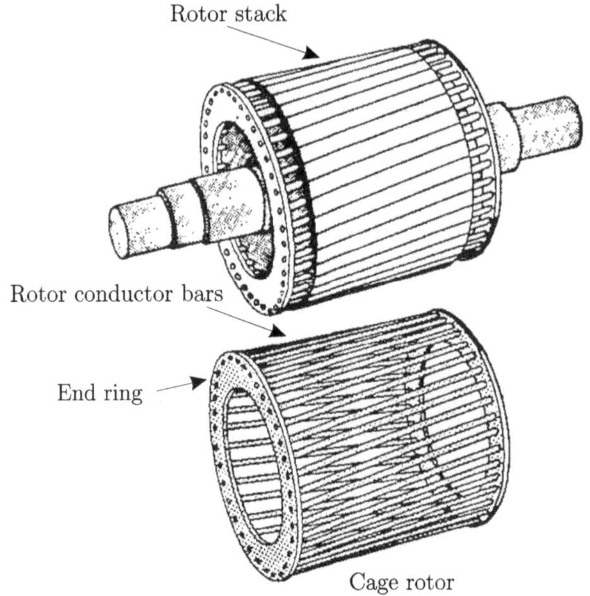

Fig. 9.1 Cage rotor for asynchronous machine

rotor stack is provided with slots around its circumference which house the rotor bars of the so-called squirrel cage. This cage, which consists of rotor bars attached to end rings, is also shown (without the rotor stack) in Fig. 9.1. The cage can be copper or die-cast in aluminum [5]. A recent development in this context has been the introduction of a copper coating on the rotor designed to replace the traditional cage concept. In some cases aluminum fan blades are attached to the end rings to serve as a fan for cooling the rotor. The cage acts as a three-phase sinusoidally distributed short-circuited winding which has a finite rotor resistance [5].

Some induction machines have a wound rotor provided with a three-phase winding where access to the three phases is provided via a brushes/sliprings (similar to the type used for the synchronous machine). This type of machine known as a "slipring" machine allows us to influence the rotor circuit, e.g., to alter the rotor resistance (and therefore the operating characteristics) by adding external rotor resistance.

The squirrel cage type rotor is very popular because of its robustness and is widely used in a range of industrial applications.

9.3 Operating Principles

The principles are again discussed with the aid of Sect. 7.3 on page 187. This type of machine has no rotor excitation hence the current source shown in Fig. 7.11 is removed and the rotor is connected to a resistance R_r. This resistance is usually the stator referred resistance of the rotor winding itself, as will become apparent in Sect. 9.5. The rotor current is in this case determined by the induced voltage $\vec{e}_{\rm m}^{\rm xy}$ and the rotor resistance $R_{\rm r}$, which with the aid of Eq. (7.20) on page 188 leads to

$$\vec{i}^{\rm xy} = \frac{{\rm j}\,(\omega_{\rm s} - \omega_{\rm m})\,\vec{\psi}_{\rm m}^{\rm xy}}{R_{\rm r}} \tag{9.1}$$

The corresponding torque produced by this machine is found using Eqs. (7.13) and (7.17) together with Eq. (9.1), which gives

$$T_{\rm e} = \frac{3\,\hat{\psi}_{\rm m}^2}{2\,R_{\rm r}}\,(\omega_{\rm s} - \omega_{\rm m}) \tag{9.2}$$

In which $\hat{\psi}_{\rm m}$ is the magnitude or peak value of the flux vector. The term $\omega_{\rm r} = (\omega_{\rm s} - \omega_{\rm m})$ is referred to as the "slip" rotational frequency. The term *asynchronous* follows from the operating condition that torque can only be produced when the rotor speed is *not* synchronous with the rotating field. This condition is readily observed from Eqs. (9.1) and (9.2) which states that there is no rotor current and hence no torque in case $\omega_{\rm r} = 0$. When the machine is operating under no-load conditions, the shaft speed will in the ideal case be equal to $\omega_{\rm s}$. When a load torque is applied the machine speed reduces and the voltage $|\vec{e}_{\rm m}^{\rm xy}|$ increases, because the slip rotational frequency $\omega_{\rm r}$ increases. A higher voltage $|\vec{e}_{\rm m}^{\rm xy}|$ leads to a higher rotor current component $i_{\rm torque}$ (see Fig. 9.2), which acts together with the flux vector to produce a torque to balance the load torque.

Additional insight into the operation principles of this machine is obtained by making use of the space vector diagram shown in Fig. 9.2. The diagram stems from the generalized diagram (see Fig. 7.12). The generalized diagram has been redrawn to correspond to the induction machine operating under motoring conditions, i.e., the rotational speed $0 < \omega_{\rm m} < \omega_{\rm s}$. Shown in Fig. 9.2 is the rotating flux vector $\vec{\psi}_{\rm m}$ which is responsible for the (in the rotor) induced voltage, represented by the

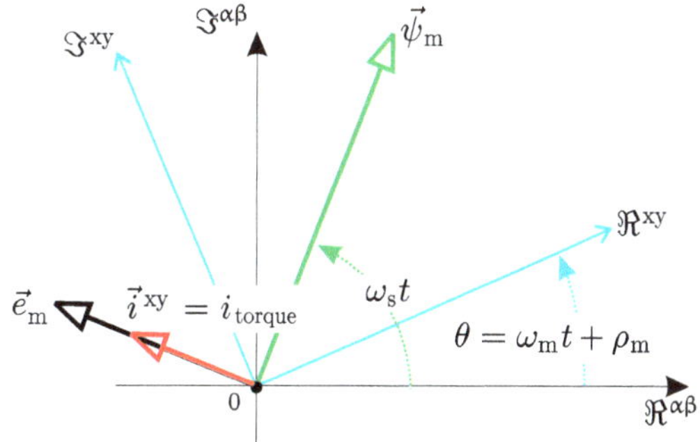

Fig. 9.2 Space vector diagram for asynchronous machine

vector $\vec{e}_{\mathrm{m}}^{\,\mathrm{xy}}$. Observe that both vectors $\vec{e}_{\mathrm{m}}^{\,\mathrm{xy}}$ and $\vec{i}$ rotate synchronous with the flux vector. Hence the current and flux vector shown in this example are orthogonal and stationary with respect to each other, which is the optimum situation in terms of torque production. It is noted that the current vector $\vec{i}^{\,\mathrm{xy}}$ rotates with respect to the rotor at a speed ω_{r}. The rotor rotates at speed ω_{m}, hence the current vector $\vec{i}$ has an angular frequency (with respect to the stator) of ω_{s}. When a higher load torque is applied the machine slows down which means that the induced voltage in the rotor increases, which in turn leads to a higher rotor current and larger torque to match the new load torque.

9.4 Zero Leakage Inductance Model Without Magnetizing Inductance

In this section a simplified model is considered where the leakage inductances $L_{\sigma\mathrm{s}}$ and $L_{\sigma\mathrm{r}}$ shown in Fig. 7.14 are set to zero which implies that the stator, magnetizing and rotor flux vectors will be equal, hence $\vec{\psi}_{\mathrm{s}} = \vec{\psi}_{\mathrm{m}} = \vec{\psi}_{\mathrm{r}}$. Furthermore the magnetizing inductance L_{m} and stator resistance R_{s} will also be ignored which implies that the model according to Fig. 7.14 may be reduced to the configuration shown in Fig. 9.3. A rotating magnetizing flux vector $\vec{\psi}_{\mathrm{m}} = \hat{\psi}_{\mathrm{m}}\,\mathrm{e}^{\mathrm{j}\omega_{\mathrm{s}}t}$ is assumed as input for our (two pole) model. This rotating field is established as a result of the machine being connected to a three-phase voltage supply with angular frequency ω_{s} (rad/s). The equation set which corresponds with Fig. 9.3 is of the form

$$\vec{i}_{\mathrm{s}}^{\,\mathrm{xy}} = \frac{1}{R_{\mathrm{r}}}\frac{\mathrm{d}\vec{\psi}_{\mathrm{m}}^{\,\mathrm{xy}}}{\mathrm{d}t} \tag{9.3a}$$

$$T_{\mathrm{e}} - T_{\mathrm{l}} = J\,\frac{\mathrm{d}\omega_{\mathrm{m}}}{\mathrm{d}t} \tag{9.3b}$$

$$\omega_{\mathrm{m}} = \frac{\mathrm{d}\theta}{\mathrm{d}t} \tag{9.3c}$$

Included (for completeness) in this equation set is the relationship between shaft torque T_{e}, load torque T_{l}, and rotor acceleration ${}^{\mathrm{d}\omega_{\mathrm{m}}}/_{\mathrm{d}t}$.

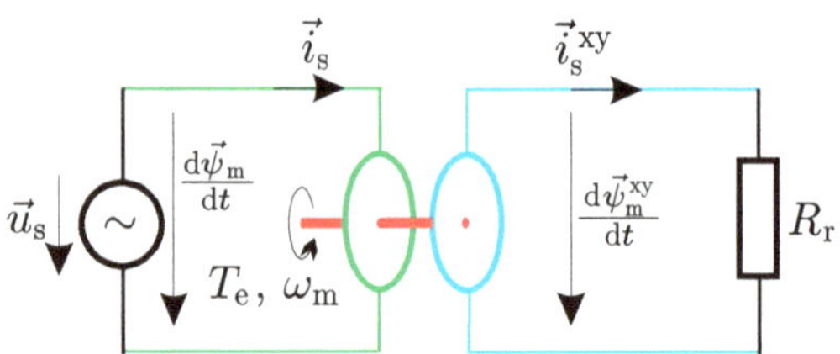

Fig. 9.3 Voltage source connected asynchronous machine, simplified version

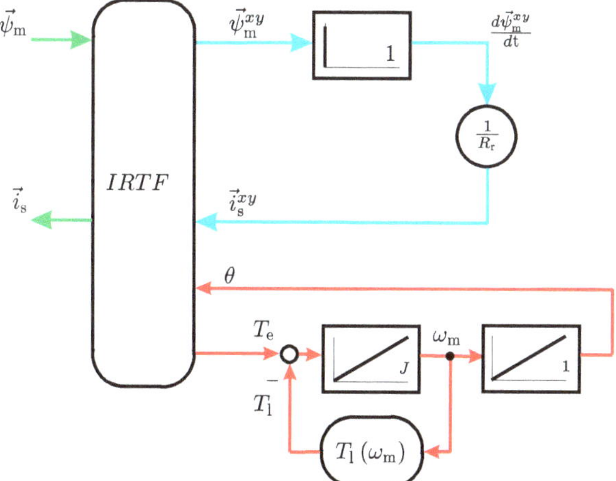

Fig. 9.4 Generic representation of simplified asynchronous machine model corresponding to Fig. 9.3 with mechanical load

9.4.1 Generic Model of a Simplified Asynchronous Machine

The generic representation of the asynchronous machine in its present form is given in Fig. 9.4. The IRTF and load torque sub-modules are identical to those used for the synchronous machine (see Fig. 8.6). This means that the IRTF model calculates the torque on the basis of the flux vector provided from the stator side and the stator current vector which has been created on the rotor side, i.e., use is made of an "IRTF-flux" module. The model according to Fig. 9.4 uses a differentiator to generate the induced voltage vector from the flux vector. From a didactic perspective this is useful as it shows the mechanism of torque production and the formation of the stator current vector. However, in simulations the use of a differentiator is not preferred given that such models are prone to numerical errors. It will be shown that we can in most cases avoid the use of differentiators. In the tutorial at the end of this chapter a steady-state type analysis will be considered which is based on Fig. 9.4. In that case the differentiator function is created with an alternative "differentiator" module (see Fig. 6.8).

9.5 Generalized Machine Model

The symbolic and generic model of the simplified machine as discussed in the previous section provides a basic understanding. Most importantly the model shows the significance of the rotor resistance $R_{\rm r}$. A more general model of the machine can be found by taking into consideration that the configuration shown in Fig. 9.5 is in

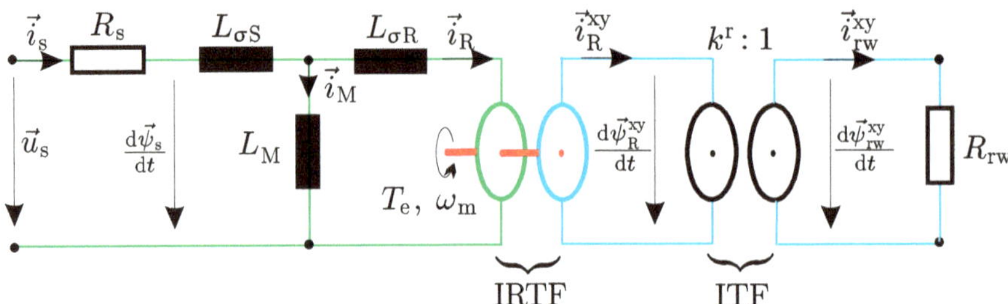

Fig. 9.5 Universal two-phase IRTF model, adapted for asynchronous machine use

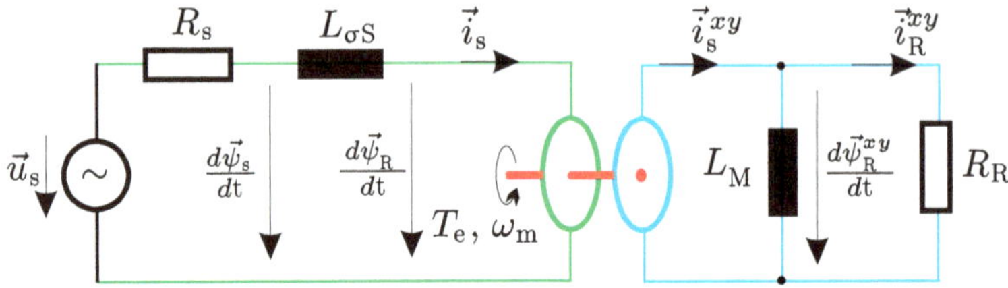

Fig. 9.6 Stator leakage inductance based asynchronous (induction) machine model

fact a special case, namely $|\vec{u}_{rw}^{xy}| = 0$ (short-circuited winding) of the universal model given in Fig. 7.14. The resistance R_{rw} shown in Fig. 9.5 represents the resistance of the squirrel cage rotor or three-phase winding for a slip-ring machine. The term "universal" underlines the fact that the transformation variable a can be chosen in such a way that the model according to Fig. 9.5 can be reduced to a two inductance configuration by setting either $L_{\sigma S}$ or $L_{\sigma R}$ to zero. A two inductance model is simpler to model and avoids the practical problem of determining values for the individual parameters $L_{\sigma S}$ and $L_{\sigma R}$ of a machine. For machine modeling purposes a universal model with variable transformation factor a is not required (other then for didactic reasons as discussed in Sect. 7.4.1). The "universal" model is, however, of prime importance for the control of induction machines [2]. For machines which are connected to a voltage source a stator based leakage inductance model is often used. The term "stator based" refers to the fact that the total leakage inductance of the machine is now represented to be a single leakage inductance which is located to the "left" of the magnetizing inductance L_M of the model. This implies that the leakage inductance $L_{\sigma R}$ is set to zero by choosing the transformation factor equal to $a = {}^{L_m}/_{L_r}$ [see Eq. (7.24)]. Note that this transformation leads to the condition $\vec{\psi}_M = \vec{\psi}_R$ which is why this type of model is also known as a *rotor flux based IRTF model*. With this choice of transformation factor the universal model given in Fig. 9.5 reduces to the configuration shown in Fig. 9.6. Note that the magnetizing inductance has been arbitrarily relocated to the rotor side of the IRTF. The rotor parameters $\vec{i}_R$ and $\vec{\psi}_R$ can be expressed in terms of the actual rotor variables using Eqs. (7.27) and (7.28) with $k^r = {}^{L_m n_s}/_{L_r n_r}$ which gives

$$\vec{i}_{\mathrm{R}} = \frac{L_{\mathrm{r}} n_{\mathrm{r}}}{L_{\mathrm{m}} n_{\mathrm{s}}} \vec{i}_{\mathrm{rw}} \tag{9.4a}$$

$$\vec{\psi}_{\mathrm{R}} = \frac{L_{\mathrm{m}} n_{\mathrm{s}}}{L_{\mathrm{r}} n_{\mathrm{r}}} \vec{\psi}_{\mathrm{rw}} \tag{9.4b}$$

$$R_{\mathrm{R}} = \left(\frac{L_{\mathrm{m}} n_{\mathrm{s}}}{L_{\mathrm{r}} n_{\mathrm{r}}} \right)^2 R_{\mathrm{rw}} \tag{9.4c}$$

Note that the turns ratio $n_{\mathrm{s}}/n_{\mathrm{r}}$ is not required when modeling squirrel cage induction machines. The reason being that the parameters R_{s}, L_{M}, $L_{\sigma\mathrm{S}}$, and R_{R} are normally found via a set of stator based measurements. Expression (9.4) becomes relevant for modeling, for example, slip-ring machines with rotor based power converters. The equation set which corresponds to Fig. 9.6 is of the form

$$\vec{u}_{\mathrm{s}} = R_{\mathrm{s}} \vec{i}_{\mathrm{s}} + \frac{\mathrm{d}\vec{\psi}_{\mathrm{s}}}{\mathrm{d}t} \tag{9.5a}$$

$$\vec{\psi}_{\mathrm{s}} = \vec{\psi}_{\mathrm{R}} + L_{\sigma\mathrm{S}} \vec{i}_{\mathrm{s}} \tag{9.5b}$$

$$\frac{\vec{\psi}_{\mathrm{R}}^{\mathrm{xy}}}{L_{\mathrm{M}}} = \vec{i}_{\mathrm{s}}^{\mathrm{xy}} - \vec{i}_{\mathrm{R}}^{\mathrm{xy}} \tag{9.5c}$$

$$0 = -\vec{i}_{\mathrm{R}}^{\mathrm{xy}} R_{\mathrm{R}} + \frac{\mathrm{d}\vec{\psi}_{\mathrm{R}}^{\mathrm{xy}}}{\mathrm{d}t} \mathrm{d}t \tag{9.5d}$$

The torque for this two-pole machine may be written as $T_{\mathrm{e}} = {}^{3}/_{2} \Im \left\{ \vec{\psi}_{\mathrm{R}}^{*} \vec{i}_{\mathrm{R}} \right\}$ [see Eq. (7.29h)] which with the aid of Eq. (9.5c) can also be written as

$$T_{\mathrm{e}} = \frac{3}{2} \Im \left\{ \vec{\psi}_{\mathrm{R}}^{*} \vec{i}_{\mathrm{s}} \right\} \tag{9.6}$$

9.5.1 Generic Induction Machine Model

The development of a generic dynamic model as shown in Fig. 9.7a on page 248 is directly based on Eq. (9.5). This generic diagram builds directly on the simplified model given in Fig. 9.4. The model shown in Fig. 9.7a also contains a differentiator model which is undesirable for dynamic simulations. An improved representation of the generic model according to Fig. 9.7a is possible which avoids the use of a numerically undesirable differentiator. This model as shown in Fig. 9.7b is preferable for dynamic simulation of induction machines for reasons mentioned above. However the model becomes untenable when considering a hypothetical machine without leakage inductance or infinite rotor resistance.

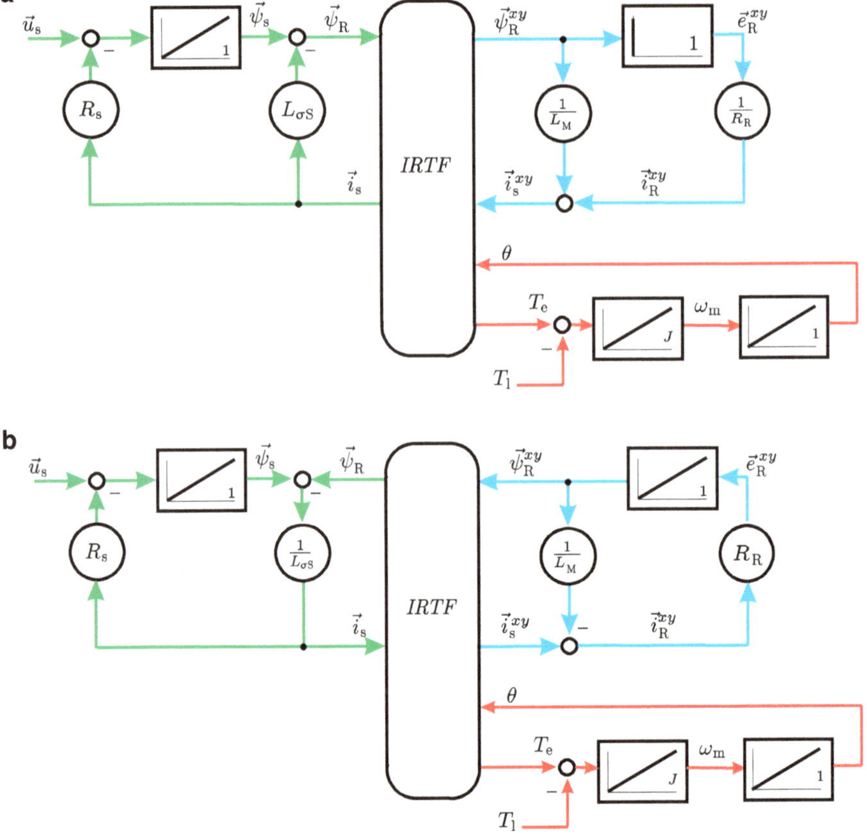

Fig. 9.7 Generic asynchronous (induction) machine dynamic models which can be used to represent the symbolic model shown in Fig. 9.6. (**a**) Differentiator based dynamic model. (**b**) Integrator based dynamic model

9.6 Steady-State Analysis

The steady-state characteristics of the asynchronous machine are studied with the aid of Fig. 9.6. The stator is again connected to a three-phase sinusoidal supply which is represented by the space vector $\vec{u}_{\mathrm{s}} = \hat{u}_{\mathrm{s}}\, \mathrm{e}^{\mathrm{j}\omega_{\mathrm{s}}t}$, hence $\underline{u}_{\mathrm{s}} = \hat{u}_{\mathrm{s}}$. The basic characteristics of the machine relate to the stator current end point locus of the machine in phasor form (when varying the shaft speed) and the torque speed curve. To arrive at a phasor type representation the space vector equations must be rewritten in terms of phasors. For example, the rotor flux on the stator and rotor side of the IRTF are of the form

$$\vec{\psi}_{\mathrm{R}} = \underline{\psi}_{\mathrm{R}}\, \mathrm{e}^{\mathrm{j}\omega_{\mathrm{s}}t} \tag{9.7a}$$

$$\vec{\psi}_R^{xy} = \underline{\psi}_R \, e^{j(\omega_s t - \theta)} \tag{9.7b}$$

where θ is equal to $\omega_m t$ (constant speed). Use of Eqs. (9.5) and (9.7) leads to the following phasor based equation set for the machine:

$$\underline{u}_s - \underline{e}_R = \underline{i}_s R_s + j\omega_s L_{\sigma S} \, \underline{i}_s \tag{9.8a}$$

$$\underline{e}_R = j\omega_s \underline{\psi}_R \tag{9.8b}$$

$$\underline{e}_R^{xy} = j\left(\omega_s - \omega_m\right) \underline{\psi}_R \tag{9.8c}$$

$$\underline{i}_R = \frac{\underline{e}_R^{xy}}{R_R} \tag{9.8d}$$

$$\underline{\psi}_R = L_M \left(\underline{i}_s - \underline{i}_R\right) \tag{9.8e}$$

Elimination of the flux phasor from Eqs. (9.8b) and (9.8c) leads to an expression for the airgap EMF on the stator and rotor side of the IRTF, namely

$$\underline{e}_R^{xy} = \underline{e}_R \, s \tag{9.9}$$

where "s" is known as the slip of the machine and is (for a two-pole machine) given by

$$s = 1 - \frac{\omega_m}{\omega_s} \tag{9.10}$$

The slip according to Eq. (9.10) is simply the ratio between the rotor rotational frequency $\omega_r = \omega_s - \omega_m$ (as apparent on the rotor side of the IRTF) and the stator rotational frequency ω_s. Three important slip values are introduced, namely

- Zero slip: $s = 0$, which corresponds to synchronous speed operation, i.e., $\omega_m = \omega_s$.
- Unity slip: $s = 1$, which corresponds to a locked rotor, i.e., $\omega_m = 0$.
- Infinite slip: $s = \pm\infty$, which according to Eq. (9.10) corresponds to infinite shaft speed or zero rotational stator frequency (DC excitation). This slip value is for $\omega_s \neq 0$ not practically achievable but this operating point is of relevance, as will become apparent at a later stage.

The rotor current phasor may be conveniently rewritten in terms of the slip parameter by making use of Eqs. (9.8d) and (9.9) which gives

$$\underline{i}_R = \frac{\underline{e}_R}{\left(\frac{R_R}{s}\right)} \tag{9.11}$$

The rotor current is according to Eq. (9.11) determined by the ratio of the airgap EMF $\underline{e}_R$ and a slip dependent resistance $\left(\frac{R_R}{s}\right)$. At slip zero its value will be ∞, which corresponds to zero current as is expected at synchronous speed, given that the rotor EMF on the rotor side $\underline{e}_{\mathrm{R}}^{\mathrm{xy}}$ will be zero. At standstill ($s = 1$) its value is simply R_{R}.

The development of the basic machine characteristics in the form of the stator current phasor as function of slip and the torque/slip curve is presented on a step by step basis. This implies that gradually more elements of the model in Fig. 9.6 are introduced in order to determine their impact on the machine characteristics. The key elements which affect the operation of the machine in steady-state are the rotor resistance R_{R} and the leakage inductance $L_{\sigma\mathrm{S}}$. We will initially consider the machine with zero stator resistance, infinite magnetizing inductance, and firstly without leakage inductance.

9.6.1 Steady-State Analysis with Zero Leakage Inductance and Zero Stator Resistance

A model with zero leakage inductance and zero stator resistance can be represented in Fig. 9.7a, assuming $L_{\sigma\mathrm{S}} = 0$ and $R_{\mathrm{s}} = 0$. Observation of Fig. 9.7a and phasor equation set (9.8), (9.9), and (9.11) shows that the stator current phasor and EMF phasor are given as $\underline{i}_{\mathrm{s}} = \underline{i}_{\mathrm{R}}$ and $\underline{e}_{\mathrm{R}} = \hat{u}_{\mathrm{s}}$, respectively. Furthermore, the stator current phasor as function of the slip may be written as

$$\underline{i}_{\mathrm{s}} = \frac{\hat{u}_{\mathrm{s}}}{\left(\frac{R_{\mathrm{R}}}{s}\right)} \tag{9.12}$$

An observation of Eq. (9.12) shows that the phasor current can be calculated using the equivalent circuit shown in Fig. 9.8. The steady-state torque may be found using Eq. (9.6) where the space vector variables are replaced by the equivalent phasor quantities. This implies that the torque is of the form

$$T_{\mathrm{e}} = \frac{3}{2}\Im\left\{\underline{\psi}_{\mathrm{R}}^{*}\underline{i}_{\mathrm{s}}\right\} \tag{9.13}$$

Use of Eqs. (9.12) and (9.8b) with (9.13) leads to the following torque slip expression for the machine in its current simplified form

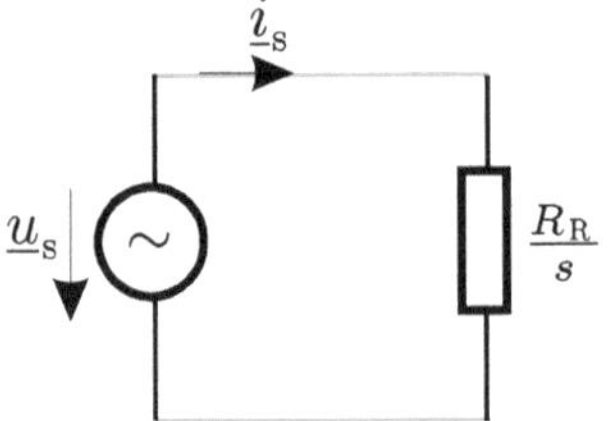

Fig. 9.8 Equivalent circuit of an asynchronous machine with zero stator impedance and voltage source in steady-state

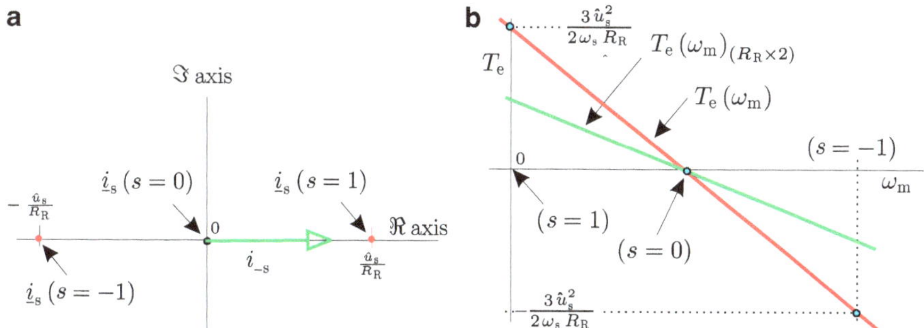

Fig. 9.9 Steady-state characteristics of voltage source connected asynchronous machine with $R_s = 0$, $L_{\sigma S} = 0$, and $L_M = \infty$, according to Fig. 9.8. (**a**) Stator current phasor. (**b**) Torque versus speed

$$T_e = \frac{3\,\hat{u}_s^2\,s}{2\,\omega_s R_R} \tag{9.14}$$

Note that this analysis makes use of the circuit condition $\underline{e}_R = \hat{u}_s$ (see Eq. (9.8a) with $R_s = 0$, $L_{\sigma S} = 0$). A graphic illustration of the current space phasor and torque versus speed characteristic is given in Fig. 9.9. The current phasor will in this case be either in phase or π rad out of phase, with respect to the supply phasor $\underline{u}_s = \hat{u}_s$. The "in phase" case will occur under motoring conditions, i.e., T_e and ω_m have the same polarity. It is noted that a torque sign change will always occur at $s = 0$, while a shaft speed reversal will take place at $s = 1$. This means that motor operation is confined to the slip range $0 \leq s \leq 1$. Note that generation (in, for example, wind turbine applications) starts when the slip becomes negative.

Shown in Fig. 9.9a are three current phasor end points which correspond to the slip conditions $s = -1, 0, 1$. The torque versus speed curve as given in Fig. 9.9b is according to Eq. (9.14) a linear function of which the gradient is determined by (among others) the rotor resistance R_R. The effect of doubling this resistance value on the torque speed curve is also shown in Fig. 9.9b by way of a "green" line. Increasing the rotor resistance leads to a torque/speed curve which is less "stiff," i.e., as a result of a certain mechanical load variation, the shaft speed will vary more.

9.6.2 *Steady-State Analysis with Leakage Inductance*

The simplified R_R-based model (Fig. 9.8) is now expanded by introducing the leakage inductance parameter $L_{\sigma S}$. Under these revised circumstances the current phasor can according to Eq. (9.8a) with $R_s = 0$, be written as

$$\underline{u}_s - \underline{e}_R = \mathrm{j}\omega_s L_{\sigma S}\,\underline{i}_s \tag{9.15}$$

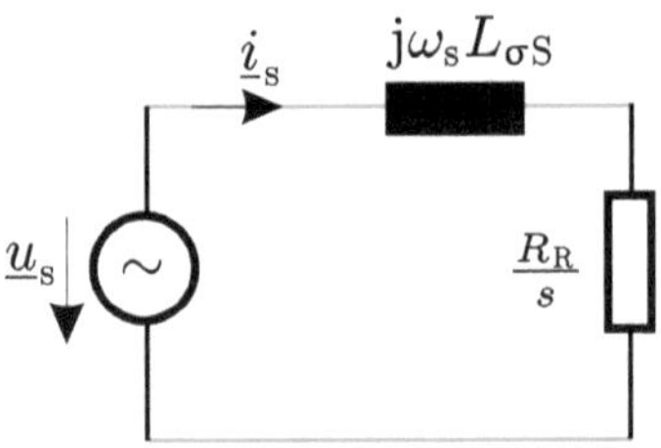

Fig. 9.10 Equivalent circuit of an asynchronous machine ($R_s = 0, L_{\sigma S} > 0$) and voltage source in steady-state

The stator current is also defined in Eq. (9.11) given that $\underline{i}_R = \underline{i}_s$. Substitution of this expression into Eq. (9.15) gives

$$\underline{i}_s = \frac{\hat{u}_s}{\frac{R_R}{s} + j\omega_s L_{\sigma S}} \tag{9.16}$$

The equivalent circuit which corresponds with Eq. (9.16) is shown in Fig. 9.10. The steady-state torque is found using Eq. (9.13) which requires access to the stator current phasor $\underline{i}_s$ as defined in Eq. (9.16). The rotor flux $\underline{\psi}_R$ is found using Eq. (9.8a) with $R_s = 0$ and Eq. (9.8b) which gives

$$\underline{\psi}_R = \frac{\hat{u}_s}{j\omega_s} - L_{\sigma S}\,\underline{i}_s \tag{9.17}$$

Subsequent evaluation of Eqs. (9.16) and (9.17) with Eq. (9.13) gives after some manipulation the following expression for the steady-state torque:

$$T_e = \frac{3\,\hat{u}_s^2}{2\,\omega_s} \frac{\frac{R_R}{s}}{\left(\frac{R_R}{s}\right)^2 + (\omega_s L_{\sigma S})^2} \tag{9.18}$$

In order to gain an understanding of the torque versus slip function it is helpful to introduce a normalized form of Eq. (9.18), namely

$$T_e^n = 2\frac{\frac{s}{\hat{s}}}{1 + \left(\frac{s}{\hat{s}}\right)^2} \tag{9.19}$$

The normalization introduced is of the form $T_e^n = T_e/\hat{T}_e$ where $\hat{T}_e = 3\hat{u}_s^2/(4\omega_s^2 L_{\sigma S})$. Furthermore, a parameter $\hat{s} = R_R/\omega_s L_{\sigma S}$ is introduced which is known as the pull-out slip value. Note that the pull-out slip $\hat{s}$ is *not* a peak value of the slip s itself, rather it refers to the slip values $s = \pm\hat{s}$, where the highest attainable torque or the so-called pull-out torque values of the machine are reached for the model discussed here. It is helpful to recall that the slip is defined as $s = \omega_r/\omega_s$, which means that the highest attainable torque and corresponding pull-out slip value is reached when the rotor angular frequency $\omega_r = R_R/L_{\sigma S}$. The extremes of Eq. (9.19) as function of the slip s may be found by differentiation of said expression with respect to the slip and zeroing the result. Alternatively, the normalized torque/slip function can be examined for two slip regions as indicated in Eq. (9.20)

$$\text{for } |s| \ll \hat{s} \quad T_{\mathrm{e}}^{\mathrm{n}} \simeq 2\frac{s}{\hat{s}} \tag{9.20a}$$

$$\text{for } |s| \gg \hat{s} \quad T_{\mathrm{e}}^{\mathrm{n}} \simeq 2\frac{\hat{s}}{s} \tag{9.20b}$$

The intersection of the two torque/slip functions indicated in Eq. (9.20) leads to the following two slip values, which correspond to a maximum and minimum torque value of $\hat{T}_{\mathrm{e}}$ and $-\hat{T}_{\mathrm{e}}$, respectively.

$$s = \pm\hat{s} \tag{9.21}$$

A normalization of the stator current phasor [Eq. (9.16)] is also helpful in terms of understanding the slip dependency. The normalized stator current phasor $\underline{i}_{\mathrm{s}}^{\mathrm{n}} = \underline{i}_{\mathrm{s}}/(\hat{u}_{\mathrm{s}}/\omega_{\mathrm{s}}L_{\sigma\mathrm{S}})$ can with the aid of Eq. (9.16) and $\hat{s} = R_{\mathrm{R}}/\omega_{\mathrm{s}}L_{\sigma\mathrm{S}}$ written

$$\underline{i}_{\mathrm{s}}^{\mathrm{n}} = \frac{\frac{s}{\hat{s}}}{1 + \mathrm{j}\frac{s}{\hat{s}}} \tag{9.22}$$

A graphical representation of the normalized stator current locus versus slip is given in Fig. 9.11a with $\hat{s} = 1/2$. It should be kept in mind that the selected pull-out slip value of 0.5 was chosen to clearly show the low and high slip regions given in Eq. (9.20). In reality the pull-out slip values of squirrel cage based machines are considerably smaller. The stator current locus is circular as determined by Heyland in 1894, hence these diagrams are commonly referred to as "Heyland diagrams." The process of verifying that the locus is a circle may be initiated by introducing the variables $x = \Re\{\underline{i}_{\mathrm{s}}^{\mathrm{n}}\}$, $y = \Im\{\underline{i}_{\mathrm{s}}^{\mathrm{n}}\}$ in Eq. (9.22) which leads to

$$x + \mathrm{j}y = \frac{\xi}{1 + \mathrm{j}\xi} \tag{9.23}$$

with $\xi = s/\hat{s}$. Equating the real and imaginary terms in Eq. (9.23) gives

$$x = \xi\,(y + 1) \tag{9.24a}$$

$$0 = y + \xi x \tag{9.24b}$$

Eliminating the slip dependent term ξ in Eq. (9.24) gives the following expression:

$$x^2 + \left(y + \frac{1}{2}\right)^2 = \left(\frac{1}{2}\right)^2 \tag{9.25}$$

which indeed represents a circle with its origin at $(0, -1/2)$ and radius $1/2$.

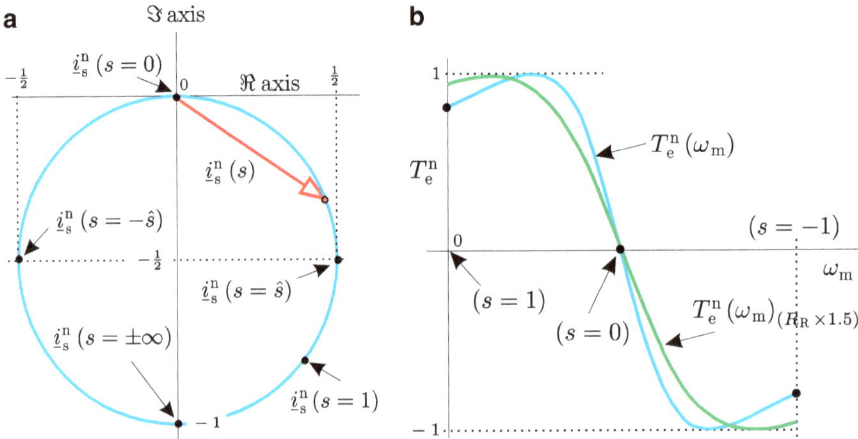

Fig. 9.11 Steady-state characteristics of voltage source connected asynchronous machine with $R_s = 0$, $L_{\sigma S} > 0$, and $L_M = \infty$, according to Fig. 9.10. (**a**) Normalized stator current phasor. (**b**) Torque versus speed

A graphical representation of the normalized torque versus shaft speed is given in Fig. 9.11b for the slip range $-1 \leq s \leq 1$ and $\hat{s} = 1/2$ ("blue" curve). A second torque/slip curve is also shown ("green" curve) which corresponds to a machine which has a rotor resistance value that is 1.5 times larger than the first. Some interesting observations may be drawn from the basic characteristics according to Fig. 9.11, namely

- The introduction of leakage inductance has a significant impact on the characteristics of the machine as may be concluded by comparing Fig. 9.11 with Fig. 9.9.
- Variations of the rotor resistance R_R affect among others the value of the pull-out slip value $\hat{s}$ and the gradient of the torque slip curve in the low slip region. The peak torque $\hat{T}_e$ is *not* affected by these changes. For a "slipring" type induction machine it is possible to vary the rotor resistance by prudently adding external resistance to improve the starting torque.
- The peak torque value is dependent on the ratio $\hat{u}_s/\omega_s$. If this ratio is kept constant the torque/shaft speed curves may be moved horizontally along the horizontal axis by changing ω_s without affecting the peak torque value. Such drives are known as "V/f" drives. Note that the pull-out slip value is inversely proportional to ω_s, which means that the peak (motoring) torque is found at an ω_m value which is $R_R/L_{\sigma S}$ (rad/s) lower than the synchronous value $\omega_m = \omega_s$. In practical V/f drives, provision needs to be made to compensate for the change in ψ_s due to the voltage drop across the stator resistance. In the present model configuration the stator resistance is set to zero.
- The Heyland diagram shows motor and generator regions. In addition it shows how the phase angle between voltage phasor $\underline{u}_s = \hat{u}_s$ and stator current phasor changes as function of the slip. For low slip operation the power factor (for the machine in its present form) approaches unity.

The machine model with leakage inductance and rotor resistance represents the basic model in terms of showing the fundamental operating principles of the machine.

9.6.3 *Steady-State Analysis with Leakage Inductance and Stator Resistance*

The machine model with leakage inductance and rotor resistance is now extended to include stator resistance. The terminal equation according to (9.15) must be revised and is now defined by expression (9.8a). Use of Eq. (9.8a) and $\underline{i}_{\mathrm{R}} = \underline{i}_{\mathrm{s}}$ with Eq. (9.11) gives

$$\underline{i}_{\mathrm{s}} = \frac{\hat{u}_{\mathrm{s}}}{\left(R_{\mathrm{s}} + \frac{R_{\mathrm{R}}}{s}\right) + \mathrm{j}\omega_{\mathrm{s}} L_{\sigma\mathrm{S}}} \tag{9.26}$$

The equivalent circuit which corresponds with Eq. (9.26) is shown in Fig. 9.12. A normalization of the stator current according to $\underline{i}_{\mathrm{s}}^{\mathrm{n}} = \underline{i}_{\mathrm{s}}/(\hat{u}_{\mathrm{s}}/\omega_{\mathrm{s}} L_{\sigma\mathrm{S}})$ as introduced to obtain Eq. (9.22) is also applied to Eq. (9.26) which gives, with pull-out slip $\hat{s}$ as defined on page 252

$$\underline{i}_{\mathrm{s}}^{\mathrm{n}} = \frac{\frac{s}{\hat{s}}}{\left(1 + r\frac{s}{\hat{s}}\right) + \mathrm{j}\frac{s}{\hat{s}}} \tag{9.27}$$

where the parameter r is introduced which is defined according to Eq. (9.28), namely

$$r = \frac{R_{\mathrm{s}}}{R_{\mathrm{R}}} \hat{s} \tag{9.28}$$

Note that Eq. (9.27) reverts back to Eq. (9.22) in case the stator resistance is set to zero. The Heyland diagram for the revised machine is again a circle as may be deduced by evaluation of Eq. (9.27) and introducing the variables $x = \Re\{\underline{i}_{\mathrm{s}}^{\mathrm{n}}\}$, $y = \Im\{\underline{i}_{\mathrm{s}}^{\mathrm{n}}\}$ as discussed for the previous case.

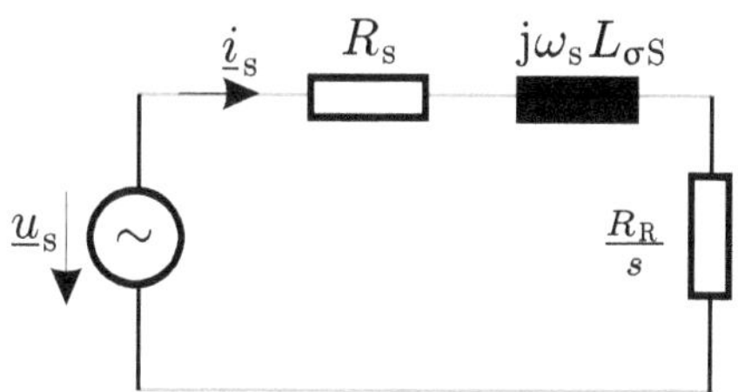

Fig. 9.12 Equivalent circuit of an asynchronous machine with both R_{s} and $L_{\sigma\mathrm{S}}$, voltage source, steady-state

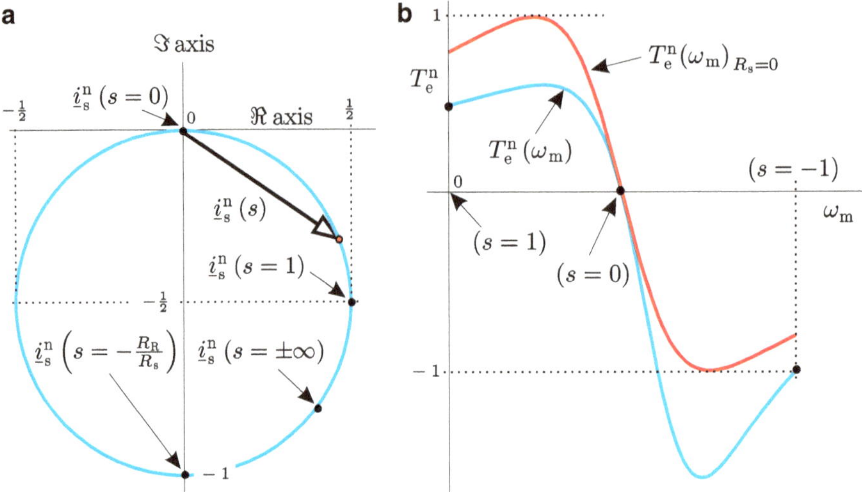

Fig. 9.13 Steady-state characteristics of voltage source connected asynchronous machine with $R_s > 0$ (in *blue*), $L_{\sigma S} > 0$, and $L_M = \infty$, according to Fig. 9.12. (**a**) Normalized stator current phasor. (**b**) Torque versus speed

In this case the circle in the complex plane is again centered at $(0, -1/2)$ and has again a radius of $1/2$ as may be observed from Fig. 9.13a. The circle has not changed in size or position relative to the origin of the complex plane when compared to the previous case (see Fig. 9.11a). The various operating regions on the circle have changed. For example, the motor operating region $0 \leq s \leq 1$ is now confined to a smaller section of the circle. In other words the non-linear slip scale along the circle has changed as a result of adding the stator resistance to the model. Furthermore, the infinite slip point has moved into the fourth quadrant. This means that the torque slip curve will no longer be symmetrical with respect to the zero slip point as will become apparent shortly.

The torque versus slip equation of the revised machine is calculated with the aid of Eqs. (9.13), (9.8b), and (9.11) with $\underline{i}_R = \underline{i}_s$. Note that this condition is still applicable, given that the magnetizing inductance L_M is still assumed to be infinite at this stage. This means that the torque can be written as $T_e = {}^{3\,|\underline{i}_s|^2}/{}_{2\,\omega_s}\, {}^{R_R}/{}_{s}$ which after substitution of Eq. (9.26) gives

$$T_e = \frac{3\,\hat{u}_s^2}{2\,\omega_s} \frac{\frac{R_R}{s}}{\left(R_s + \frac{R_R}{s}\right)^2 + (\omega_s L_{\sigma S})^2} \tag{9.29}$$

Normalization of Eq. (9.29) as undertaken for the previous case [Eq. (9.19)] leads to

$$T_e^n = 2\frac{\frac{s}{\hat{s}}}{\left(1 + r\frac{\hat{s}}{s}\right)^2 + \left(\frac{\hat{s}}{s}\right)^2} \tag{9.30}$$

A graphical representation of Eq. (9.30) as function of shaft speed ω_m with $R_s/R_R = 1$ (which to $r = 1/2$) with $\hat{s} = 1/2$ as used previously is given in Fig. 9.13b. The torque characteristic shown in Fig. 9.13b has as may be expected a maximum and minimum value which correspond to the slip values

$$s = \pm \frac{\hat{s}}{\sqrt{1+r^2}} \tag{9.31}$$

It is emphasized that the slip values which correspond to the peak torque values [as given in Eq. (9.31)] are in this new model no longer equal to the pull-out slip values $\hat{s}$. The new pull-out slip values according to Eq. (9.31) are found by differentiation of Eq. (9.30) with respect to the slip and setting the result to zero. The peak torque values which correspond to the pull-out slip values given in Eq. (9.31) are equal to

$$T^{n}_{e+} = \frac{1}{\sqrt{1+r^2}+r} \tag{9.32a}$$

$$T^{n}_{e-} = \frac{-1}{\sqrt{1+r^2}-r} \tag{9.32b}$$

Equation (9.32) confirms the earlier statement that the two peak torque values indicated in Fig. 9.13b will be unequal when the value of the stator resistance is non-zero.

9.6.4 *Steady-State Analysis with Leakage Inductance, Stator Resistance, and Finite Magnetizing Inductance*

The development of the asynchronous machine model is completed by adding the magnetizing inductance L_M. The Heyland diagram for this revised model is found by making use of Eq. (9.8) which leads to the following expression for the phasor based stator current:

$$\underline{i}_s = \frac{\hat{u}_s\left(\frac{R_R}{s} + j\omega_s L_M\right)}{j\omega_s L_M \frac{R_R}{s} + (R_s + j\omega_s L_{\sigma S})\left(\frac{R_R}{s} + j\omega_s L_M\right)} \tag{9.33}$$

The equivalent circuit which corresponds with Eq. (9.33) is shown in Fig. 9.14. A normalization of expression (9.33) is again introduced which is again of the form $\underline{i}^n_s = \underline{i}_s/(\hat{u}_s/\omega_s L_{\sigma S})$. Furthermore, the parameters $\hat{s} = R_R/\omega_s L_{\sigma S}$, $r = R_s/R_R\,\hat{s}$ again introduced together with a new parameter $l = L_{\sigma S}/L_M$. Use of these parameters with Eq. (9.33) leads to

$$\underline{i}^n_s = \frac{\frac{s}{\hat{s}} - jl}{\left(1 + r\frac{s}{\hat{s}} + l\right) + j\left(\frac{s}{\hat{s}} - lr\right)} \tag{9.34}$$

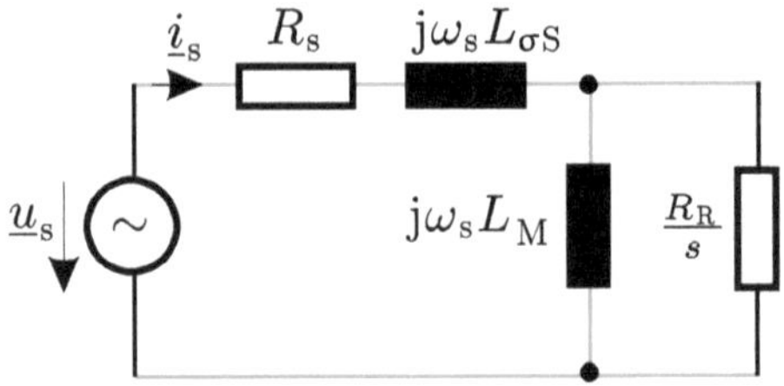

Fig. 9.14 Equivalent circuit of an asynchronous machine with $L_{\sigma S}$, R_s, and L_M, voltage source, steady-state version of dynamic model in Fig. 9.6

An indication of its validity can be obtained by setting $l = 0$ which corresponds to the case $L_M \rightarrow \infty$, in which case Eq. (9.34) is reduced to expression (9.27). A similar exercise can be undertaken with Eq. (9.34) for the case $r = 0$, $l = 0$, which corresponds to infinite magnetizing inductance and zero stator resistance. The Heyland diagram is found by introducing the variables $x = \Re\left\{\underline{i}_s^n\right\}$, $y = \Im\left\{\underline{i}_s^n\right\}$ in Eq. (9.34) and eliminating the slip dependency. Subsequent mathematical handling along the lines indicated in the previous section shows that the Heyland diagram is a circle with midpoint coordinates (x_c, y_c) and radius r_c which are defined as follows:

$$x_c = \frac{r\, l}{1 + l\,(1 + r^2)}$$

$$y_c = -\frac{\left(\frac{1}{2} + l\right)}{1 + l\,(1 + r^2)}$$

$$r_c = \frac{\frac{1}{2}}{1 + l\,(1 + r^2)}$$

An example of the Heyland diagram for the complete machine model is shown in Fig. 9.15a for the cases $\hat{s} = 1/2$, $r = 1/2$, and $l = 1/5$. The parameters values $\hat{s}$ and r are identical to the values introduced previously for the sake of comparison. The value $l = 1/5$ is chosen considerably higher than those normally encountered (typical values for l would be in the order of 0.05 or smaller) in machines in order to demonstrate the impact on the diagram when compared to the cases $r = 0$, $l = 0$ and $r > 0$, $l = 0$ which are also shown in Fig. 9.15a for comparative purposes.

Also indicated in Fig. 9.15a are the normalized stator current phasor end points which correspond with the three key slip points: $s = 0$, $s = 1$, $s = \pm\infty$ as calculated using Eq. (9.34) with the present choice of parameters $\hat{s}$, r, and l. A comparison between the two Heyland circles shows that the introduction of magnetizing inductance reduces the radius of the circle and also causes its midpoint to move to the right and downwards. At synchronous speed ($s = 0$) the stator current will in this case no longer be zero but is instead determined by the stator resistance R_s and the sum (which is the stator reactance $\omega_s L_s$) of the magnetizing reactance $\omega_s L_M$ and leakage reactance $\omega_s L_{\sigma S}$, i.e., under these circumstances $R_R/s \rightarrow \infty$.

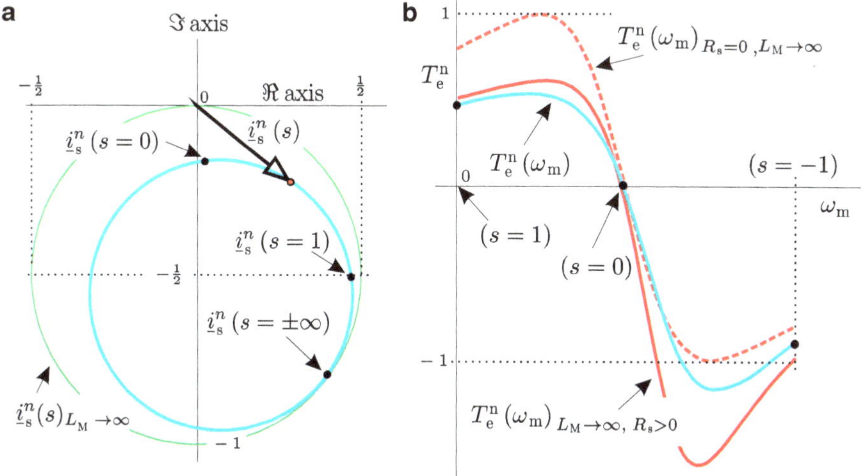

Fig. 9.15 Steady-state characteristics of voltage source connected asynchronous machine, complete model (in *blue*) according to Fig. 9.14. (**a**) Normalized stator current phasor. (**b**) Torque versus speed

The task of finding the torque versus shaft speed characteristic is initiated by making use of Eqs. (9.11) and (9.8) which leads to an expression for the flux phasor $\underline{\psi}_R$, namely

$$\underline{\psi}_R = \frac{\underline{i}_s \frac{R_R}{s} L_M}{\frac{R_R}{s} + j\omega_s L_M} \tag{9.35}$$

Use of Eq. (9.13) with expression (9.35) gives the following steady-state torque equation:

$$T_e = \frac{3}{2}\left|\underline{i}_s\right|^2 \frac{\omega_s L_M^2 \frac{R_R}{s}}{\left(\frac{R_R}{s}\right)^2 + (\omega_s L_M)^2} \tag{9.36}$$

Substitution of the stator current expression (9.33) into Eq. (9.36) gives after some considerable mathematical manipulation the following normalized torque expression:

$$T_e^n = 2\frac{\frac{s}{\hat{s}}}{\left(1 + r\frac{s}{\hat{s}} + l\right)^2 + \left(\frac{s}{\hat{s}} - l\,r\right)^2} \tag{9.37}$$

The normalization introduced is identical to that used in the previous cases, namely $T_e^n = T_e/\hat{T}_e$, with $\hat{T}_e = 3\hat{u}_s^2/4\omega_s^2\hat{s}$. Furthermore, the parameters $\hat{s}$, r, and l, as defined for Eq. (9.34) are also introduced in Eq. (9.37). The torque versus speed characteristic as shown ("blue" curve) in Fig. 9.15b has a maximum and minimum value which correspond to the slip values

$$s = \pm \hat{s}\sqrt{\frac{(l\,r)^2 + (1+l)^2}{1+r^2}} \tag{9.38}$$

The peak torque values which correspond to the pull-out slip values given in Eq. (9.38) are of the form

$$T^{\mathrm{n}}_{\mathrm{e+}} = \frac{1}{\sqrt{\left((lr)^2 + (1+l)^2\right)(1+r^2)} + r} \tag{9.39a}$$

$$T^{\mathrm{n}}_{\mathrm{e-}} = \frac{-1}{\sqrt{\left((lr)^2 + (1+l)^2\right)(1+r^2)} - r} \tag{9.39b}$$

Two additional torque speed curves have been added to Fig. 9.15b which correspond to the cases $r = 0, l = 0$, and $r > 0, l = 0$. It is noted that in most practical cases the influence of magnetizing inductance on the Heyland diagram and torque speed curve is marginal, given the relatively small value of l, i.e., the magnetizing inductance L_{M} is much larger than the leakage inductance $L_{\sigma\mathrm{S}}$.

9.7 Tutorials

9.7.1 *Tutorial 1: Grid Connected Simplified Induction Machine*

This tutorial considers a three-phase IRTF based asynchronous (induction) machine in its simplified form, i.e., no stator resistance R_{s}, magnetizing inductance L_{M}, or leakage inductance $L_{\sigma\mathrm{S}}$. The aim is to build a PLECS model of this machine, which is in accordance with the generic model given in Fig. 9.7a (without L_{M}, R_{s}, or $L_{\sigma\mathrm{S}}$). An implementation of the PLECS model in its present form is given in Fig. 9.16.

The model in its present form is *not* designed for dynamic analysis, given that the so-called alternative differentiator module (see Fig. 6.8) has been introduced in the simulation model which are only usable for quasi-steady-state conditions. The aim of this example is to examine the steady-state characteristics, hence the use of numerical display modules as well as RMS and real power modules introduced in earlier tutorials. The machine in question has an inertia of $J = 0.001\,\mathrm{kg\,m^2}$.

A rotating stator flux vector $\vec{\psi}_{\mathrm{s}}$ with angular frequency $\omega_{\mathrm{s}} = 100\pi$ rad/s and amplitude $\hat{\psi}_{\mathrm{s}} = 1.0$ Wb is assumed as an input to this model. This means that the integrator with output $\vec{\psi}_{\mathrm{s}}$ (see Fig. 9.7a) can be omitted in this tutorial.

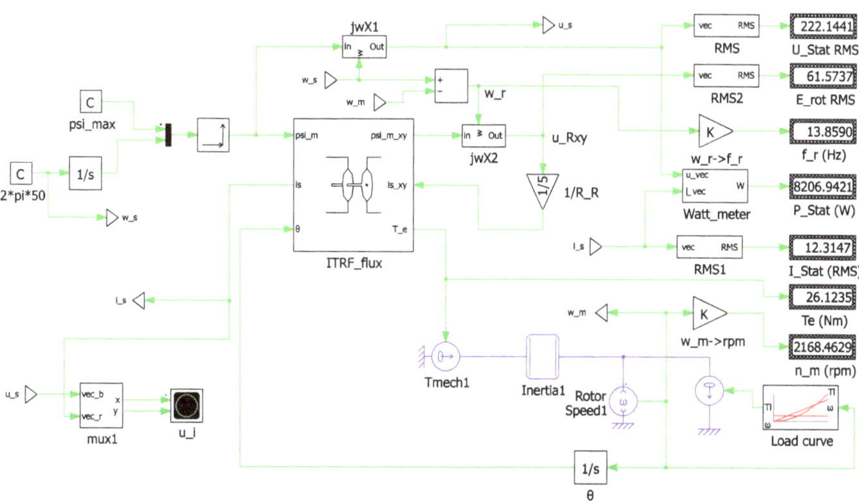

Fig. 9.16 PLECS model of (simplified) asynchronous machine

The generation of the vector $\vec{\psi}_{\mathrm{s}}$ in PLECS is described in Sect. 8.7.1. The machine is connected to a mechanical load with a quadratic torque speed curve, where we are able to vary the load torque (so we can achieve motor operation) value. Add a XY Plot module to observe the stator voltage and stator current vectors at the end of the simulation.

Under "simulation parameters" set the simulation run time to 0.1 s and observe the results by way of the values on the display modules.

The operation of the model is to be examined for two values of the rotor resistance, namely $R_{\mathrm{R}} = 5\,\Omega$ and $R_{\mathrm{R}} = 10\,\Omega$. Set the reference speed $\omega^{\mathrm{ref}} = 100\pi$ rad/s in the load torque module. Change the load torque T^{ref} (in the load torque module) in five steps in the range 0–30 Nm and record (after doing a simulation run for each *step*) the data from the display modules. Note that the machine shaft speed will be at that point where the load torque speed curve $T_{\mathrm{l}}(\omega_{\mathrm{m}})$ and machine torque speed curve $T_{\mathrm{e}}(\omega_{\mathrm{m}})$ intersect. An example of the results which should appear on the displays is given in Table 9.1. The load torque reference value T^{ref} used with the load torque sub-module is also given in Table 9.1. Note that RMS stator voltage reading will remain constant at $U_{\mathrm{s}} = 222.14$ V. Rerun your simulation for the case where the rotor resistance is doubled, i.e., $R_{\mathrm{R}} = 10\,\Omega$. For this example change the load torque T^{ref} (in the load torque module) in five steps in the range 0–50 Nm. The results which should appear on the display are given in Table 9.2. An example of the vector plots which should appear on the XY Plot module `u_i` at the end of the simulation is given in Fig. 9.24. The results given represent the attenuated stator voltage vector and current vector at a shaft speed of 2200 rpm for rotor resistance values of $5\,\Omega$ (subplot (a)) and $10\,\Omega$ (subplot (b)), respectively. Clearly noticeable from these vectors is that the voltage and current vectors are aligned, hence the stator

Table 9.1 Simulation results asynchronous machine $R_R = 5\,\Omega$: load→no-load

T^{ref} (Nm)	n_m (rpm)	T_e (Nm)	I_s (A)	P_s (W)	E_R (V)	f_r (Hz)
30	2392.60	19.08	8.99	5994.74	44.97	10.12
20	2542.67	14.36	6.77	4513.57	33.86	7.62
10	2735.36	8.31	3.91	2611.79	19.59	4.41
5	2855.77	4.53	2.13	1423.39	10.67	2.40
0	3000.00	0	0	0	0	0

Table 9.2 Simulation results asynchronous machine $R_R = 10\,\Omega$: load→ no-load

T^{ref} (Nm)	n_m (rpm)	T_e (Nm)	I_s (A)	P_s (W)	E_R (V)	f_r (Hz)
50	1823.70	18.47	8.71	5804.78	87.10	19.60
30	2081.01	14.43	6.80	4535.01	68.04	15.31
20	2270.61	11.45	5.40	3599.39	54.00	12.15
10	2542.67	7.18	3.38	2256.78	33.86	7.62
5	2735.36	4.15	1.95	1305.89	19.59	4.41
0	3000.00	0	0	0	0	0

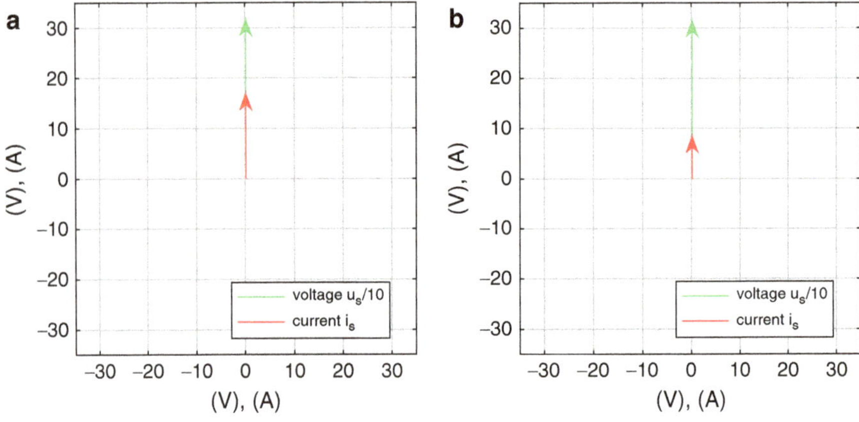

Fig. 9.17 Vector plot: scaled stator voltage vector $\vec{u}_s/10$ ("*green*") and stator current vector $\vec{i}_s$ ("*red*") for shaft speed $n_m = 2200$ rpm. (**a**) $R_R = 10\,\Omega$, $T_e = 25$ Nm. (**b**) $R_R = 5\,\Omega$, $T_e = 12.5$ Nm

reactive power is zero. Consequently, the real input power to the machine is the sum of the shaft power $P_m = T_m.^{2*\pi*n_m}/60$ and dissipated power $P_r = 3.I_R^2\,R_R$, where I_R is the RMS current in the rotor (which in this case is equal to I_s). Furthermore, recall that the stator flux lags the stator vector by 90°, hence the stator current is perpendicular to the flux, which is the most favorable orientation for efficient torque production. Note that doubling the rotor resistance halves the stator current (as may be observed from Fig. 9.17) and consequently the torque.

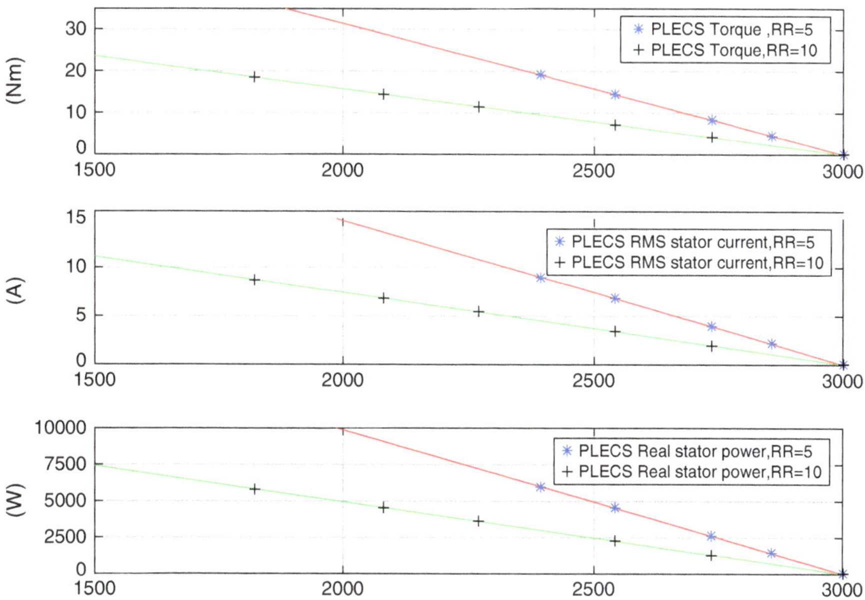

Fig. 9.18 PLECS/MATLAB result: $T_e\,(n_m)$, $I_s\,(n_m)$, $P_s\,(n_m)$, $R_R = 5,\ 10\,\Omega$

Build an M-file which will display the data from your simulations in the form of three subplots $T_e\,(n_m)$, $I_s\,(n_m)$, $P_s\,(n_m)$. In addition, undertake a phasor analysis of the drive and add these results to the PLECS generated data. An example of the results which should appear is given in Fig. 9.18. Also shown in Fig. 9.18 (continuous lines) are the results obtained via a steady analysis of the model in its current form. The M-file which contains the steady-state analysis and required code to plots the results obtained from the PLECS model is as follows:

M-Code

```
%tutorial 1, chapter 9
close all
clear all
%steady-state analysis
psis_ph=-j*1.0;                     % stator flux
                                      vector
ws=100*pi;%stator frequency rad/s
us_ph=j*ws*psis_ph;                 % voltage phasor
                                     (assumed real)
Us=abs(us_ph)/sqrt(2);              % stator voltage
                                      RMS
nm=[0:10:3000];                     % selected speed
                                      range
wm=2*pi*nm/60;                      % shaft speed
```

```
                                                    rad/s
slip=(ws-wm)/ws;                                  % slip
                                                    calculation
wr=ws-wm;                                         % slip frequency
fr=wr/(2*pi);                                     % rotor freq Hz
P_a=[]; Te_a=[]; I_sa=[];
for j=1:2
RR=5*j;
is_ph=us_ph./(RR./slip);                       % stator current
                                                 phasor
I_s=abs(is_ph)/sqrt(2);                        % RMS value phase
                                                 current
I_sa=[I_sa;I_s];
P=3/2*real(us_ph*conj(is_ph));                   % real stator
                                                   power
P_a=[P_a;P];
Te=3/2*imag(conj(psis_ph)*is_ph);                % torque
Te_a=[Te_a;Te];
end
%show PLECS results
%with RR=5,
nm1=[2392.66 2542.67 2735.36 2855.77 3000];% shaft speed
Te1=[19.08 14.36 8.31 4.53 0];                    % torque
Is1=[8.99 6.77 3.91 2.13 0];                      % RMS stator
                                                    current
Ps1=[5994.74 4513.57 2611.79 1423.39 0];          % real power
%%%%%%%%%%%%%%%%%%%%%%%%%
%with RR=10,
nm2=[1823.7 2081.01 2270.61 2542.67 2735.36 3000];
Te2=[18.47 14.43 11.45 7.18 4.15 0];
Is2=[8.71 6.80 5.40 3.38 1.95 0];
Ps2=[5804.78 4535.01 3599.39 2256.78 1305.89 0];
%plot PLECS results
figure(1)
subplot(3,1,1)
plot(nm1,Te1,'*')
grid
hold on
plot(nm2,Te2,'+')
legend('PLECS Torque ,RR=5','PLECS Torque,RR=10')
plot(nm,Te_a(1,:),'r')
plot(nm,Te_a(2,:),'k')
axis([1500 3000 0 35])
ylabel('(Nm)')
subplot(3,1,2)
plot(nm1,Is1,'*')
grid
hold on
plot(nm2,Is2,'+')
legend('PLECS RMS stator current,RR=5','PLECS RMS stator
  current,RR=10')
plot(nm,I_sa(1,:),'r')
plot(nm,I_sa(2,:),'k')
axis([1500 3000 0 15])
```

```
ylabel('(A)')
subplot(3,1,3)
plot(nm1,Ps1,'*')
grid
hold on
plot(nm2,Ps2,'+')
legend('PLECS Real stator power,RR=5','PLECS Real stator
  power,RR=10')
plot(nm,P_a(1,:),'r')
plot(nm,P_a(2,:),'k')
axis([1500 3000 0 10000])
ylabel('(W)')
```

9.7.2 Tutorial 2: Grid Connected Simplified Induction Machine with Leakage Inductance

The purpose of this tutorial is to modify the PLECS model shown in Fig. 9.16 to include leakage inductance $L_{\sigma \mathrm{S}}$. A suitable starting point for this task is the model shown in Fig. 9.6 which for this tutorial has been simplified to include the leakage inductance and rotor resistance. Furthermore a rotating flux vector $\vec{\psi}_{\mathrm{s}} = \hat{\psi}_{\mathrm{s}}\, \mathrm{e}^{\mathrm{j}\omega_{\mathrm{s}}t}$ as introduced in the previous tutorial is to be used here. The aim is to use the revised PLECS model to plot the torque, real/reactive power, and RMS current versus shaft speed characteristics so that the reader can ascertain the significance of adding leakage inductance.

A solution to this problem may be found by making use of the IRTF model given in Fig. 9.6 AND realizing that such models allow inductive elements to be moved to either stator or rotor side without impunity. Hence in this case it is convenient to move the leakage inductance to the rotor side as shown in Fig. 9.19. In which case the rotor resistance as used in the previous PLECS tutorial (see Fig. 9.16) must be replaced by a set of control blocks that represent a series network formed by the elements R_{R} and L_{σ}^{s}. Furthermore, the excitation term $\vec{u}_{\mathrm{s}}^{\mathrm{xy}} = {\mathrm{d}\vec{\psi}_{\mathrm{s}}^{\mathrm{xy}}}/{\mathrm{d}t}$ can be simplified given the use of a constant amplitude $\hat{\psi}_{\mathrm{s}} = 1.0\,\mathrm{Wb}$ rotating stator flux vector and the IRTF transformation $\vec{\psi}_{\mathrm{s}}^{\mathrm{xy}} = \vec{\psi}_{\mathrm{s}}\, \mathrm{e}^{\mathrm{j}\omega_{\mathrm{m}}t}$. Evaluation of the differential term gives: $\vec{u}_{\mathrm{s}}^{\mathrm{xy}} = \mathrm{j}\,(\omega_{\mathrm{s}} - \omega_{\mathrm{m}})\, \vec{\psi}_{\mathrm{s}}^{\mathrm{xy}}$, which can be readily implemented using an

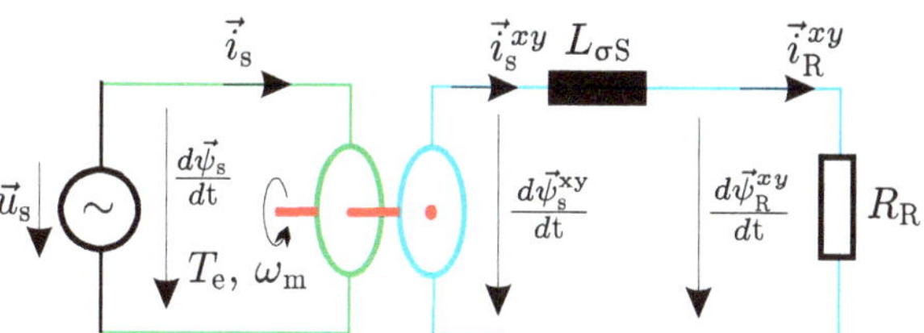

Fig. 9.19 Simplified IRTF based model of asynchronous machine with leakage inductance

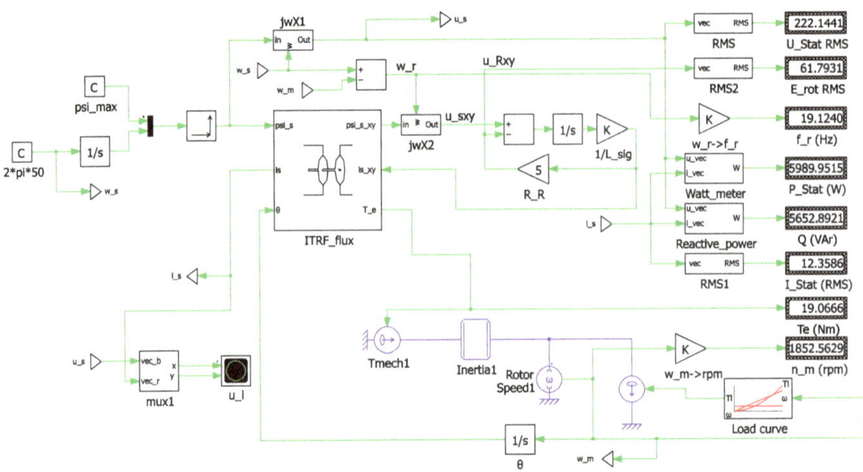

Fig. 9.20 PLECS model of (simplified) asynchronous machine with leakage inductance

'alternative differentiator function' as introduced earlier (see Fig. 6.8). The revised PLECS model as given in Fig. 9.20 shows this alternate differentiator module as `jwX2`. A second module `jwX1` is used to generate the stator voltage vector $\vec{u}_{\mathrm{s}}$ in line with the approach used in the previous tutorial. Note that the rotor EMF vector $\vec{u}_{\mathrm{R}}^{\mathrm{xy}}$ is now found by evaluating the term $\vec{u}_{\mathrm{R}}^{\mathrm{xy}} = \vec{i}_{\mathrm{R}}^{\mathrm{xy}} R_{\mathrm{R}}$. A set of control modules `\RR`, `L_sig`, and integrator are used to model the rotor resistor/leakage inductance network. A set of vector to RMS conversion models and real/reactive power modules with corresponding numerical displays have been added to analyses steady-state operation. Furthermore, a XY plot module `u_i` has been added to observe the stator voltage and stator current vectors at the end of the simulation. The mechanical part of the PLECS model remains unchanged when compared to the previous tutorial.

Repeat the simulation exercise described in tutorial 1 with ten load reference torque T^{ref} steps in the range of $1000 \rightarrow 0\,\mathrm{Nm}$. An example of the data which should appear after running this simulation (after each load step) is given in Table 9.3. An example of the vector plots which should appear on the XY Plot module `u_i` at the end of the simulation is given in Fig. 9.21. The results given represent the attenuated stator voltage vector and current vector for two different operating points with rotor resistance and leakage inductance values of $5\,\Omega$ and $\pi/80\,\mathrm{H}$, respectively. Clearly noticeable from these vectors is the impact of the leakage reactance to rotor resistance ratio: $\omega_{\mathrm{r}} L_{\sigma}^{\mathrm{s}}/R_{\mathrm{R}}$. For low slip (subplot (b)) the rotor frequency ω_{r} is relatively low as is the induced rotor voltage E_{R}. Consequently the phase relationship between the stator voltage and current is dominated by the rotor resistance term. Under high slip conditions slip frequency and induced voltage are high but the leakage reactance dominates the rotor resistance hence the increased lag between the two vectors as observed in subplot (a).

On the basis of the data given in Table 9.3 generate four subplots which represent the relationships $T_{\mathrm{e}}(n_{\mathrm{m}})$, $I_{\mathrm{s}}(n_{\mathrm{m}})$, $P_{\mathrm{s}}(n_{\mathrm{m}})$, $Q_{\mathrm{s}}(n_{\mathrm{m}})$. In addition plot the

Table 9.3 Simulation results asynchronous machine $R_R = 5\,\Omega$, $L_{\sigma S} = \pi/80$ H: load→no-load

T^{ref} (Nm)	n_m (rpm)	T_e (Nm)	I_s (A)	P_s (W)	Q_s (VA)	E_R (V)
1000	369.48	14.51	16.35	4561.17	9898.17	81.76
200	859.14	16.40	15.65	5153.05	9073.43	78.28
100	1271.95	17.97	14.72	5647.42	8026.46	73.63
70	1555.47	18.81	13.77	5911.99	7023.86	68.87
60	1690.23	19.04	13.19	5983.43	6445.60	65.98
50	1852.56	19.06	12.35	5989.95	5652.89	61.79
30	2257.59	16.98	9.38	5337.29	3258.96	46.91
20	2484.84	13.72	7.02	4310.56	1826.39	35.12
10	2724.03	8.24	3.98	2590.20	587.89	19.92
5	2853.89	4.52	2.14	1421.51	170.82	10.74
0	3000.00	0.00	0.00	0.00	0.0	0.00

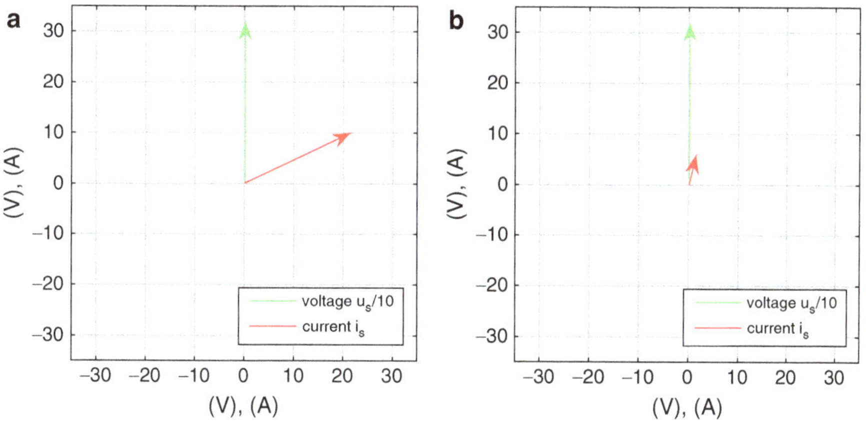

Fig. 9.21 Vector plot: scaled stator voltage vector $\vec{u}_s/10$ ("*green*") and stator current vector $\vec{i}_s$ ("*red*") for two slip conditions. (**a**) High slip: $T_e = 14.5$ Nm, $n_m = 369$ rpm. (**b**) Low slip: $T_e = 8.2$ Nm, $n_m = 2724$ rpm

Heyland diagram for the machine under motor operation. An example of the results which should appear is given in Fig. 9.22. Also shown in Fig. 9.22, in the form of continuous lines are the results from the steady-state analysis. The M-file given at the end of this tutorial shows the steady-state calculations, together with the MATLAB code required to plot the results from Table 9.3.

A Heyland diagram of the machine which shows the stator current phasor end point versus slip can be made by using the data given in Table 9.3. This requires use of the real and reactive power table entries and stator current data. On the basis of this data the angle between the voltage and current phasor is found using: $\rho_s = -\arctan\left(Q_s/P_s\right)$ (the negative sign is introduced because the phasor current lags the voltage phasor). The current phasor magnitude $|\underline{i}_s|$ is found using $|\underline{i}_s| = I_s\sqrt{2}$.

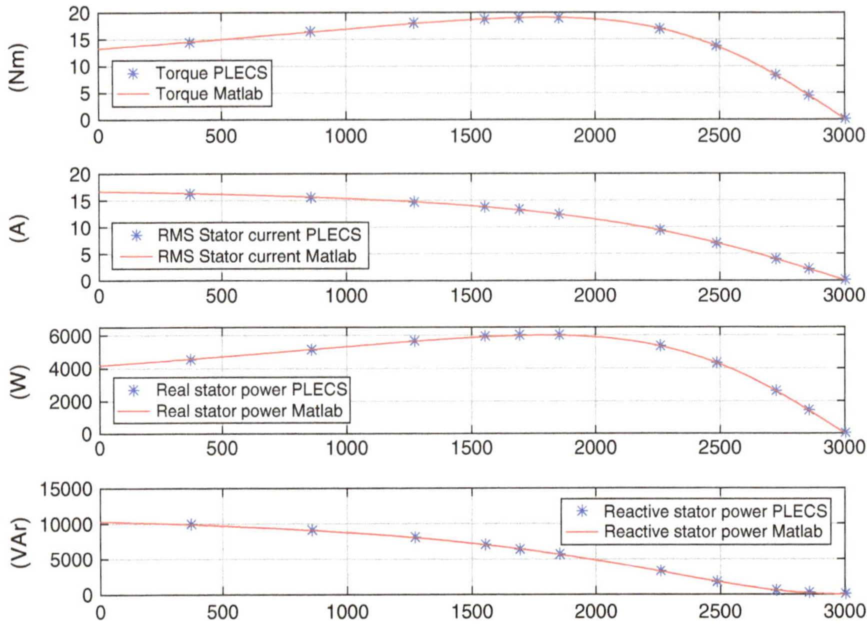

Fig. 9.22 Simulink/MATLAB result: $T_e\,(n_m)$, $I_s\,(n_m)$, $P_s\,(n_m)$, $Q_s\,(n_m)$, $R_R = 5\,\Omega$, $L_{\sigma S} = \pi/80\,\mathrm{H}$

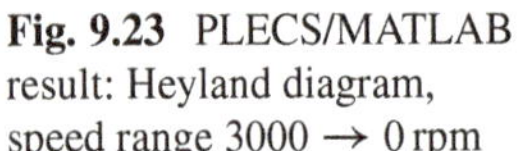

Fig. 9.23 PLECS/MATLAB result: Heyland diagram, speed range 3000 → 0 rpm

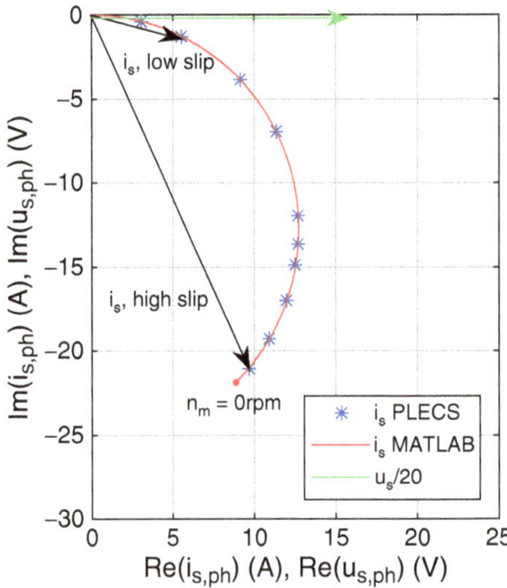

The result which should appear is given in Fig. 9.23 together with the supply voltage phasor $\underline{u}_s$. Shown in this Heyland diagram are the low and high slip operating point also shown in the XY plot, see Fig. 9.21. The operating points shown in the Heyland diagram are located on a circle which is according to the theory discussed in this chapter (see Fig. 9.11). The M-file given below shows the code that undertakes the steady-state phasor analysis and plotting of the results for the desired speed range (standstill to synchronous speed).

M-Code

```
%tutorial 2 chapter 9
close all
clear all
psis_ph=-j*1.0;                          % stator flux vector
ws=100*pi;%stator frequency rad/s
us_ph=j*ws*psis_ph;                      % voltage phasor
                                           (assumed real)
Us=abs(us_ph)/sqrt(2);                   % stator voltage RMS
RR=5;                                    % rotor resistance
L_sig=pi/80;                             % leakage inductance
nm=[0:10:3000];                          % selected speed range
wm=2*pi*nm/60;                           % shaft speed rad/s
slip=(ws-wm)/ws;                         % slip calculation
wr=ws-wm;                                % slip frequency
fr=wr/(2*pi);                            % rotor freq Hz
is_ph=us_ph./(RR./slip+j*ws*L_sig);      % stator current phasor
I_s=abs(is_ph)/sqrt(2);                  % RMS value phase
  current
P=3/2*real(us_ph*conj(is_ph));               % real stator
  power
Q=3/2*imag(us_ph*conj(is_ph));               % reactive stator
                                               power
Te=3/2*imag(conj(psis_ph)*is_ph);            % torque
psiM_ph=psis_ph-L_sig*is_ph;             % magnetizing flux
e_ph=j*wr.*psiM_ph;                      % e_rot
E=abs(e_ph)/sqrt(2);                     % RMS value e_rot
%
%with RR=5 and leakage inductance Lsig=pi/80
nm3=[369.48 859.14 1271.95 1555.47 1690.23 1852.56 2257.59
  2484.84 ...
2724.03 2853.89 3000];
Te3=[14.51 16.40 17.97 18.81 19.04 19.06 16.98 13.72 8.24
  4.52 0];
Is3=[16.35 15.65 14.72 13.77 13.19 12.35 9.38 7.02 3.98 2.14
  0];
Ps3=[4561.17 5153.05 5647.42 5911.99 5983.43 5989.95 5337.29
  4310.56 ...
2590.20 1421.51 0];
Qs3=[9898.17 9073.43 8026.46 7023.86 6445.6 5652.89 3258.96
  1826.39 ...
587.89 170.82 0];
%%%%%%%%%%%%%%%%%%%%%%%%%%%%%%%%%%%%%%%%%%%%%%%%%%%%%%%%%%%%%%%%%%%%%%%%
%plot PLECS results
subplot(4,1,1)
plot(nm3,Te3,'*')
grid
```

```
hold on
plot(nm,Te,'r')
ylabel('(Nm)')
legend('Torque PLECS','Torque Matlab ',0)
subplot(4,1,2)
plot(nm3,Is3,'*')
grid
hold on
plot(nm,I_s,'r')
ylabel('(A)')
legend('RMS Stator current PLECS','RMS Stator current
  Matlab',0)
subplot(4,1,3)
plot(nm3,Ps3,'*')
grid
hold on
plot(nm,P,'r')
ylabel('(W)')
legend('Real stator power PLECS','Real stator power
  Matlab',0)
subplot(4,1,4)
plot(nm3,Qs3,'*')
grid
hold on
plot(nm,Q,'r')
ylabel('(VAr)')
legend('Reactive stator power PLECS','Reactive stator power
  Matlab',0)
%%%%%%plot Heyland diagram
figure(2)
rhos=-atan(Qs3./Ps3);
isR=sqrt(2)*Is3.*cos(rhos);
isI=sqrt(2)*Is3.*sin(rhos);
plot(isR,isI,'*')
hold on
plot(real(is_ph),imag(is_ph),'r')
axis equal
grid
axis ([0 20 -30 0])
xlabel('Re(is_{ph}) (A)')
ylabel('Im(is_{ph}) (A)')
legend('PLECS','Matlab')
```

9.7.3 Tutorial 3: Asynchronous Machine Connected to a Three-Phase Source

The purpose of this tutorial is to consider the operation of a general purpose asynchronous machine model under load and connected to a three-phase sinusoidal voltage source. Central to the development of this simulation diagram is the

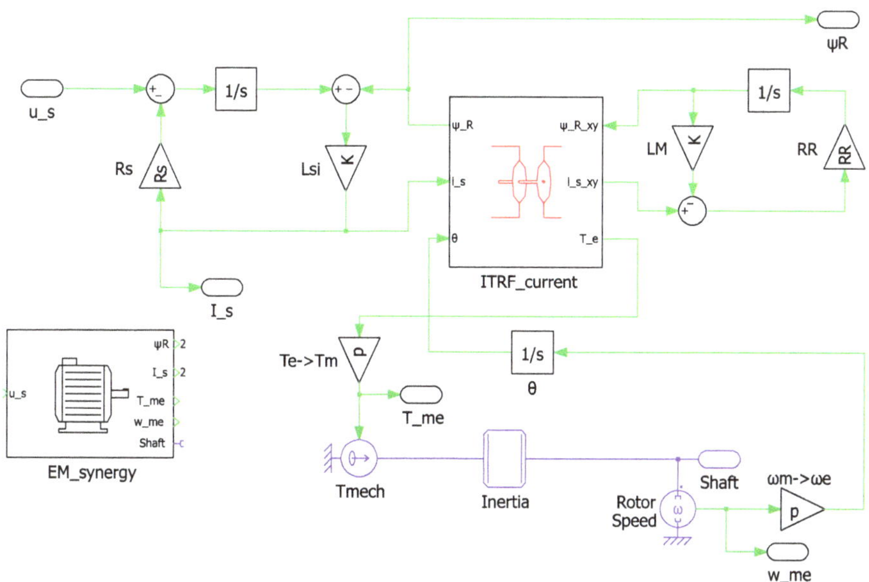

Fig. 9.24 PLECS model: asynchronous machine

Table 9.4 Parameters Texas Instruments LVACIMTR (EMsynergy) induction machine

Parameters		Value
Leakage inductance	$L_{\sigma S}$	6.22 mH
Stator resistance	R_s	1.84 Ω
Rotor resistance	R_R	1.08 Ω
Magnetizing inductance	L_M	30.0 mH
Inertia	J	0.07 10^{-3} kg m^2
Pole pairs	p	2
Initial rotor speed	ω_m^0	0 rad/s

generic model given in Fig. 9.7b. The IRTF module shown in this diagram is a "IRTF-current" type unit, given that the stator current vector is selected as an input on the stator side. Furthermore, it is instructive to expand this model to suit multi-pole machine operation as discussed in Sect. 7.2.1. A PLECS model interpretation along the lines discussed above is given in Fig. 9.24, which shows the basic IRTF module with (two pole) torque output `T_e`. A set of mechanical blocks is to used to implement the relationship between shaft T_m, load torque T_l, inertia J, and shaft speed ω_m. For this tutorial a Texas Instruments LVACIMTR (EMsynergy) 13.6 W, 4-pole machine [10] is used with the a set of parameters given in Table 9.4. This machine, which is shaft coupled to the Teknic PM machine introduced in the previous chapter, will also be deployed for de "demonstration" lab at the end of this section, hence its inclusion here. Note that the inertia given in Table 9.4 is the combined value of the induction machine and the attached PM (Teknic) machine as shown in Fig. 8.29. Said model is to be connected to a three-phase 7.0 V RMS/50 Hz

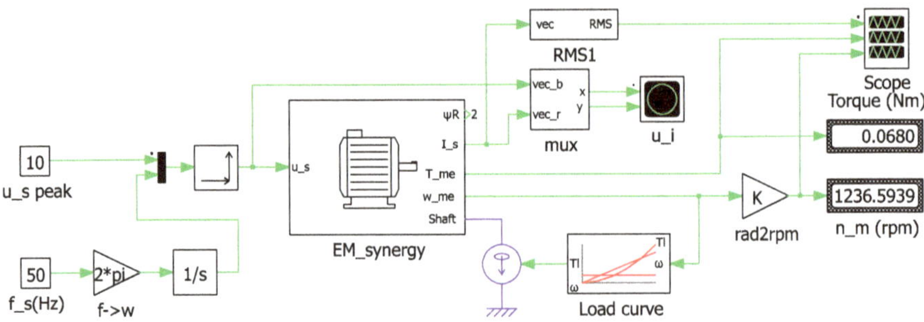

Fig. 9.25 PLECS model: asynchronous machine connected to a three-phase voltage source

supply. Hence a 3000 rpm rotating vector with an amplitude of 10.0 V must be generated as an input for this model. Connect a load module [see Eq. (8.28)] to the machine model. This module must be configured with a quadratic load speed curve and parameters $T^{\mathrm{ref}} = 100\,\mathrm{mNm}$, $\omega^{\mathrm{ref}} = 2\pi 1500/60\,\mathrm{rad/s}$. A PLECS implementation of the machine connected to the required mechanical load and three-phase supply is given in Fig. 9.25. Run the simulation for 0.5 s and add a scope model which shows the RMS phase current, shaft torque, and shaft speed. In addition, add a Vector plot module to show the stator voltage (attenuated by a factor 10) and stator current space vectors. Observe these vectors and the end of the simulation period under load AND no-load conditions. The results from your simulation Matlab processed should be in accordance with those given in Fig. 9.26. A set of numerical display modules has also been added which show the steady-state torque and shaft speed at the end of the simulation interval. The required vector plots of the machine under no-load (reference torque in load module `load curve` set to zero for this simulation example) and load are shown in Fig. 9.27. Readily observable from the waveforms shown in Fig. 9.26 is the presence of a transient on the current and torque during start up. Current during this type of start up can typically be three times the rated value. In this case the effect is less pronounced during to the presence of a relatively large stator resistance (typical for small electrical machines).

9.7.4 Tutorial 4: Steady-State Analysis of an Asynchronous Machine Connected to a Three-Phase Supply

It is instructive to examine the steady operation of the IM machine by way of a phasor analysis according to the theory presented in Sect. 9.6.4. More specifically it is considered important to understand the relationship between current and voltage phasors under steady-state operation as the machine moves from no-load to load conditions. Construct a Heyland diagram for the EMsynergy machine as presented in the previous tutorial using the steady conditions discussed and present in this diagram:

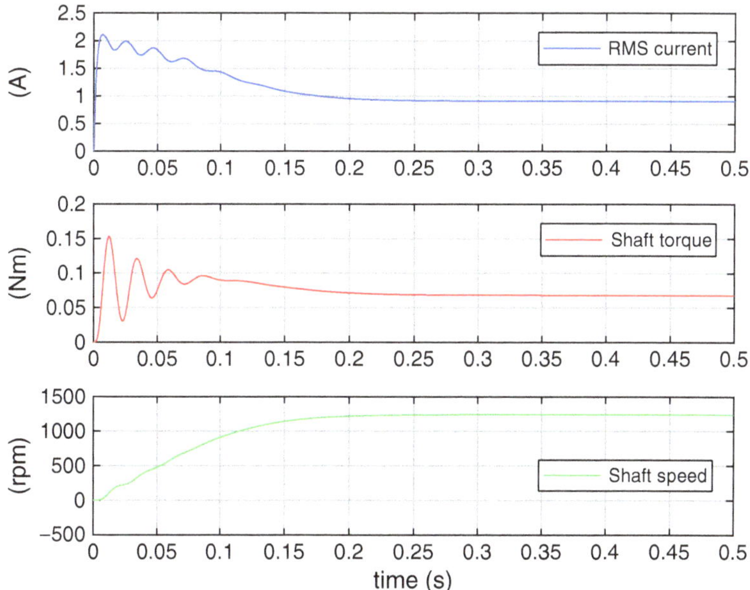

Fig. 9.26 Startup sequence: LVACIMTR (EMsynergy) asynchronous machine connected to a three-phase voltage source

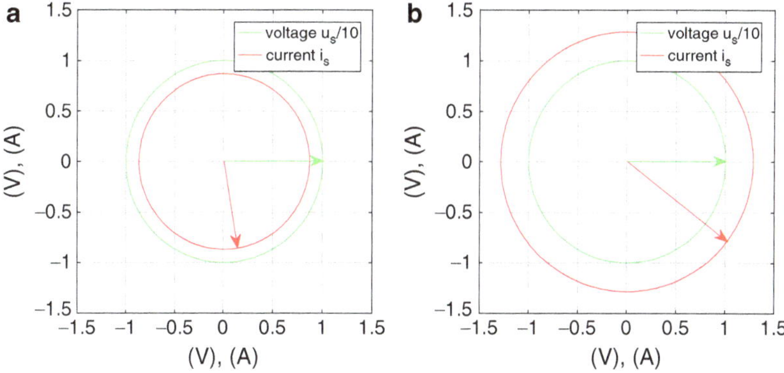

Fig. 9.27 Vector plot: scaled stator voltage vector $\vec{u}_s/10$ ("*green*") and stator current vector $\vec{i}_s$ ("*red*") for two slip conditions: $s = 0$, $n_m = 1500$ rpm and $s = 0.176$, $T_m = 68 \cdot 10^{-3}$ Nm, $n_m = 1236$ rpm. (**a**) No-load ($s = 0$). (**b**) load ($s = 0.176$)

- the operating trajectory for motor/generator operation
- the operating points which corresponds to standstill, no-load operation, and operation with the steady-state torque and speed value found in the previous tutorial at the end of the simulation interval.

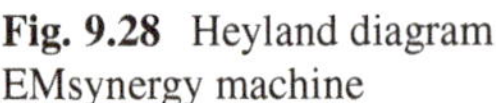

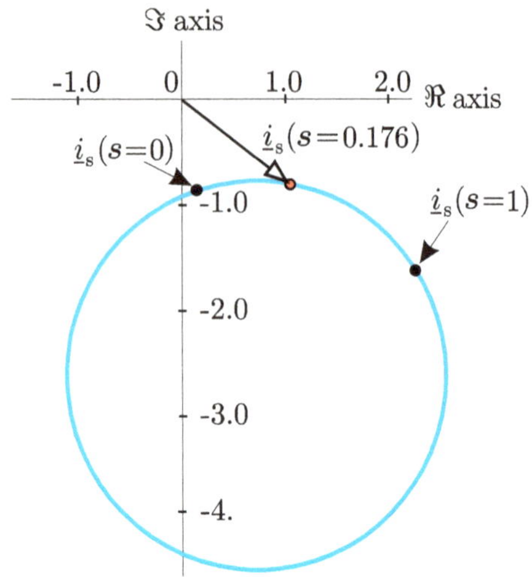

Fig. 9.28 Heyland diagram: EMsynergy machine

Show your calculations in the form of an M-file. In addition, use this M-file to plot the shaft torque versus speed curve for the machine in use over the speed range $0 \rightarrow 3000$ rpm. Also add the load torque versus speed characteristic and confirm that the steady-state operating torque/shaft speed match those found in the previous tutorial.

An example of the results obtained from the M-file is shown in Fig. 9.28 in the form of a Heyland diagram. Central to this figure is the Heyland circle, which shows the required operating points under no-load $s = 0$, standstill $s = 1$, and under load conditions with $T_m = 68 \cdot 10^{-3}$ Nm, shaft speed 1236 rpm, which corresponds to a slip value of $s = 0.176$. This diagram clearly shows how the relationship between voltage and current phasors changes as the machine operating point moves from $s = 0 \rightarrow s = 0.176$ and conforms with the results shown in the vector plot for both load conditions (see Fig. 9.25). The shaft torque versus speed curve over the required speed range can be found with the aid of Eq. (9.37) and the machine parameters given in Table 9.4. Said equation must be scaled as shown in Sect. 9.6.4, given the need to present actual shaft torque and speed variables as may be observed from Fig. 9.29. Also shown in this figure is the quadratic load versus shaft speed as used in the previous tutorial. The intersection of the machine and load torque curves represents the steady-state operating point, which matches the torque and speed value found on the numerical displays of the previous tutorial (see Fig. 9.25). An example of an M-file which shows the required calculations for this tutorial is given below:

M-Code

```
%tutorial 4 chapter 9
%parameters machine
```

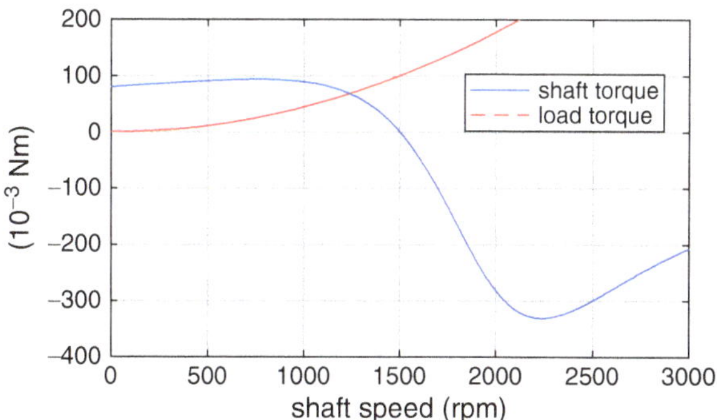

Fig. 9.29 MATLAB results: shaft torque versus slip curve, EMsynergy machine, and load torque/speed characteristic

```
Rs=1.84;          %stator resistance
RR=1.08;          % Rotor resistance
Lsig=6.22e-3;     % leakage inductance
LM=30e-3;         % magnetizing inductance
p=2;              % pole pair number
%%%%%%%
%excitation in use
us_hat=10;        %peak voltage (V)
fs=50;            % stator frequency (Hz)
ws=2*pi*fs;       % stator freq (rad/s)
%%%
%calculate Heyland circle  see eqn (9.34)
s_p=RR/(ws*Lsig); % peak slip frequency
l=Lsig/LM;
r=Rs/RR*s_p;
%scale factor
Isc=us_hat/(ws*Lsig); %scaling factor current
%%%center circle
xc=r*l/(1+l*(1+r^2));
yc=-(0.5+l)/(1+l*(1+r^2));
rc=0.5/(1+l*(1+r^2));
Xc=xc*Isc;
Y=yc*Isc;
R=rc*Isc;
%%%%%%%%%%%%
%determine the current phasor
%location for a given slip(s) using eqn (9.34)
s=(1500-1236)/1500; %equal s value
isn_ph=(s/s_p-j*l)/((1+r*(s/s_p)+l)+j*(s/s_p-l*r));
rho_ph=180/pi*angle(isn_ph); % angle between voltage and
  current (deg)
is=abs(isn_ph)*Isc;              % current amplitude (A)
%%%add torque/slip curve
Te_p=3*us_hat^2/(4*ws^2*Lsig) %peak electrical torque
sl=1:-0.01:-1;
Te_n=2*(sl./s_p)./((1+r*sl./s_p+l).^2+(sl./s_p-l*r).^2);
```

```
  %perunit torque
%calculated using eqn(9.38)
T_m=Te_n*Te_p*p;
nmm=(1-sl)*1500;
plot(nmm,T_m*1000);
hold on %plot load curve
Tl=nmm.^2*100/1500^2;
plot(nmm,Tl,'--');
grid on
legend('shaft torque', 'load torque')
ylabel('milli-Nm')
xlabel('shaft speed (rpm)')
axis([0 3000 -400 200])
```

9.7.5 Demo Lab 2: Voltage/Frequency IM Drive

For this demonstration the LVACIMTR (EMsynergy) asynchronous machine which is the motor used in the previous two tutorials is to be connected to the aft (furthest away from the USB connector) BOOSTXL-DRV8301 module shown in Fig. 8.30. A voltage/frequency controller is used [10] which can generate a three-phase sinusoidal supply with the required voltage amplitude and frequency. This controller implemented in embedded software VisSim [14] is shown in Fig. 9.30. The LVSERVOMTR (Teknic) PM machine is connected to the forward converter and the resulting drive functions as a dynamometer for the induction machine. The drive setup was used to measure the variables u_{α} and i_{α} (which is also the phase current) under no-load and identical load conditions (see Fig. 9.31) as discussed in

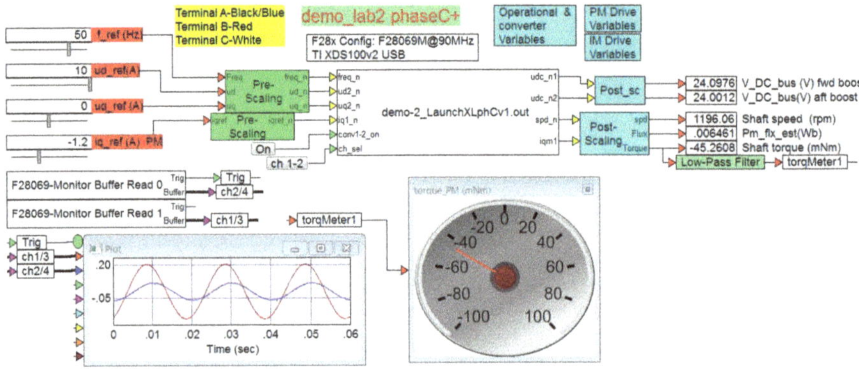

Fig. 9.30 VisSim [14] based V/f drive Demonstration controller [10]

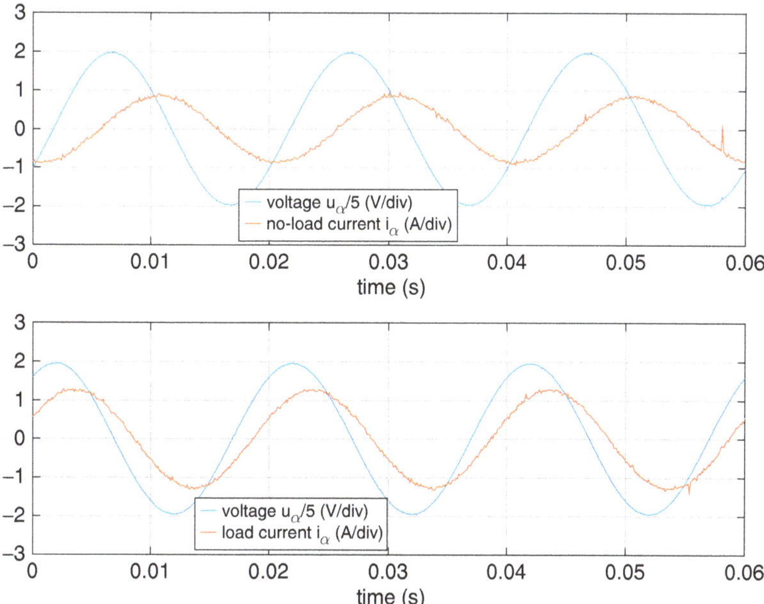

Fig. 9.31 Matlab processed scope results VisSim [14] based V/f drive: no-load ($s = 0$) and load ($s = 0.716$) condition [10]

the previous tutorial. The results derived from the demonstration setup have been processed to show actual values. The following critical observations of the results according to Fig. 9.31 can be made in conjunction with the results from the vector plot in Fig. 9.27.

- The voltage waveform is 20 ms, which corresponds with a rotating voltage vector of 3000 rpm, i.e., a excitation frequency of 50 Hz.
- The drive was set to operate with voltage space vector $\vec{u}_s = u_{s\alpha} + j\,u_{s\beta}$ with amplitude $|\vec{u}_s| = 10$ V. The amplitude of the attenuated real component $u_{s\alpha} = {}^{u_{s\alpha}}/_{5}$ is indeed 2.0 V.
- Inspection of the no-load current amplitude $|i_{s\alpha}|$ and the phase angle relative to the attenuated voltage waveform $u_{s\alpha}$ shows that these are equal to $|i_{s\alpha}| = 0.86$ A and 71°, respectively. This compares favorably with the current vector amplitude and phase angle between the voltage and current vectors shown in Fig. 9.27 ("left" subplot). The no-load vector plot shows a current amplitude of 0.9 A and phase angle between the two vectors of $\approx$80°.
- Inspection of the load (with the same load conditions used in the previous tutorial) current amplitude $|i_{s\alpha}|$ and the phase angle relative to the attenuated voltage waveform $u_{s\alpha}$ shows that these are equal to $|i_{s\alpha}| = 1.26$ A and 27.9°, respectively. This compares favorably with the current vector amplitude and

phase angle between the voltage and current vectors shown in Fig. 9.27 ("right" subplot). The load vector plot shows a current amplitude of 1.25 A and phase angle between the two vectors of $\approx 36°$.

An overall conclusion is that the results derived with the DEMO lab confirm the theoretical and simulated results shown and most importantly shown that the theory introduced in this chapter is applicable to actual machines.

Chapter 10
Direct Current Machines

10.1 Introduction

The genesis of the ideas required to build an electrical machine can be traced back to the discovery of electromagnetism by the Danish scientist Oersted in 1819–1820. Oersted discovered that a current in a wire could deflect a compass needle. Thus the connection between a current carrying conductor, a magnetic field, and a mechanical movement was established. A German chemist named Schweigger, who studied Oersted's experiment, found that if the wire carrying the current was wound into a coil then the deflection of the magnet was greatly increased. The Professor of Chemistry at Cambridge, Cumming, coined the term "Galvanometer" for this configuration and used it as a current detector. At around the same time Ampère developed a theory to support the observations made about current carrying coils in wire. In 1825 Sturgeon found that putting an iron core in the coil increased the magnetic field strength considerably for the same current.

Meanwhile, a laboratory assistant by the name of Faraday, working for the Royal Institution in England, developed what could be called the first electrical motor in September 1821. It was a crude device that caused a wire suspended in a basin of mercury with a vertical magnet, to rotate around the magnet when current flowed through the wire and mercury. It produced continuous motion.

In 1837 Davenport filed a patent for an electro-magnetic engine. It incorporated a crude commutator and a multi-pole stator with electromagnets rather than permanent magnets. He could use this machine to drill 6 mm holes in wood and steel. Gramme introduced a ring type rotor configuration in 1870, which in turn led to a rotor concept introduced by Von Hefner in 1873, which is still in use today.

Electronic supplementary material The online version of this chapter (doi: 10.1007/978-3-319-29409-4_10) contains supplementary material, which is available to authorized users.

© Springer International Publishing Switzerland 2016

A. Veltman et al., *Fundamentals of Electrical Drives*, Power Systems,
DOI 10.1007/978-3-319-29409-4_10

The direct current (DC) machine is still in use given that it has certain advantages in terms of controllability and low manufacturing costs. In household appliances and automotive applications series DC machines or universal machines (these run on both DC or AC) are often used for the reasons mentioned above. However, the development of adjustable speed drives has demonstrated that the same type of performance can also be achieved with, for example, an asynchronous machine in combination with a power electronic converter. Asynchronous machines are considerably cheaper to manufacture at most power levels and very much easier to maintain than a DC machine.

In this chapter we will consider the various DC machine configurations in use, which also including those which make use of permanent magnets. The operating principle of these machines as well as the steady-state characteristics will be discussed. In addition, a dynamic model of this machine will be introduced, which will be verified against a "demonstration" example in the trial section of this chapter.

10.2 Machine Configuration

A simple two pole stator of a DC machine is shown in Fig. 10.1. The field winding consists of an N turn concentrated winding of which each half is wrapped around a pole. The two parts of the winding are connected in series and attached to a DC power supply. The frame is in this case also the yoke-part of the magnetic flux path. The use of a field winding gives us the ability to control the magnetic flux in the circuit in terms of amplitude and polarity. Permanent magnets are often used to replace the field winding which leads to a more compact and efficient machine. However, we loose in practical terms one degree of freedom, as we are now unable to

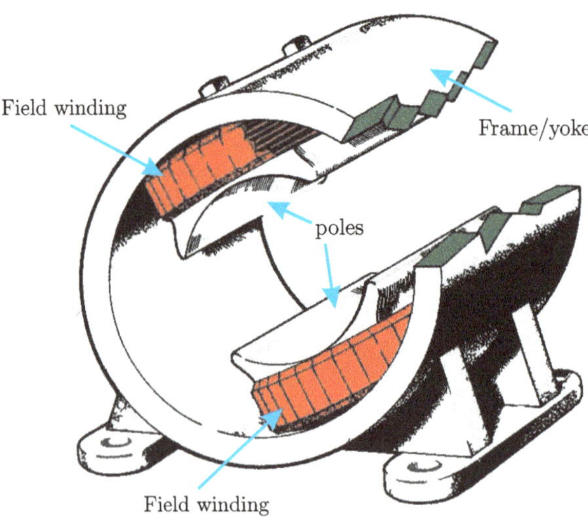

Fig. 10.1 DC machine stator

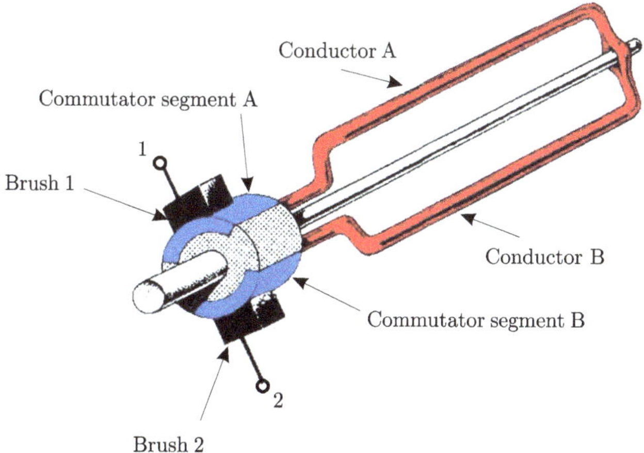

Fig. 10.2 Two segment/conductor DC machine armature

control (in electrical terms) the flux magnitude during operation. We also loose the potential of operating the machine on an AC source, therefore universal machines (AC/DC) always have a field winding and no permanent magnets.

A very simple example of a DC rotor, more commonly referred to as the "armature", is given in Fig. 10.2. Said figure shows the same single turn winding introduced for the synchronous machine. In this case, the slipring/brush combination is replaced by the so-called commutator. This commutator consists, for this simple rotor (armature), of two brushes and two commutator segments. Segment 1 and 2 are connected to coil which is formed by conductors A and B. The "conductor" part of the coil is exposed to the magnetic field due to the field winding (or magnets) and is therefore instrumental for torque production. The brushes are connected to a direct current (DC) power supply. The purpose of the commutator is to reverse the current polarity in the armature winding every half revolution in this case. For example, in the case shown, conductor A is connected via segment 1 and a brush to terminal 1. Likewise side B is connected via segment 2 and a brush to terminal 2. When we rotate the rotor by 180°, conductor A will be electrically connected to terminal 2 and conductor B to terminal 1. This means that a current reversal in the winding will take place twice during one period.

10.3 Operating Principles

The operating principles of this machine are discussed with the aid of Fig. 10.3, which is based on the armature model shown in Fig. 10.2. Furthermore a set of magnetic poles have been added to represent the magnetic field ψ_{m} due to a field winding or permanent magnetics. A current source with amplitude i_{a} is connected to the armature via the brushes. Also shown in Fig. 10.3 are two conductors (located on the rotor) which are exposed to a magnetic field that is represented by the space

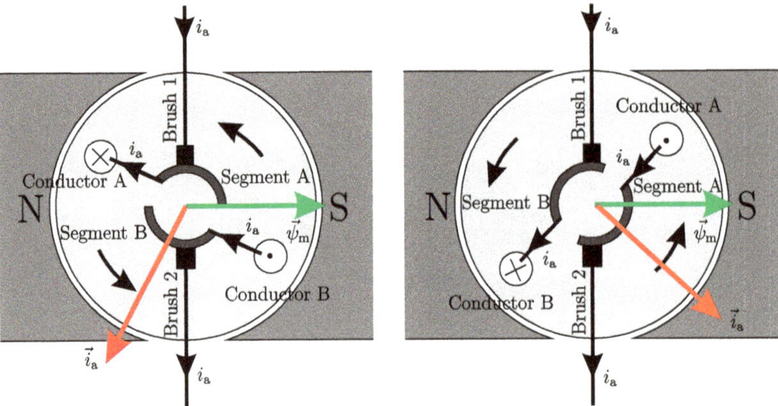

Fig. 10.3 Simplified DC machine with two segments/conductors

vector $\vec{\psi}_m$. The flux vector is aligned with the real axis of an αβ coordinate frame. Both conductors carry a current i_a which due to the presence of the magnetic field causes Lorentz forces (as shown in Fig. 1.8) to rotate in a clockwise direction. Torque production can also be shown with aid of current and flux space vectors (see Chap. 7), hence the current in the two conductors can be represented by the current vector $\vec{i}_a$, in which case the torque is defined in Eq. (7.13), with $\vec{i} = \vec{i}_a$. Furthermore the factor 3/2 in this expression must be removed, because we are NOT dealing with a three-phase system. Consequently, the torque for this simplified machine with a 2 segment commutator can be written as

$$T_e = -\psi_m \, i_a \, \sin \rho_i \tag{10.1}$$

where ψ_m and i_a represent the amplitudes of the flux vectors $\vec{\psi}_m$ and $\vec{i}_a$, respectively. Furthermore, the angle between said vectors is defined as ρ_i which is tied directly to the orientation of the rotor. Maximum torque occurs with $\rho_i = -90°$, i.e., with the current vector oriented along the negative β axis. Zero torque occurs when flux and current vectors are aligned ($\rho_i = 0°$, $\rho_i = -180°$), which is also the ideal moment to commutate the current in the armature as may be observed from Fig. 10.3. Observation of the commutation process shown in Fig. 10.3 shows that the current polarity in the conductors changes due to the presence of the brush/commutator assembly. This implies that the current space vector will stay in quadrants 3 and of the αβ coordinate system. Hence the vector will rotate clockwise starting at $\rho_i = -180°$ to $\rho_i = 0°$ at which point commutation will ensure the (near instantaneous) transition $\rho_i = 0 \rightarrow -180°$. The torque associated with this process, as defined in Eq. (10.1) is shown in Fig. 10.4, subplot (a), with $\psi_m = 1.0\,\text{Wb}$ and

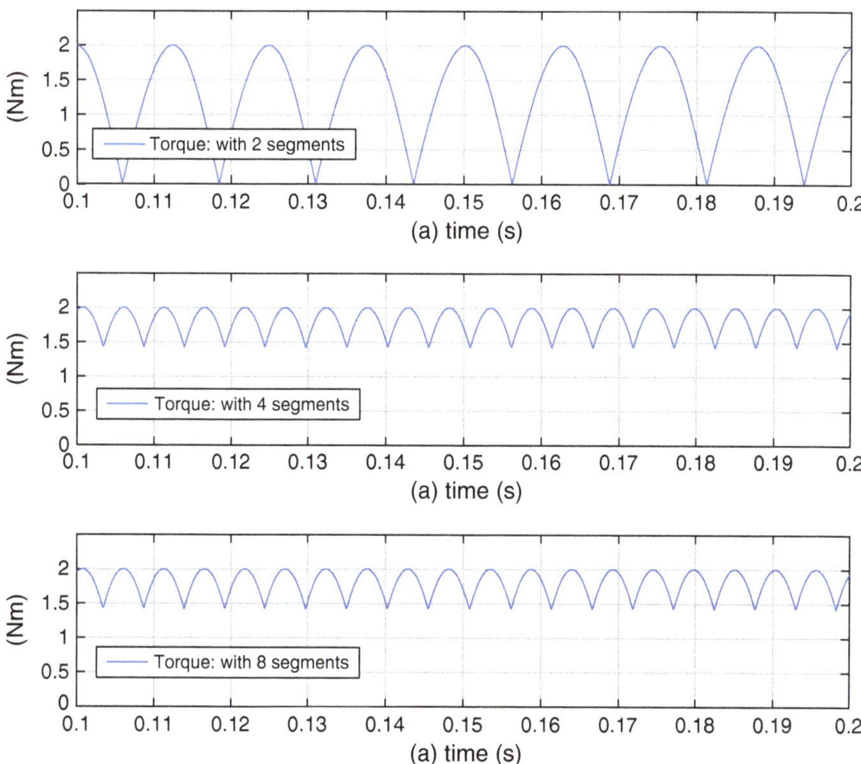

Fig. 10.4 Torque production with 2, 4, and 8 segmented commutator

$i_a = 2.0\,\mathrm{A}$. Clearly observable from Fig. 10.4, subplot (a) is that the torque is not constant. By adding more conductors and segments the effect of commutation of any two conductors becomes less influential in which case the armature current vector will deviate less from the negative β axis. The two DC machine configurations shown in Fig. 10.5 both utilize a 18 segment commutator. Both machines use the same armature but excitation is provided by either a field winding or permanent magnets. Readily observable is that a more compact machine can be realized when permanent magnets are used, however, the disadvantage is that the flux ψ_m cannot be changed. Torque ripple will also reduce as segment numbers are increased as may also be observed from Fig. 10.4, subplots (b) and (c), where use is made of a commutator with 4 respectively eight segments. With sufficient segments present the current vector will remain firmly aligned with the negative β axis, i.e., $\rho_i = -90°$ in which case Eq. (7.13), with $\vec{i} = \vec{i}_a$ (without the factor $3/2$) is reduced to

$$T_e = \psi_m \, i_a \tag{10.2}$$

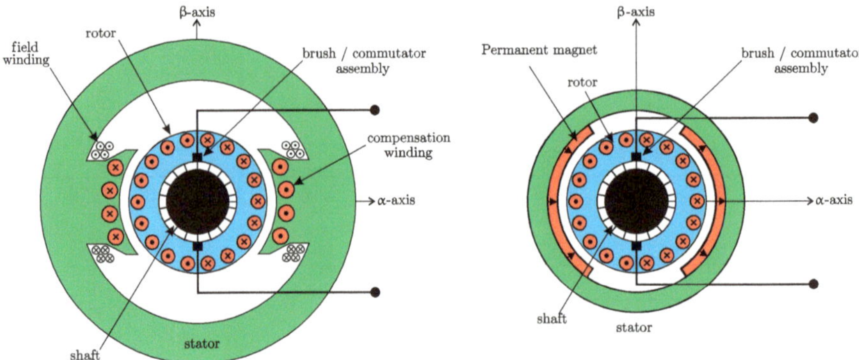

Fig. 10.5 DC machine with field winding (*Left*) and permanent magnet excitation

which for the chosen flux and current value used to derive Fig. 10.4 corresponds to $T_e = 2.0\,\text{Nm}$. Equation (10.1) shows that the torque produced by the DC machine can be independently controlled by the flux DC current amplitude i_a. For example, a torque step of the machine can be realized by an appropriate armature current step.

In the sequel to this section the reader is alerted to the presence of the so-called compensation winding winding in Fig. 10.5, that is, located within the stator poles and also carries the armature current i_a. The current space vector due to the compensation winding is purposely chosen to be in the OPPOSITE direction of the space vector due to the armature. The reason for this is that the armature, which also has inductance L_a, will produce a flux $\vec{\psi}_a$ (which also point in the negative β direction) that must be added to the flux vector $\vec{\psi}_m$. The effect of this so-called armature reaction is that a slight misalignment of the main flux vector $\vec{\psi}_m$ relative to the α axis will occur, that can negatively impact the commutation process, i.e., brush sparking can take place. The presence of the compensation winding, which is only found on relatively large machines, produces a magnetic flux component that reduces the armature flux component. This ensures that the main flux vector $\vec{\psi}_m$ remains aligned with the α axis, even under large armature current conditions, which in turn insures that the commutation process is not hindered.

10.4 Armature Based Voltage Source Model

In this section we will consider the development of a dynamic model of either machine shown in Fig. 10.5. A suitable starting point for this analysis is Fig. 10.6 that shows a symbolic model of the machine, which is connected to a current source with magnitude i_a. Motor operation is assumed, with the machine operating at a speed ω_m, with a torque of T_e. The armature of the machine shown is assumed to have zero resistance and inductance. Under these conditions an induced voltage

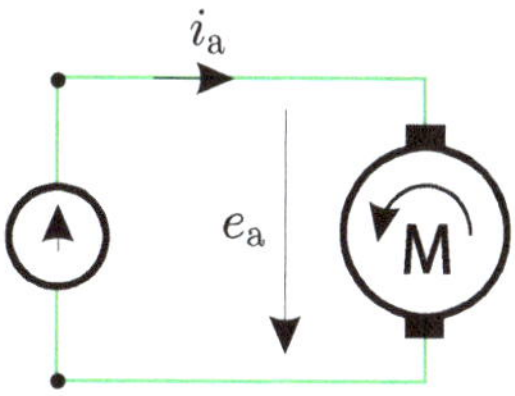

Fig. 10.6 Symbolic armature model of machine connected to a current source DC

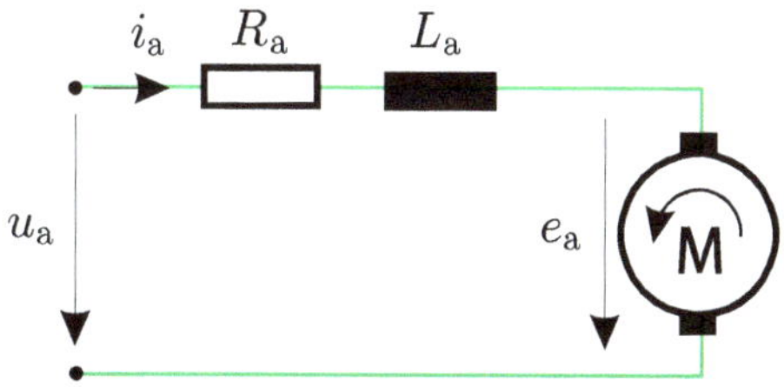

Fig. 10.7 DC machine model with armature resistance R_a and armature inductance L_a

e_a will be present across the brush terminals, which can be determined using the input/output power balance for this machine. Electrical input power to this machine is equal to $p_{in} = e_a i_a$, whereas the output power is given as $p_{out} = T_e \omega_m$. Subsequent evaluation of the power balance using the variables introduced gives

$$e_a = \frac{T_e \omega_m}{i_a} \tag{10.3}$$

which after substitution of expression (10.2) leads to

$$e_a = \psi_m \omega_m \tag{10.4}$$

Hence the EMF generated by the machine is proportional to the flux-linkage ψ_m (due to the field winding or permanent magnets) and the shaft speed ω_m. In reality the armature has resistance R_a and inductance L_a which must be added to the symbolic armature model introduced in Fig. 10.6. The full symbolic model of the machine is shown in Fig. 10.7. On the basis of Fig. 10.7 the terminal equation for the machine connected to a voltage source u_a can be derived, which is of the form

$$u_a = R_a i_a + L_a \frac{di_a}{dt} + e_a \tag{10.5}$$

The corresponding generic model of the DC machine as shown in Fig. 10.7 is directly based on the use of Eqs. (10.5), (10.4), and (10.2), respectively. Also added to the generic module is the mechanical equation set (7.14) and a load torque/speed module $T_l(\omega_m)$. The flux-linkage ψ_m input for this machine can be provided by either a field winding or permanent magnet. Both these options will be entertained in the next section.

10.5 Steady-State Characteristics

It is instructive to consider the steady-state characteristics of the machine. The variables which exist under steady circumstances are given in the form $\bar{x}$. For example, the steady-state speed would appear as $\bar{\omega}_m$. The steady equation set is directly derivable from the armature terminal equation (10.5), torque equation (10.2), EMF equation (10.4), and mechanical equation (7.14a) which leads to

$$\bar{u}_a = R_a\bar{i}_a + \psi_m\bar{\omega}_m \tag{10.6a}$$

$$\bar{T}_e = \bar{i}_a\psi_m \tag{10.6b}$$

$$\bar{T}_e = \bar{T}_l \tag{10.6c}$$

The equation set emphasizes that under steady-state conditions the electro-magnetic torque produced by the machine is equal to the load torque. The equation set is given for the case that the field is provided by a permanent magnet. However, as pointed out earlier, the magnet can and often is replaced by a field winding. Hence from the point of considering the characteristics we can assume that the flux ψ_m can be set to different values.

The torque versus speed characteristic can be found with the aid of Eq. (10.6) which after some manipulation gives

$$\bar{T}_e = \frac{\psi_m}{R_a}\bar{u}_a - \frac{\psi_m^2}{R_a}\bar{\omega}_m \tag{10.7}$$

Observation of Eq. (10.7) shows that the torque speed curves are in fact linear functions with a gradient given by $-\psi_m^2/R_a$. Furthermore, the no-load shaft speed is given by $\bar{\omega}_{mo} = \bar{u}_a/\psi_m$.

In terms of changing the shaft speed it is possible to resort to the so-called armature voltage control. In this case the voltage $\bar{u}_a$ is varied. The effect of changing the armature voltage is shown in Fig. 10.8.

From Fig. 10.8 it can be observed that the shaft speed is reduced as the load torque is increased. The sensitivity of speed variations to load changes is dependent on the gradient of the torque speed curve. Machines with a lower armature resistance will be less sensitive to this effect. Note that an increase in torque will also

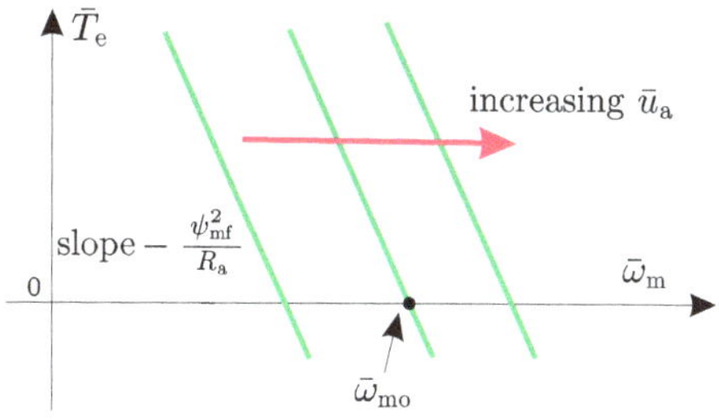

Fig. 10.8 Steady-state torque speed curves with armature voltage control

Fig. 10.9 Steady-state torque speed curves with field flux control

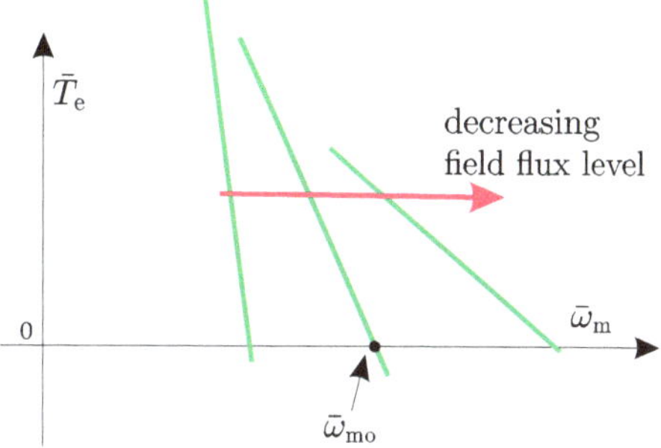

increase the armature current given that $\bar{i}_a = \bar{T}_e/\psi_m$. The maximum load is therefore constrained by the condition $\bar{i}_a \leq \bar{i}_a^R$, where $\bar{i}_a^R$ represents the rated armature current.

Replacing the permanent magnet with a field winding provides an extra degree of freedom which comes at the price of a larger volume of the machine (when compared to a permanent magnet machine of the same rating) and the need to excite the field winding.

The effect of a variable field flux level on the torque speed may be again analyzed with the aid of Eq. (10.7), when we vary ψ_m and keep $\bar{u}_a$ constant. Figure 10.9 shows how the torque speed curves are effected when controlling the speed of the machine with the aid of a field winding.

The machine configuration which we have discussed up to now is referred to as "separately excited" given that the field flux can be chosen (or varied) independently of the armature variables. For machines which utilize a field winding it is possible to electrically connect the latter in series or parallel with the armature winding. If the field winding is connected in parallel the machine is referred to as "shunt-wound." The third option, which is commonly used for traction drives and starter motors for combustion engines, has the field winding in series with the armature winding in which case the machine is referred to as "series-wound." The field flux level is then proportional to the armature current, hence $\psi_m = k_f \bar{i}_a$ where k_f is a constant which is largely determined by the winding arrangement of the machine. In a series-wound machine the field winding has a relatively low number of turns, but uses a larger wire diameter (to carry the armature current) when compared with the configurations indicated above. The torque for this type of machine is of the form

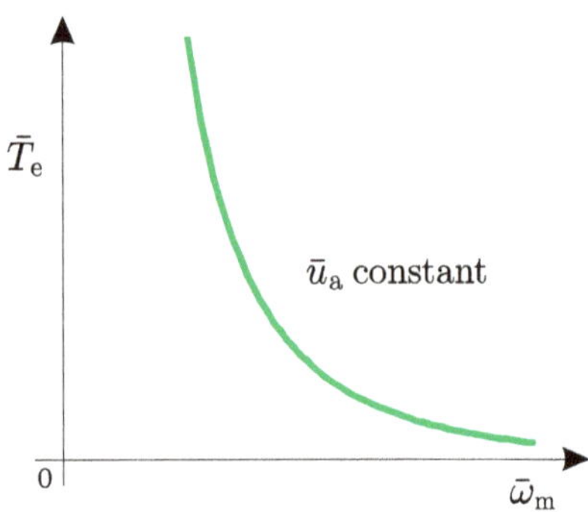

Fig. 10.10 Steady-state torque speed curves with series connected field winding

$$\bar{T}_e = k_f \,\bar{i}_a^2 \tag{10.8}$$

The torque speed curve for the voltage fed series connected machine can be found by substitution of Eq. (10.8) into Eq. (10.6), in which ψ_m is replaced with $\psi_m = k_f \bar{i}_a$ which gives

$$\bar{T}_e = k_f \frac{\bar{u}_a^2}{(R_a + k_f \bar{\omega}_m)^2} \tag{10.9}$$

A typical torque speed for this type of machine is given in Fig. 10.10. As with the previous cases the torque upper limit is determined by the rated current $\bar{i}_a^R$ of the machine. It is emphasized that motor operation with the series-wound machine is advantageous for traction drives, given that low speed operation yields a high starting torque. However, care should be taken to ensure that the load is not removed from this machine given that there is no finite "no-load" operating speed (see Fig. 10.10). The series machine is also commonly used in electrical appliances where it is often referred to as a "universal machine" given the fact that the supply can be either AC or DC. The use of an AC machine is possible because the supply voltage is proportional to the square of the voltage [see Eq. (10.9)]. This implies that, for example, a sinusoidal supply voltage will provide a non-zero torque value.

10.6 Tutorials

10.6.1 Tutorial 1: PLECS Based Model of a Separately Excited DC Machine

The aim of this tutorial is to build a dynamic, PLECS based, DC model which can be connected to a voltage source. The model is to be evaluated for transient and steady state conditions.

The motor module parameters are set to the values indicated in Table 10.1 given that these correspond to the data of the machine used in the "demonstration lab" introduced at the end of this section. A PLECS sub-module is to be build on the basis of the generic diagram for this machine as given in Fig. 10.11. An example of a possible implementation is given in Fig. 10.12. The dynamic simulation is to be undertaken with the aid of the PLECS model given in Fig. 10.13. The machine is to be studied by considering a start up sequence from zero shaft speed with a mechanical load connected. The load torque/speed curve is assumed to be "constant" (hence speed independent) function, which can be set in the load module. The load module settings must be set to produce a load torque of $80 \cdot 10^{-3}$ Nm. Electrically the machine is to be connected to a DC voltage supply source of 20 V at $t = 0$. Add a set of display modules as indicated in Fig. 10.13, so that we are able to view the shaft speed (in rpm), shaft torque, and armature current at the end of the simulation. Run the simulation for 0.5 s and add a "Scope" module as indicated in Fig. 10.12 so that the results can be observed and processed in MATLAB. In addition, two numerical display modules should be provided so that the torque and speed at the end of the simulation (when the drive is operating under steady-

Table 10.1 DC machine parameters

Parameters		Value
Armature inductance	L_a	7.35 mH
Armature resistance	R_a	5.0 Ω
Field flux (due to PM)	ψ_m	95.0 mWb
Inertia (estimated)	J	80.0 μkg m^2
Initial rotor speed	ω_m^0	0 rad/s

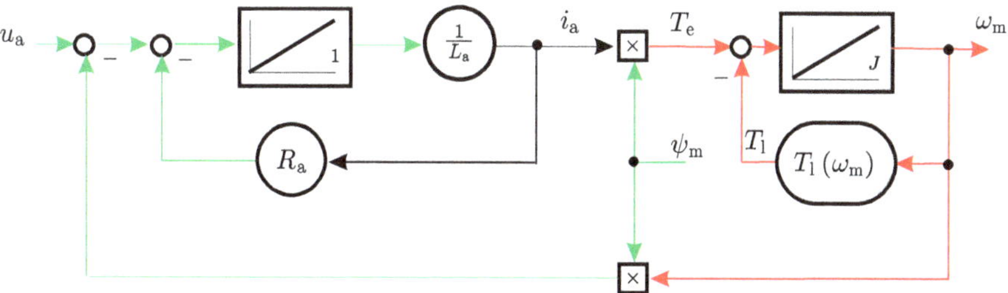

Fig. 10.11 Generic dynamic DC machine model, with mechanical load

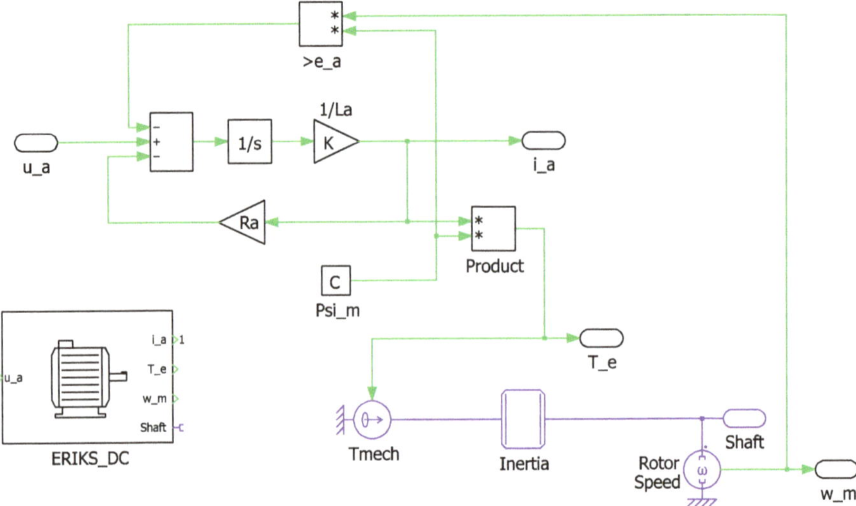

Fig. 10.12 PLECS model: separately excited brushed DC machine model

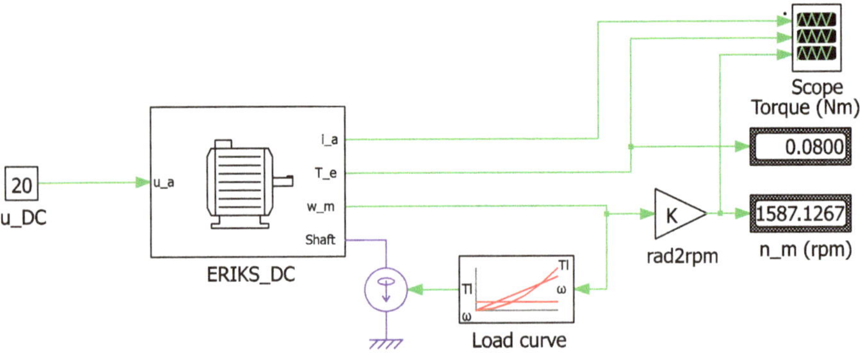

Fig. 10.13 PLECS model: dynamic simulation with general machine

state conditions) can be derived. An example of the Scope results which should appear with the chosen armature voltage, field flux (due to the magnets in this case), and load module settings is given in Fig. 10.14. Observation of the results given in Fig. 10.14 shows that a peak transient current of approximately 3.8 A appears during startup. Accordingly, a torque peak will also be present given that torque is directly proportional to current and flux [see Eq. (10.2)]. The two numerical displays show the torque and shaft speed present at the end of the simulation, i.e., machine operation under steady-state conditions.

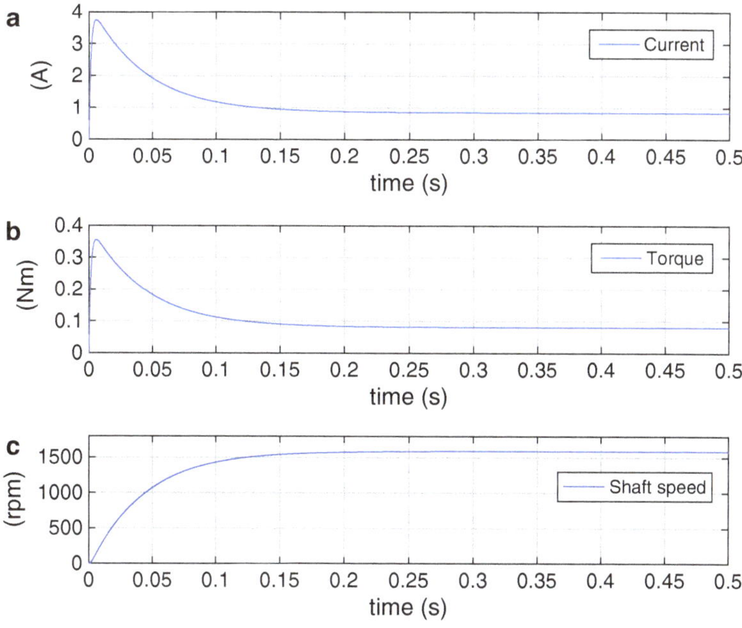

Fig. 10.14 Direct-on-line (DOL) start of a separately excited DC machine

10.6.2 Tutorial 2: Separately Excited DC Machine with Variable Load, Supply Conditions

To test the dynamic model according to tutorial 1 with respect to its correct functioning we can observe the display module values given in the PLECS model (see Fig. 10.13) at the end of a simulation run. Set the run time for your simulation to 0.5 s, to ensure that steady-state operation is reached at the end of each simulation run. Vary the load reference torque level from $80 \cdot 10^{-3}$ to $0 \cdot 10^{-3}$ Nm in five steps and record the display readings. Rerun your simulation after each reference torque change. Redo this tutorial for two different armature voltages, namely: $u_a = 17.5$ V and $u_a = 15.0$ V. An example of the steady-state results obtained with the PLECS simulation is given in Table 10.2.

The data obtained from these simulation runs should be compared against the results calculated with the aid of the steady state analysis given in Sect. 10.5. Calculate and plot the torque versus speed curves for the machine over a speed range from $0 \rightarrow 2100$ rpm with an armature voltage of $u_a = 20$ V, $u_a = 17.5$ V, and $u_a = 15$ V, respectively. Build an M-file to show these performance curves and add the steady-state torque versus speed data obtained from your PLECS simulations.

Figure 10.15 shows how your results should appear for the three armature voltage levels chosen. Also shown in this figure by way of discrete data points, are the results obtained via the PLECS model. The results confirm the qualitative analysis as

Table 10.2 Steady-state results PLECS model with variable armature voltage

		n_{m} (rpm)		
T^{ref} (10^{-3} Nm)	T_{e} (10^{-3} Nm)	$U_{\mathrm{a}} = 20.0$ V	$U_{\mathrm{a}} = 17.5$ V	$U_{\mathrm{a}} = 15.0$ V
80	80	1587.12	1335.83	1084.53
60	60	1692.93	1441.64	1190.34
40	40	1798.74	1547.44	1296.15
20	20	1904.55	1653.25	1401.96
0	0	2010.36	1759.06	1507.77

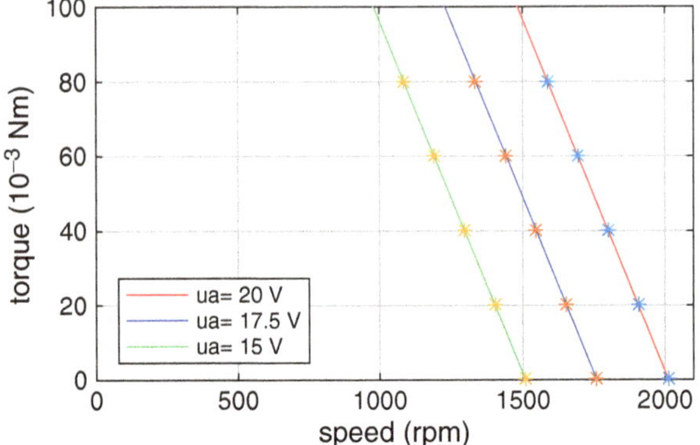

Fig. 10.15 MATLAB results: torque/speed curves, varying armature voltage

indicated in Fig. 10.8. Observe that the gradient of the torque/speed characteristics is NOT affected by changes to the armature voltage. The reason for this is that said gradient is proportional to the square of the field flux and inversely proportional to the armature resistance [see Eq. (10.7)].

An example of an M-file which can produce the results shown in Fig. 10.15 is as follows:

M-file Code

```
%Tutorial 2, chapter 10
close all
clear all
%results from PLECS model (constant load)
Tem=[80 60 40 20 0];%mill-Nm
nm1=[1587.12 1692.93 1798.74 1904.55 2010.36]; %Ua=20
nm2=[1335.83 1441.64 1547.44 1653.25 1759.06]; %Ua=17.5
nm3=[1084.53 1190.34 1296.15 1401.96 1507.77]; %Ua=15
%steady-state analysis DC machine
%machine parameters
Ra=5;                                    % armature resistance
La=7.35e-3;                              % armature inductance
```

```
psif=95e-3;                              %  PM field flux value
%%%%%%%%%%%%%%%%%%%%%%%%%%%5
ua=[20 17.5 15];                         % armature voltages
ns=[0:100:2100];                         % speed points for plot
TeA=[];
for i=1:3,
ws=2*pi*ns/60;                           % speed in rad/s
Te=psif/Ra*ua(i)-psif^2/Ra*ws;           % torque data
TeA=[TeA;Te];
end
%plot torque speed curves
plot(ns,TeA(1,:)*1000,'r')
grid
hold on
plot(ns,TeA(2,:)*1000,'b')
plot(ns,TeA(3,:)*1000,'g')
legend('ua= 20 V','ua= 17.5 V','ua= 15 V')
plot(nm1,Tem,'*')
plot(nm2,Tem,'*')
plot(nm3,Tem,'*')
axis([0 2100 0 100])
xlabel('speed (rpm)')
ylabel('torque (mill-Nm)')
```

10.6.3 Separately Excited Variable Field Conditions

For machines which carry a field winding the option is present to change the field flux-linkage flux ψ_f by altering the field current i_f. The machine used in this tutorial section has a field which is provided by permanent magnets. Nevertheless it is instructive to consider how the torque versus speed curves of this machine will change as a result of increasing or decreasing the size of the magnets. The latter implies that the flux-linkage ψ_f will also change accordingly.

For this tutorial keep the armature voltage u_a=17.5 V constant and vary the field flux value. Consider three cases, namely $\psi_f = \psi_m * 1.2$, $\psi_f = \psi_m$, and $\psi_f = 0.8 * \psi_m$, respectively, where ψ_m represents the PM flux of the "original" machine as introduced in the previous two tutorials. An example of the steady-state results obtained with the PLECS model is given in Table 10.3.

The torque versus speed curves for this part of the tutorial should again be calculated using the theory presented in Sect. 10.5. The results from these calculations should be of the form given in Fig. 10.16. Also shown in this figure are the data points obtained via the PLECS model. The results according to Fig. 10.16 reinforce the quantitative analysis given in Sect. 10.5 and Fig. 10.9 in particular. Clearly observable from Fig. 10.16 is that weakening the flux of the machine allows operation at higher speeds for a given supply voltage.

The M-file which corresponds to this part of the tutorial is as follows:

Table 10.3 Steady-state results PLECS model with variable field flux

T^{ref} (10^{-3} Nm)	T_e (10^{-3} Nm)	n_m (rpm)		
		$\psi_f = 1.2 * \psi_m$	$\psi_f = \psi_m$	$\psi_f = 0.8 * \psi_m$
80	80	1171.98	1335.83	1536.55
60	60	1245.46	1441.64	1701.77
40	40	1318.94	1547.44	1867.00
20	20	1392.42	1653.25	2032.22
0	0	1459.90	1759.06	2197.44

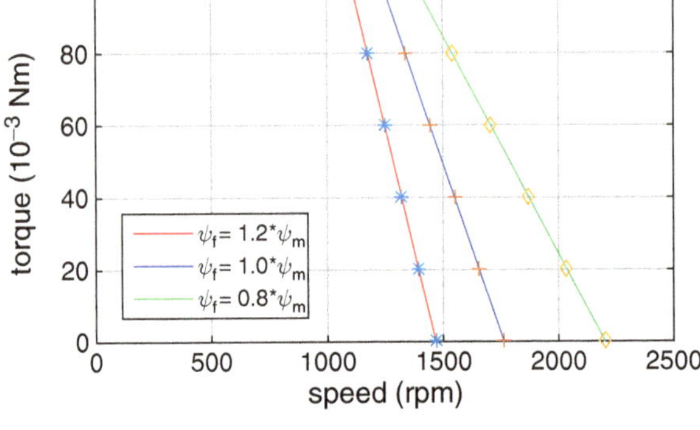

Fig. 10.16 MATLAB results: torque/speed curves, varying field flux

M-file Code

```
%Tutorial 3, chapter 10
close all
clear all
%results from PLECS model (constant load)
Tem=[80 60 40 20 0];%mill-Nm
nm1=[1171.98 1245.46 1318.94 1392.42 1465.90]; %psi_f=1.2*psi_m
nm2=[1335.83 1441.64 1547.44 1653.25 1759.06]; %psi_f=1.0*psi_m
nm3=[1536.55 1701.77 1867.00 2032.22 2197.44]; %psi_f=0.8*psi_m
%steady-state analysis DC machine
%machine parameters
Ra=5;                                % armature resistance
La=7.35e-3;                          % armature inductance
psim=95e-3;                          %  PM field flux value
%%%%%%%%%%%%%%%%%%%%%%%%%%%%5
ua=17.5;                             % armature voltage
psif=[1.2 1.0 0.8]*psim;             % flux values
ns=[0:100:2500];                     % speed points for plot
TeA=[];
for i=1:3,
ws=2*pi*ns/60;                       % speed in rad/s
Te=psif(i)/Ra*ua-psif(i)^2/Ra*ws;    % torque data
TeA=[TeA;Te];
end
```

```
%plot torque speed curves
plot(ns,TeA(1,:)*1000,'r')
grid
hold on
plot(ns,TeA(2,:)*1000,'b')
plot(ns,TeA(3,:)*1000,'g')
legend('\psi_f= 1.2*\psi_m','\psi_f= 1.0*\psi_m','\psi_f= 0.8*\psi_m')
plot(nm1,Tem,'*')
plot(nm2,Tem,'+')
plot(nm3,Tem,'d')
axis([0 2500 0 100])
xlabel('speed (rpm)')
ylabel('torque (mill-Nm)')
```

10.6.4 Demo Lab 3: Voltage Controlled DC Drive

For this demonstration the LVACIMTR (EMsynergy) asynchronous machine which was connected to the aft (furthest away from the USB connector) BOOSTXL-DRV8301 module shown in Fig. 8.29 is replaced by the DC machine [11] introduced in the tutorials above. A voltage controller is used [10] which can generate a variable DC voltage supply. For this purpose, phases A and B of the aft boost converter are combined to form the so-called H-bridge [10], which is also discussed in the next chapter. This controller implemented in embedded software VisSim [14] is used in Fig. 10.17, where the user is able to set the armature voltage via slider `uA_ref`. The LVSERVOMTR (Teknic) PM machine is connected to the forward boost converter and the resulting drive functions as a dynamometer for the DC machine. The torque of this machine is controlled via slider `Iq_ref`. The drive setup was used to measure the variables i_{Arm} (armature current) and n_{m} (shaft speed) under load conditions as discussed in tutorial 2 (see Sect. 10.6.2). The results derived from the demonstration setup have been processed to show the shaft torque versus speed for DC supply voltages of 20 V and 15 V, respectively. For comparison purposes

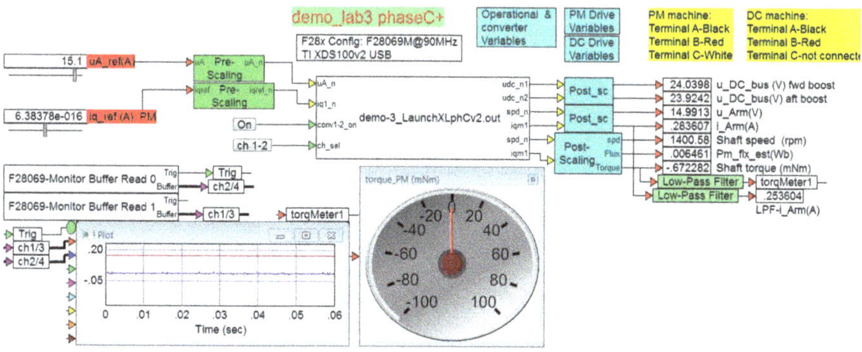

Fig. 10.17 VisSim [14] based "H-drive" Demonstration controller [10]

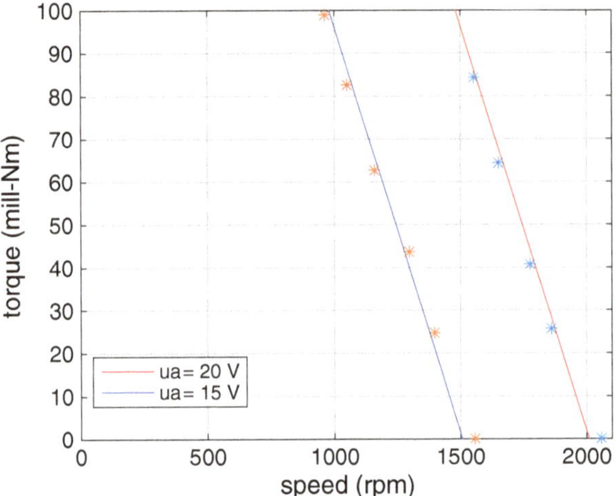

Fig. 10.18 MATLAB results: shaft torque versus shaft speed characteristics

the theoretical steady-state torque/speed curves as calculated in tutorial 2 have also been added. Note that the experimental drive has mechanical friction. This implies that the DC machine must provide the torque needed to overcome this friction in addition to any load imposed by the PM machine. The magnitude of the friction torque present can be found by using the PM machine as a motor and adjusting the slider `Iq_ref` to a value where the armature current reduces to zero. Under these (constant speed) conditions the torque meter will show the friction torque which is in this case approximately $20 \cdot 10^{-3}$ Nm.

The following comments/observations of the results according to Fig. 10.18 can be made:

- There is good agreement between the measured results (shown by discrete data points "$*$") and those calculated using a steady-state analysis.
- The "measured" torque was calculated using the measured armature current which was then multiplied by the known magnet flux ψ_{m} of the machine.

An overall conclusion is that the results derived with the DEMO lab confirm the theoretical and simulated results shown and most importantly shown that the theory introduced in this chapter is applicable to actual machines.

Chapter 11
Pulse Width Modulation and Current Control for DC Drives

11.1 Introduction

In this chapter we will look at some basic drive implementations with a DC machine as discussed in Sect. 10.4. Our aim is to arrive at generic models of all major drive components (excluding the DC machine which has already been discussed) which we can then transpose to a PLECS type environment in the tutorials at the end of this chapter. A central topic of this chapter is the so-called pulse width modulation (PWM) which is used to realize uni-polar and bipolar voltage control of converters.

In the sequel of the chapter a so-called model based current control algorithm will be presented [2, 10, 12] which will allow precise torque control of the DC machine used. The techniques described here are fundamental not only to the DC machine but also to all the machines discussed in this book.

11.2 Single Phase Uni-Polar "Drive" Circuit

An elementary drive model as shown in Fig. 11.1 has almost the same structure as the general drive model given in Fig. 1.2 on page 4. In this example the mechanical "load" module has been removed. Furthermore, the power source is now shown in a two-wire configuration (+, −) which is helpful here because a symbolic implementation of the "converter" module is shown. The DC motor is represented as an "R-L-e_a" network as was discussed in Sect. 10.4 on page 285. The purpose of the modulator is to control the converter switch shown in Fig. 11.1 on the basis of a set-point given by the controller. A simple controller structure is also

---

Electronic supplementary material The online version of this chapter (doi: 10.1007/978-3-319-29409-4_11) contains supplementary material, which is available to authorized users.

© Springer International Publishing Switzerland 2016

A. Veltman et al., *Fundamentals of Electrical Drives*, Power Systems,
DOI 10.1007/978-3-319-29409-4_11

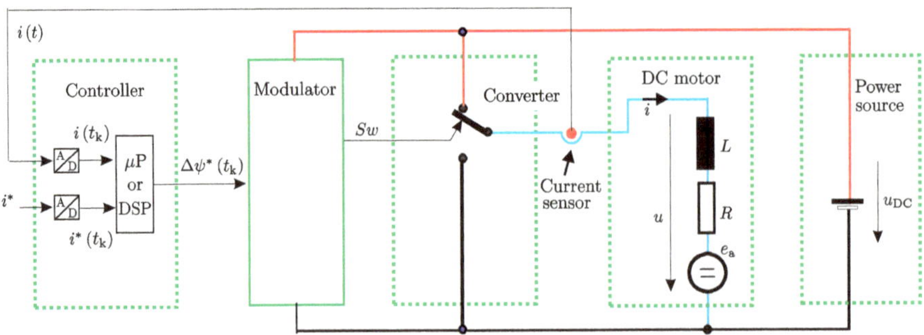

Fig. 11.1 Basic electrical drive

given in Fig. 11.1 which shows a micro-processor (μP or DSP), which is a digital computational element that implements the control algorithm of the drive. The input to the control module is the load current $i\,(t)$ which is obtained via a current sensor which measures the load current, i.e., the armature current of the DC machine in this case. A user input value i^* is also shown which represents the reference current level.

The aim of this drive circuit is to control the current in the motor in such a way that the reference current value matches the actual load current under all circumstances, i.e., transient as well as in steady state. A typical situation to be discussed is to apply a step change to the reference current and our aim is to ensure that the load current will match this step change, within the limits of the system. To achieve this aim we will need to initially discuss in some detail the functioning of the modules shown in Fig. 11.1. Afterwards we will develop a control structure which can be implemented in the micro-processor (μP or DSP) as to realize our task.

11.2.1 Power Source

A DC voltage source is assumed here which has a value of u_{DC}. The bottom side is set to 0 V which means that the upper wire (red) shown in Fig. 11.1 has a potential of u_{DC}. The voltage source is "uni-polar" which means that there is only one voltage level other than zero. At a later stage in this chapter we will replace the power source by a bipolar power source which gives us a positive and a negative supply voltage level with respect to 0 V. The term "uni-polar drive" reflects the ability to operate with a variable but single positive supply value.

11.2.2 Converter Module

The converter module shown in Fig. 11.1 consists of a single two-way switch. In reality such a switch is formed by two switches as shown in Fig. 11.2 which also

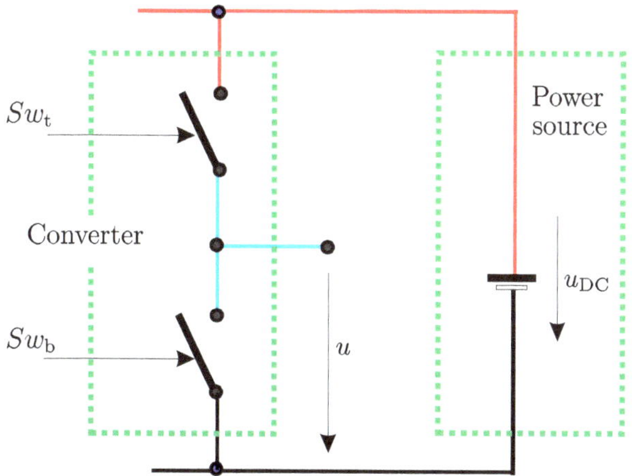

Fig. 11.2 Half-bridge converter with single voltage supply

gives the power source module. The switches are controlled by two logic signals $\mathrm{Sw_t}$ and $\mathrm{Sw_b}$ where logic 1 corresponds to a "closed" switch state and logic 0 to an "open" switch state. In this case there are four possible switch combinations of which states: $\mathrm{Sw_t} = 1$ and $\mathrm{Sw_b} = 1$ (both switches closed; "shoot-through" mode, this state should always be avoided) and $\mathrm{Sw_t} = 0$ and $\mathrm{Sw_b} = 0$ (both switches open; "idle" mode, normally used to disable the inverter output), both are not considered in the following part. The remaining two states are: $\mathrm{Sw_t} = 1$ and $\mathrm{Sw_b} = 0$ (top switch closed/bottom switch open) and $\mathrm{Sw_t} = 0$ and $\mathrm{Sw_b} = 1$ (top switch open/bottom switch closed). In the first case ($\mathrm{Sw_t} = 1$ and $\mathrm{Sw_b} = 0$), the converter output is connected to the positive supply line, i.e., $u = u_{\mathrm{DC}}$, while in the second case $\mathrm{Sw_t} = 0$ and $\mathrm{Sw_b} = 1$, the output line is connected to the lower supply line, i.e., $u = 0$. The two switches can therefore in symbolic form be replaced by a single two-way switch as shown in Fig. 11.1, where the logic signal Sw is used to control its state. The state $\mathrm{Sw} = 1$ corresponds to the switch in the "up" state, i.e., the output voltage is given as $u = u_{\mathrm{DC}}$. As expected the switch state $\mathrm{Sw} = 0$ corresponds to the switch in the "down" state, i.e., the output voltage is given as $u = 0$.

11.2.3 Controller Module

Today, the controller is in most cases digital. This means that the analog input variables, here in the form of the measured current $i(t)$ and user defined reference current $i^*(t)$, need to be converted to a digital form. We have therefore introduced in Fig. 11.1 a new building block in the form of an "analog-digital" (A/D) converter.

The function of the unit is readily shown with the aid of Fig. 11.3. Figure 11.3 shows an input function $x(t)$ to the A/D converter. The diagram shows an example

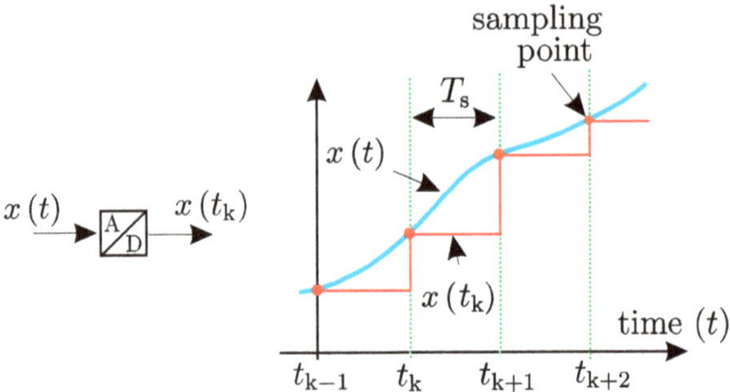

Fig. 11.3 A/D converter unit with example input/output waveforms

waveform together with a set of discrete time points t_{k-1}, t_k, and t_{k+1} where k can be any integer value. The difference in time between any two time points is constant and equal to the "sampling interval period" T_s. For drive systems the sampling time is in the order of 100-μs, 1 ms. The output of the converter module is such that at these time points the input is "sampled," i.e., the output is then set equal to the input value. Hence, at these "sampling points" the output changes to match the instantaneous value found at the input of the converter. The output is therefore held constant during the sampling time. For example, the output $x(t_k)$ represents the value of the input variable as sampled at the time mark t_k.

The A/D units are used to sample the measured and reference current values. These inputs, at for example t_k, are then used by the micro-processor or DSP to calculate an output variable known as the "reference average voltage per sample $U^*(t_k)$," which acts as an input to the modulator. We will define the variable $U^*(t_k)$ in the next section.

11.2.4 Modulator Module

The basic task of the modulator module is to control the switch or switches of the converter module in such a way that the condition according to Eq. (11.1) is met (within the constraint of this unit) for each sampling interval.

$$U^*(t_k) = U(t_k) \tag{11.1}$$

where $U(t_k)$ is known as the average voltage per sample value which is defined as

$$U(t_k) = \frac{1}{T_s}\int_{t_k}^{t_k+T_s} u(t)\,\mathrm{d}t \tag{11.2}$$

The term $u(t)$ shown in Eq. (11.2) represents the instantaneous voltage across the load (output from the converter) within a sample period in this case between sample points t_k, t_{k+1}. We have in the past [see Eq. (2.7)] commented on the fact that it is the incremental flux-linkage which controls the current in an inductance. For electrical loads where the time constant $\sigma = {}^{L}/_{R}$ of the load is deemed to be relatively large (as is normally the case for electrical machines) the incremental current is given as

$$\Delta i(t_k) \cong \frac{U(t_k)\,T_s}{L} \tag{11.3}$$

Expression (11.3) is significant as it shows that in regularly sampled systems with sampling time T_s the current change per sample is defined by the average voltage per sample value $U(t_k)$. Condition (11.1) in fact states that the modulator should set the converter switches during each sampling interval in such a way as to ensure that the reference average voltage per sample value at, for example, time t_k (as provided by the controller) matches the converter average voltage per sample value [as defined by Eq. (11.2)].

We will now consider two basic "single edged" modulation strategies by examining the converter average voltage per sample as a function of the switch on/off time within a sample interval $t_k \ldots t_{k+1}$. We will in the first instance make use of the converter configuration as shown in Fig. 11.1.

The so-called rising edge type modulation strategy calls for the switch Sw to be placed in the "up" (switch logical control level 1) position after a time t^r measured from the start of the sampling interval. The switch is placed in the "down" (switch logical control level 0) position at the end of each sampling interval. An example of the output voltage waveform which appears as a result of this modulation strategy is given in Fig. 11.4 for the sampling interval $t_k \ldots t_{k+1}$. Also shown in Fig. 11.4 is the average voltage per sample value $U(t_k)$ as a function of the rise time t^r and the switch state Sw of the converter. The variable t^r can change between zero and T_s. For a particular value t^r we can evaluate the average voltage per sample value by making use of Eq. (11.2) which in this case gives the function $U(t^r) = u_{DC}\left(1 - {}^{t^r}/_{T_s}\right)$, which is illustrated in Fig. 11.4. It is noted that this function repeats each sample interval and this gives us the possibility to find the required t^r value for a given average voltage per sample reference value.

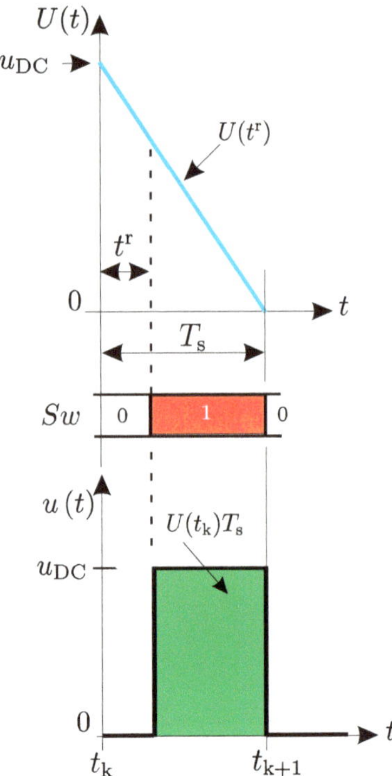

Fig. 11.4 Average voltage per sample and output waveforms: rising edge modulation

The basic algorithm for finding the rise time t^{r} is based on the use of Eq. (11.1). Basically, we compare for each sample interval the required reference value with the average voltage per sample function $U(t^{\mathrm{r}})$ and move the converter switch to the "up" position when the condition $U^* \geq U(t^{\mathrm{r}})$ is met. An example as given in Fig. 11.5 shows two consecutive sampling intervals where the reference average voltage per sample levels (as provided by the controller) is taken to be $U^*(t_{k-1})$ and $U^*(t_k)$, respectively. The switching point for the converter (which corresponds to the required t^{r} value) is identified by comparison with the $U(t^{\mathrm{r}})$ function for each sample interval. We note that the converter will provide the correct average voltage per sample value which means that the modulator will achieve its aim.

A further two observations of Fig. 11.5 are of interest. Firstly, this modulator–converter combination will provide an average voltage per sample value between zero and u_{DC}, which is why this converter topology is "uni-polar." Secondly, the required average voltage per sample value produced by the converter is realized by adjusting the *width* of the output voltage pulse for each sample interval. This is why this modulation strategy is known as PWM. So far, we have discussed a "single rising edge" PWM scheme which operates with a uni-polar converter.

The so-called falling edge type PWM modulation strategy calls for the switch Sw to be placed in the "up" (switch logical control level 1) position at the start of each sampling interval and moved to its "down" (switch logical control level 0) position

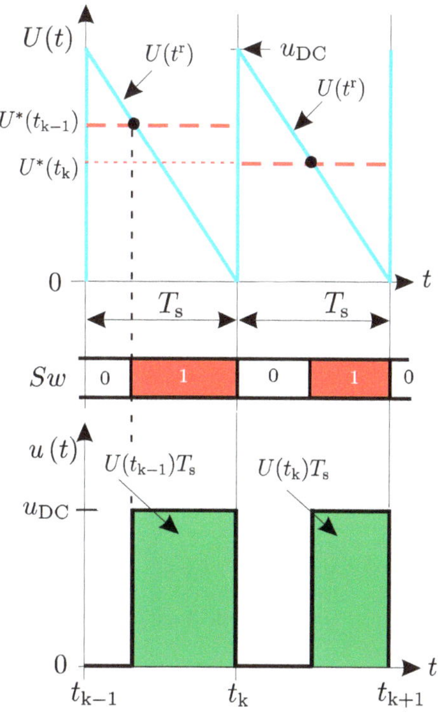

Fig. 11.5 Switch algorithm for rising edge modulation

after a time t^{f} (measured from the start of the sampling interval). An example of the output voltage waveform which appears as a result of this modulation strategy is given in Fig. 11.6 for the sampling interval $t_k \ldots t_{k+1}$. Also shown in Fig. 11.6 is the average voltage per sample value as a function of the fall time t^{f}. This variable can change between zero and T_{s}. For a particular value t^{f} we can evaluate the average voltage per sample by making use of Eq. (11.2), which in this case gives the function $U\left(t^{\mathrm{r}}\right) = u_{\mathrm{DC}}\, {}^{t^{\mathrm{f}}}\!/_{T_{\mathrm{s}}}$ which is also illustrated in Fig. 11.6 together with the switch Sw sequence.

The algorithm for finding the time t^{f} is again based on the use of Eq. (11.1). Basically, we compare for each sample interval the required reference value with the average voltage per sample function $U\left(t^{\mathrm{f}}\right)$ and move the converter switch to the "down" position when the condition $U^* < U\left(t^{\mathrm{f}}\right)$ is met. How this is achieved is illustrated in Fig. 11.7, which shows two consecutive sampling intervals where the reference average voltage per sample values (as provided by the controller) is taken to be $U^*\left(t_{k-1}\right)$ and $U^*\left(t_k\right)$, respectively. The switching point for the converter (which corresponds to the required t^{f} value) is identified by comparison with the $U\left(t^{\mathrm{f}}\right)$ function for each sample interval. We note that the converter will provide the correct average voltage per sample value which means that the modulator will achieve its aim. Note that in this case we have discussed a "single falling edge" PWM scheme which operates with a uni-polar converter. A generic implementation of a "falling edge" PWM strategy is given in Fig. 11.8. Shown in Fig. 11.8 are two

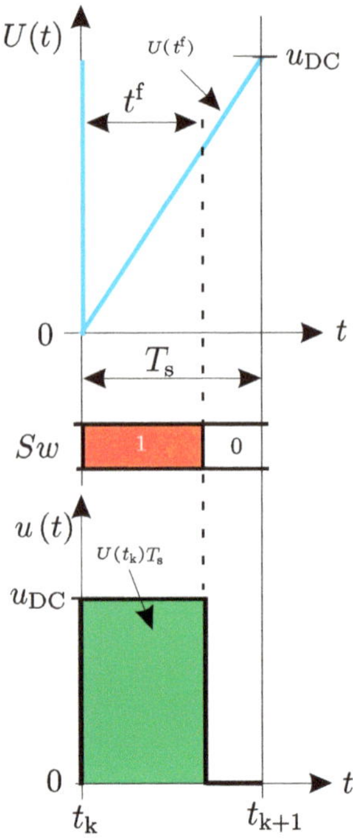

Fig. 11.6 Average voltage per sample and output waveforms: falling edge modulation

A/D modules which take the average voltage per sample reference value (from the controller) and the measured u_{DC} value from the converter module. We in fact use this value as to allow us to adjust the converter switch or switches as to accommodate voltage changes as will be discussed shortly. The sampled DC voltage is multiplied by a "saw tooth" function which is in fact the $U\left(t^{f}\right)$ function with the maximum value set to 1. We could have implemented this function directly in the "function generator" module shown. However, the chosen implementation allows us to take into consideration changes to the supply voltage u_{DC}. A summation module compares the voltage values and its output ε is used by a so-called comparator module. This is a new addition to our building block library and its function is given by Eq. (11.4).

$$\text{if } \varepsilon > 0 \quad \text{comparator output} = 1 \tag{11.4a}$$

$$\text{if } \varepsilon \leq 0 \quad \text{comparator output} = 0 \tag{11.4b}$$

In our case the output of the comparator is known as Sw and drives the converter switch. A logical level Sw $=$ 1 moves the converter switch to the "up" position and Sw $=$ 0 sets it to the "down" position. It is left to the reader to consider how Fig. 11.8 should be changed when a rising edge PWM strategy is to be implemented.

The modulator generic structure as given by Fig. 11.8 was specifically chosen to allow changes to the supply voltage which (within limits) will *not* affect the ability

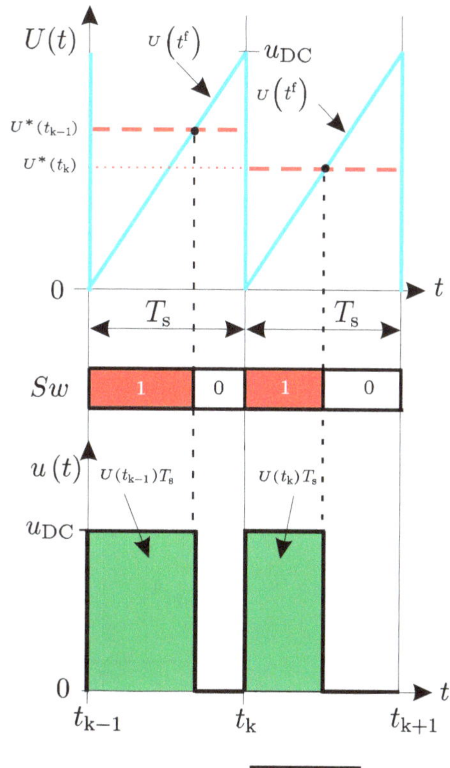

Fig. 11.7 Switch algorithm for falling edge modulation

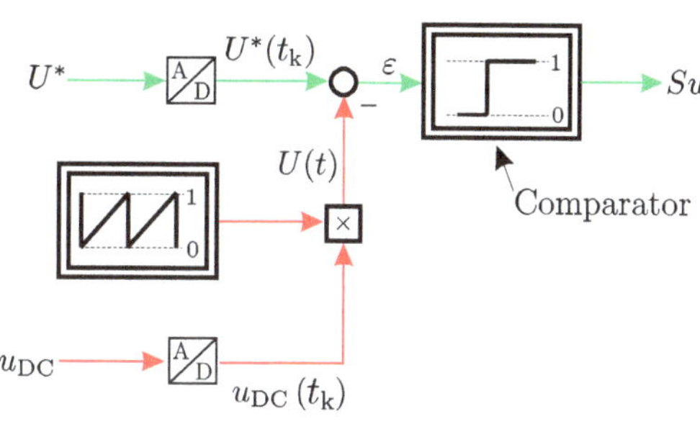

Fig. 11.8 Generic model of falling edge PWM

of the modulator/converter combination to meet condition (11.1). With the aid of Fig. 11.9, we will show how the modulator/converter combination is able to cope with a change in supply voltage. In this example, the average voltage per sample reference U^* is held constant. The supply voltage has been changed with time and this is reflected by the sampled DC bus voltage, which in the second sample is arbitrarily taken to be lower than in the first sample. The immediate effect is that the average voltage per sample function $U\left(t^{\mathrm{f}}\right)$ in the second sample will have a *lower* gradient than in the first. The consequence of this is that the fall time for the second sample is *increased*. Furthermore, the output waveform voltage in the second sample is reduced (because the converter supply was lowered). What in fact

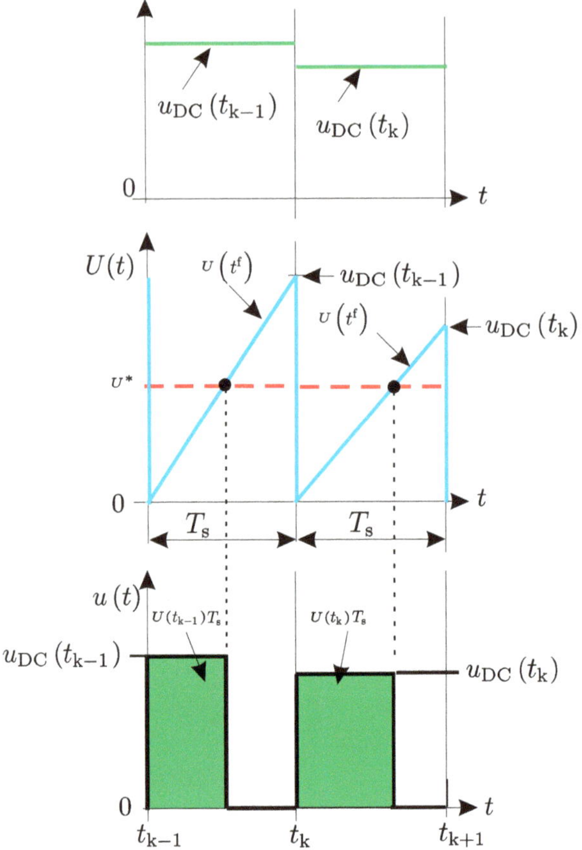

Fig. 11.9 Falling edge PWM, with change of supply voltage

has occurred is that the modulator has increased the fall time in the second sample to offset the reduced output voltage level of the converter. This means that the average voltage per sample value delivered to the load remains *unaffected* by the supply voltage change as long as the supply voltage is of sufficient magnitude.

At the conclusion of the discussion on single edged PWM we will look at an important modulation strategy known as "double edged" PWM. This modulation strategy combines the rising and falling edged PWM strategies into one. Basically this new modulation strategy alternates between the two single edged PWM options for every sample. For example, in the first sample we use rising edge PWM and in the next we use falling edge PWM. How this operates in practice is shown in Fig. 11.10. Two sample intervals are shown and in the first sample interval a rising edged PWM strategy (as given in Fig. 11.5) is shown. In the second sample a falling

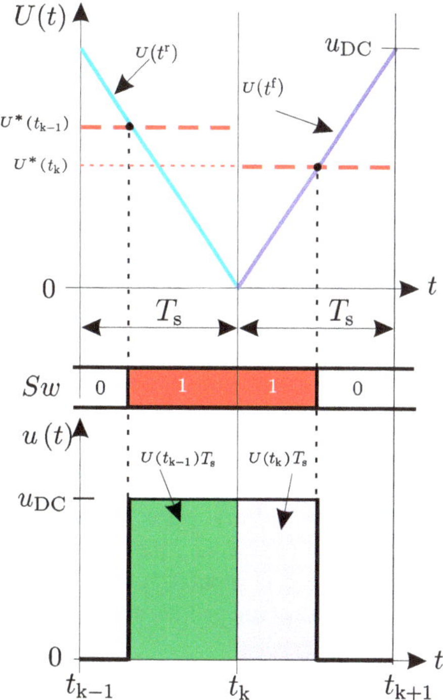

Fig. 11.10 Double edged PWM strategy, with a uni-polar converter

edge strategy is shown (as given earlier in Fig. 11.7). The reference average voltage per sample value for the second sample was arbitrarily set lower than the first sample. Also shown in Fig. 11.10 is the switch Sw sequence for the double edged PWM strategy. The generic module according to Fig. 11.8 is readily modified given that the "saw tooth" generator is now replaced by a triangular function (of unity amplitude). Several important observations are to be made with respect to this modulation strategy.

Firstly, the fundamental frequency f_{mod} of the new modulator is now changed from $1/T_s$ to $1/2T_s$, i.e., it is halved. Secondly, each so-called modulator period or carrier-wave period $1/f_{\mathrm{mod}}$ consists of two samples. This type of PWM strategy is in this text referred to as double edged PWM. However a variety of other names are found in the literature, for example, "center-aligned PWM with double update," "symmetric PWM" (referring to the symmetrical carrier wave) and also "asymmetrically sampled PWM" (referring to the sampling), or "sine-triangle modulation." It has been shown [4] that this type of modulation is superior to single edged PWM strategies, due to its double update frequency and the fact that the pulse's center of gravity does not depend on its width (for a DC set-point).

11.3 Half-Bridge Single Phase Bipolar Converter

The converter module shown in Fig. 11.1 consists of a single two-way switch, which in reality is formed by two switches as was shown in Fig. 11.2, which also gives the power source module.

The problem with the uni-polar circuit discussed in the previous section is that we cannot change the sign of the average voltage per sample quantity delivered (each sample) to the load. We can improve the situation by splitting the supply source u_{DC} and rearranging the two new supply sources as shown in Fig. 11.11. The load is in this case connected between the converter midpoint, i.e., to both switches and the 0 V (neutral) point shown in Fig. 11.11. The switches are controlled by two logic signals $\mathrm{Sw_t}$ and $\mathrm{Sw_b}$ where logic 1 corresponds to a closed switch state, similar to the half bridge in Fig. 11.2. The only difference with the explanation in Sect. 11.2.2 is a shift in output voltage by $u_{DC}/2$. For the half-bridge converter in Fig. 11.11, the state $\mathrm{Sw} = 1$ corresponds to the switch in the "up" state, i.e., the output voltage is given as $u = u_{DC}/2$. As expected the switch state $\mathrm{Sw} = 0$ corresponds to the switch in the "down" state, i.e., the output voltage is given as $u = -u_{DC}/2$. A modulator module with user input U^*, which is the commanded average voltage per sample value, controls both switches as will be discussed in the next section.

11.3.1 Modulation Strategy for the Half-Bridge Converter

Similar to the uni-polar modulator we can make a choice as to whether we use a single or double edged PWM strategy. The double edged strategy is preferable and will therefore be applied to this case. Note that the output waveforms will now change during each sample as shown in Fig. 11.12. Furthermore, the average voltage per sample versus rise and fall time must in this case be recalculated using

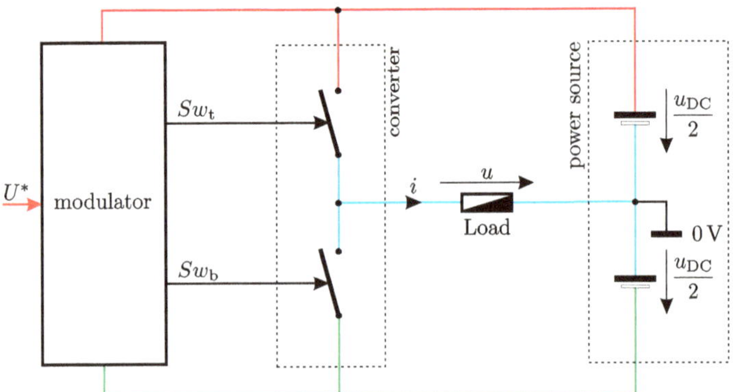

Fig. 11.11 Two-switch "half bridge" bipolar converter with power source and modulator

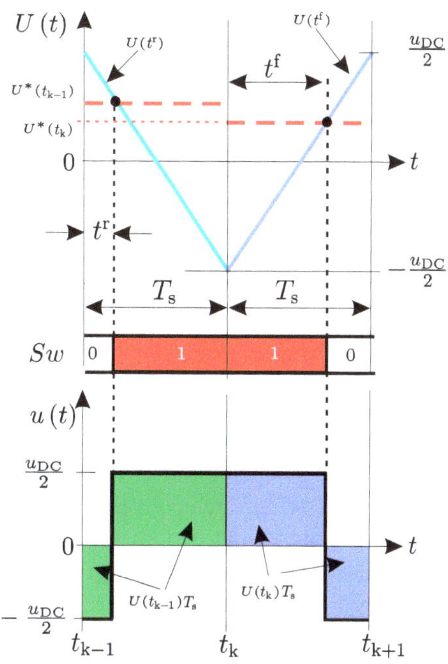

Fig. 11.12 Double edged PWM strategy, with half-bridge converter

Eq. (11.2). Figure 11.12 shows two samples, which correspond to one modulator period of operation. The first sample shows rising edge modulation and this means that the converter sets the load voltage to $u = {-u_{DC}}/2$ at the beginning of the sampling interval and after a time interval t^r the output voltage is changed to $u = {u_{DC}}/2$. Use of Eq. (11.2) allows us to find the average voltage per sample value versus rise time function, which is in this case of the form $U(t^r) = {u_{DC}}/2 - {u_{DC}\, t^r}/{T_s}$. In the second sample a falling edge PWM strategy is used which means that the converter switches the load voltage to $u = {u_{DC}}/2$ at the beginning of the sample interval and after a time interval t^f it is switched to $u = {-u_{DC}}/2$. Use of Eq. (11.2) allows us to find the average voltage per sample value versus fall time function which is in this case of the form $U(t^f) = {-u_{DC}}/2 + {u_{DC}\, t^f}/{T_s}$. The combination of both rising and falling single edged PWM strategies gives us the double edged PWM strategy. Also shown in Fig. 11.12 is the switch Sw state for the operating sequence shown. A comparison between the reference incremental flux value and the flux versus rise or fall time gives us the correct value for t^r and t^f, as to meet the condition as specified by Eq. (11.1) for each sample. Bipolar output voltage capability enables negative output voltages, but reduces the maximum positive output voltage in comparison with the uni-polar case (assuming the same total DC supply voltage). One main reason not to use bipolar half-bridge converters for low-frequency applications such as DC motors is the fact that energy is not only moved from the supply to the load, but also from the upper supply to the lower supply (the so-called supply-pumping). The always present supply capacitors can handle the effect of an AC output voltage and current, but low-frequency or DC output cannot be maintained by a half-bridge converter with a practical supply. For this reason most converters intended to drive DC loads are of the full-bridge (H-bridge) type. The two DC supplies in the bipolar converter needs to be able to handle bidirectional current flow especially when DC or low-frequency outputs levels are required. Such supplies are uncommon, hence mostly full-bridge or H-bridge converters are used.

11.4 Full-Bridge Single Phase Bipolar Converter

The problem with the half-bridge circuit discussed in the previous section is the presence of a split DC power supply. We can improve the situation by combining two half-bridge converters, which results in a so-called full- or H-bridge configuration as shown in Fig. 11.13. The load is in this case connected between the two half-bridge converter midpoints, i.e., between both switches. The switches of each half bridge are controlled by two logic signals $\mathrm{Sw}_{\mathrm{t}}^{\mathrm{a,b}}$ and $\mathrm{Sw}_{\mathrm{b}}^{\mathrm{a,b}}$ where logic 1 corresponds to a closed switch state, similar to the half bridge in Fig. 11.2. For each half-bridge converter in Fig. 11.13, a logic variable $\mathrm{Sw}^{\mathrm{a,b}}$ is introduced which is linked to the switch states. For example, $\mathrm{Sw}^{\mathrm{a}} = 1$ corresponds with switch state $\mathrm{Sw}_{\mathrm{t}}^{\mathrm{a}} = 1$ (hence closed) and switch state $\mathrm{Sw}_{\mathrm{b}}^{\mathrm{a}} = 0$ (hence open). Likewise, $\mathrm{Sw}^{\mathrm{b}} = 1$ corresponds with switch state $\mathrm{Sw}_{\mathrm{t}}^{\mathrm{b}} = 1$ (hence closed) and switch state $\mathrm{Sw}_{\mathrm{b}}^{\mathrm{b}} = 0$ (hence open). When the states $\mathrm{Sw}^{\mathrm{a}} = 1$, $\mathrm{Sw}^{\mathrm{b}} = 0$ are active the output voltage is equal to $u = u_{\mathrm{DC}}$. Alternatively states $\mathrm{Sw}^{\mathrm{a}} = 0$, $\mathrm{Sw}^{\mathrm{b}} = 1$ will ensure an output voltage of $u = -u_{\mathrm{DC}}$ hence bipolar operation is also possible with this converter, without the need for a split DC supply. Finally the voltage across the load will be zero for switch state combinations $\mathrm{Sw}^{\mathrm{a}} = 0$, $\mathrm{Sw}^{\mathrm{b}} = 0$ and $\mathrm{Sw}^{\mathrm{a}} = 1$, $\mathrm{Sw}^{\mathrm{b}} = 1$, respectively. The two modulators (shown as separate modules for convenience) control the half bridges and the average voltage per sample reference values U_{a}^{*}, U_{b}^{*} must be appropriately chosen to ensure that the required load average voltage per sample value U^{*} is realized.

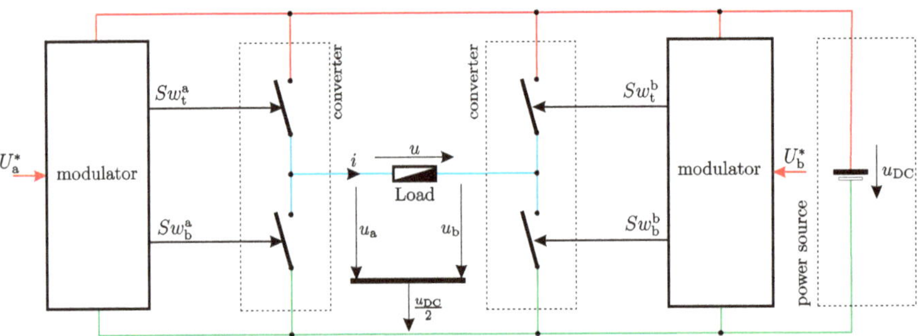

Fig. 11.13 Four switch "full-bridge" bipolar converter with power source and modulators

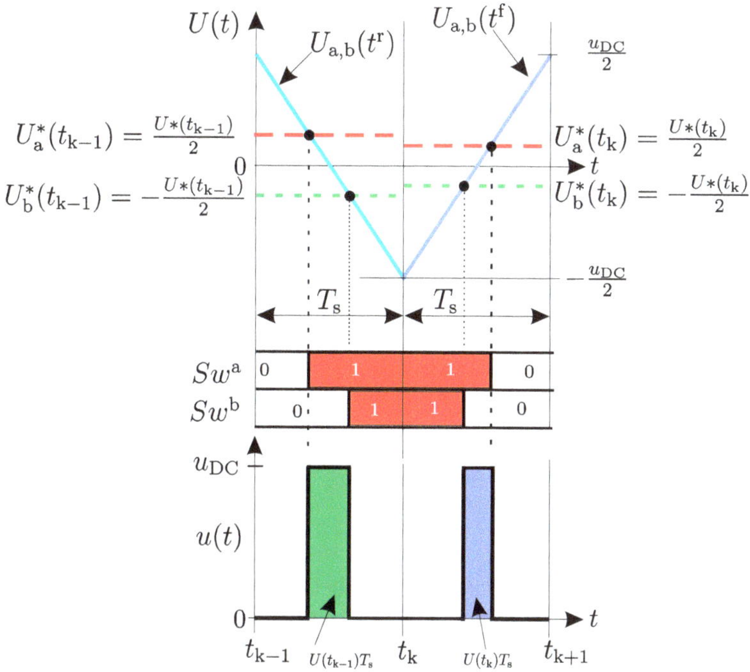

Fig. 11.14 Double edged PWM strategy, for full-bridge converter

11.4.1 *Modulation Strategy for the Full-Bridge Converter*

To understand the modulation of the full-bridge converter it is helpful to initially consider one half bridge, whilst the other is idle (both switches open). For this discussion will be assume the "left" half-bridge converter as active, whilst the right converter is "idle." A double edged modulation strategy is used, for which the average voltage per sample functions must be recalculated using Eq. (11.2). The half-bridge converter output voltages $u_{a,b}$ are defined relative to a DC voltage $u_{DC}/2$ for reasons which will become apparent at a later stage. Figure 11.14 shows two samples, which correspond to one modulator period of operation. The first sample shows rising edge modulation and this means that the half-bridge converter sets the voltage to $u_a = -u_{DC}/2$ at the beginning of the sampling interval and after a time interval t^r (when $U^*_a(t_{k-1}) = U_a(t^r)$) the output voltage is changed to $u_a = u_{DC}/2$. Use of Eq. (11.2) allows us to find the average voltage per sample value versus rise time function, which is in this case of the form $U_a(t^r) = u_{DC}/2 - u_{DC}\, t^r/T_s$. In the second sample a falling edge PWM strategy is used which means that the half-bridge converter switches the output voltage to $u_a = u_{DC}/2$ at the beginning of the sample interval and after a time interval t^f ($U^*_a(t_k) = U_a(t^f)$) it is switched to $u_a = -u_{DC}/2$. Use of Eq. (11.2) allows us to find the average voltage per sample value versus fall time function which is in this case of the form $U_a(t^f) = -u_{DC}/2 + u_{DC}\, t^f/T_s$.

The combination of both rising and falling single edged PWM strategies gives us the double edged PWM strategy. Also shown in Fig. 11.12 is the switch Sw^{a} state for the half-bridge operating sequence discussed. A comparison between the reference average voltage per sample value and the average voltage per sample function $U_{a}(t^{r})$, $U_{a}(t^{f})$ gives the correct values for t^{r} and t^{f}, required to meet the condition specified by Eq. (11.1) for each sample. This analysis undertaken with the "left" half-bridge converter active can readily be undertaken for the case where the "right" half-bridge converter is active and left converter "idle." A subsequent analysis of this scenario leads to the average voltage per sample functions $U_{b}(t^{r})$, $U_{b}(t^{f})$ and corresponding switch state Sw^{b} as shown in Fig. 11.14.

When considering the operation of both half bridges together the question arises how to choose the reference voltage per sample values U_{a}^{*}, U_{b}^{*} relative to the reference voltage per sample value U^{*} that must be applied to the load shown in Fig. 11.13. This issue may be addressed by considering the voltage across the load u, which can be written as

$$u = u_{a} - u_{b} \tag{11.5}$$

where u_{a}, u_{b} are the output voltages of the half-bridge converters. The corresponding reference average voltage per sample output voltage U^{*} can, with the aid of Eqs. (11.5), (11.3), and (11.2), be written as

$$U^{*} = U_{a}^{*} - U_{b}^{*} \tag{11.6}$$

where U_{a}^{*}, U_{b}^{*} are the reference voltage per sample value of the left and right half-bridge converters. A detailed discussion on choosing the most appropriate values for U_{a}^{*}, U_{b}^{*} relative to the output average voltage per sample value U^{*} is shown in our book "Advanced Electrical Drives" [2], which leads to the following guidelines:

$$U_{a}^{*} = \frac{1}{2}U^{*} \tag{11.7a}$$

$$U_{b}^{*} = -\frac{1}{2}U^{*} \tag{11.7b}$$

In the modulation example given in Fig. 11.14 the half-bridge reference value was precisely chosen according to Eq. (11.7). Note also that the limit average voltage per sample outputs of a full-bridge converter is $\pm u_{DC}$, which is double the value possible with a single half-bridge converter. Furthermore, the voltage across the load is now equal to $\pm u_{DC}$ pending the polarity of the reference average voltage per sample value U^{*}. When positive (as shown in Fig. 11.14) the output pulses will be u_{DC}, vice versa when a negative pulse amplitude $-u_{DC}$ will appear. Finally if the reference average voltage per sample value U^{*} is set to zero, the average voltage per sample of both half-bridge converters will be equal to $u_{DC}/2$, which was the reason for choosing the half-bridge converter output voltages u_a, u_b relative to said value (see Fig. 11.13). Under these conditions the output voltage across the load will be zero.

11.5 Current Control Algorithm

For regularly sampled control systems as considered here we are able to derive a simple controller structure which can be used to control the current in the DC machine (in this case represented as an R-L-e_a circuit). Central to our proposed controller is the use of a modulator/converter combination which is designed to deliver (in a given sampling interval T_s) an average voltage per sample quantity $U(t_k)$, which corresponds to the set-point (reference) $U^*(t_k)$ value provided by the current control module. The task of the current control module is thus reduced to determining this set-point (reference) average voltage per sample value on the basis of the measured discretized load current $i(t_k)$ and the set-point current value i^*.

The nature of this control philosophy is shown with the aid of Fig. 11.15. Shown in this figure are the non-sampled reference current $i^*(t)$ and typical (for PWM based control) converter current $i(t)$. These currents are sampled by the controller at time marks $0, t_1, t_2, \ldots$, etc. At time $t = t_1$ a current error (between the sampled reference and sampled converter current) $i^*(t_1) - i(t_1)$ is present and the objective of the control approach is to determine the required average voltage reference value needed to zero said error. This leads to the condition $i^*(t_1) = i(t_2)$.

The control objective aimed at driving the current error to zero during each sample interval may be written as

$$i(t_{k+1}) = i^*(t_k) \tag{11.8}$$

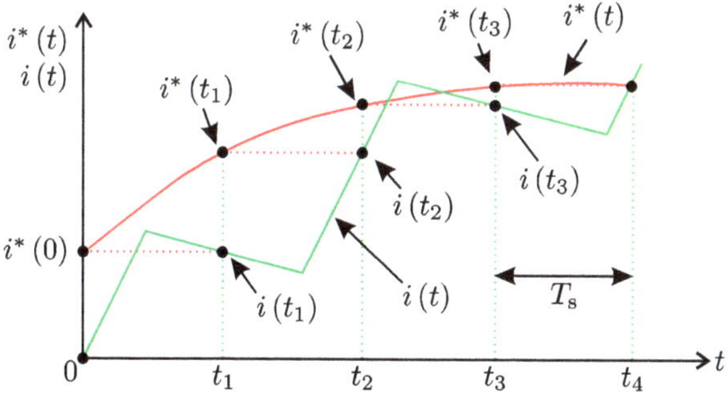

Fig. 11.15 Model based current control [2]

for a regularly sampled system with sampling time T_s. The modulator converter module will ensure that the following condition is satisfied

$$U^*(t_k) = \int_{t_k}^{t_{k+1}} u(\tau)\,d\tau \tag{11.9}$$

where u represents the voltage across the load [see Fig. 11.1]. An observation of Fig. 11.1 shows that the load voltage may be expressed as

$$u = Ri + L\frac{di}{dt} + e_a \tag{11.10}$$

which is precisely the differential equation used to represent the armature based machine model (see Eq. (10.5)). For simplicity the armature "a" subscripts have been removed and the variable $e_a = \psi_{mf}\omega_m$ is introduced. Use of Eq. (11.10) with Eq. (11.9) allows the latter to be written as

$$U^*(t_k) = \frac{R}{T_s}\int_{t_k}^{t_k+T_s} i(\tau)\,d\tau + \frac{L}{T_s}\int_{i(t_k)}^{i(t_k+T_s)} di + \frac{1}{T_s}\int_{t_k}^{t_k+T_s} e_a(\tau)\,d\tau \tag{11.11}$$

Equation (11.11) forms the basis for determining a generic control structure that is able to calculate the required average voltage per sample quantity capable of satisfying condition (11.8). It is noted that this set-point value can only be determined on the basis of a detailed knowledge of the load parameters R and L.

For a discrete (use of sampled data for processing via a micro-processor) type controller considered here discretization of Eq. (11.11) is required. A first order approximation technique can be used, provided that the sampling time is sufficiently small, in which case the resistive term of (11.11) is reduced to

$$\frac{R}{T_s}\int_{t_k}^{t_{k+1}} i(\tau)\,d\tau \cong \frac{R}{2}\left(i(t_{k+1}) + i(t_k)\right) \tag{11.12}$$

The inductive term present in Eq. (11.11) can be directly evaluated which gives

$$\frac{L}{T_s}\int_{i(t_k)}^{i(t_{k+1})} di = \frac{L}{T_s}\left(i(t_{k+1}) - i(t_k)\right) \tag{11.13}$$

The third term which contains the induced voltage (back-emf) is reduced to the form

$$\frac{1}{T_s}\int_{t_k}^{t_{k+1}} e_a(\tau)\,d\tau \cong e_a(t_k) \tag{11.14}$$

where use is made of the fact that the voltage variations of $e_a(t)$ are relatively small within one sampling period. The reason being that such variations are linked to the mechanical time constant of the machine.

The resultant reference average voltage per sample value is found by adding together the three terms given in Eqs. (11.12)–(11.14) which upon use of (11.8) gives

$$U^*(t_k) \cong \frac{R}{2}\left(i^*(t_k) + i(t_k)\right) + \frac{L}{T_s}\left(i^*(t_k) - i(t_k)\right) + e_a(t_k) \tag{11.15}$$

A further simplification is possible by rewriting the resistive term in Eq. (11.15) as

$$\frac{R}{2}\left(i^*(t_k) + i(t_k)\right) = \frac{R}{2}\left(i^*(t_k) - i(t_k)\right) + R\,i(t_k) \tag{11.16}$$

in which case the set-point average voltage per sample value is given as

$$U^*(t_k) \cong R\,i(t_k) + \left(\frac{L}{T_s} + \frac{R}{2}\right)\left(i^*(t_k) - i(t_k)\right) + u_e(t_k) \tag{11.17}$$

A control structure based on Eq. (11.17) is basically a so-called proportional type controller. In practice, a so-called proportional-integral type structure is preferable and this may be achieved by rewriting the term $R\,i(t_k)$ as

$$R\,i(t_k) \cong R\sum_{j=0}^{j=k-1}\left(i^*(t_j) - i(t_j)\right) \tag{11.18}$$

which means that the current $i(t_k)$ is composed of a series of difference terms as can be observed from Fig. 11.15. Use of Eq. (11.18) with (11.17) leads to

$$U^*(t_k) \cong R\sum_{j=0}^{j=k-1}\left(i^*(t_j) - i(t_j)\right) + \left(\frac{L}{T_s} + \frac{R}{2}\right)\left(i^*(t_k) - i(t_k)\right) + e_a(t_k) \tag{11.19}$$

In practical terms some further simplification can be applied by assuming that the gain term $(L/T_s + R/2)$ can be reduced to $\approx L/T_s$. Furthermore Eq. (11.19) can also be expressed in terms of a gain $K_p = \approx L/T_s$, bandwidth $\omega_i = R/L$, and current error term $\epsilon(t_k) = (i^*(t_k) - i(t_k))$ which leads to

$$U^*(t_k) \cong K_p\left(\epsilon(t_k) + \omega_i \sum_{j=0}^{j=k-1}\epsilon(t_k)\right) + e_a(t_k) \tag{11.20}$$

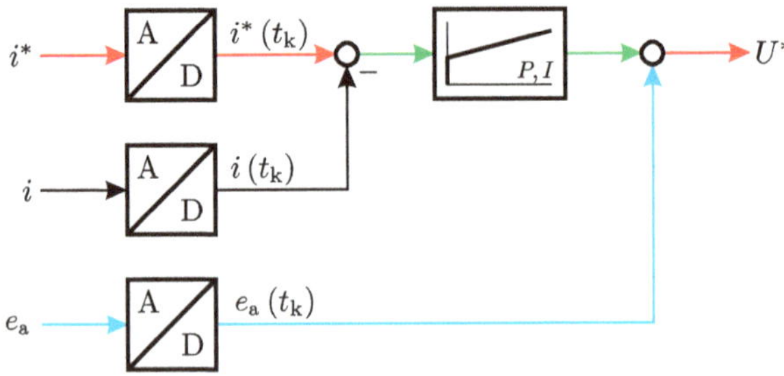

Fig. 11.16 Model based current controller structure

The generic structure which corresponds to Eq. (11.20) as shown in Fig. 11.16 contains a so-called PI (proportional-integral) controller and "feed-forward" term $e_a(t_k)$. In practical implementations the PI controller in its current form is prone to "windup" which occurs when the limits of a linear system are reached. Windup can occur in this case when the reference average voltage value generated by the PI controller exceeds the maximum value which can be delivered by the converter. Under such circumstances a current error occurs at the input of the controller which will cause the integrator output to ramp up or down. Practical controllers have a "anti-windup" feature which limits integrator action when the user defined limits are reached. In the tutorial section controllers with and without "anti-windup" are introduced so that the reader can examine the impact of "anti-windup." A disadvantage of model based control is that a priori knowledge of the load is required, which is not the case for the so-called hysteresis type current controllers which are discussed (among others) in our text "Advanced Electrical Drives" [2].

11.6 Tutorials

11.6.1 Tutorial 1: "Rising" Edge PWM with a Uni-Polar Converter

This tutorial considers a simple example which consists of a modulator, uni-polar converter, and load in the form of an ideal inductance. A single "rising" edge type modulator will be considered. The average voltage per sample value is taken to be the input to the modulator. A sampled (discrete) system is assumed with a sampling time $T_s = 1$ ms. The supply voltage u_{DC} is taken to be 300 V. Furthermore, the load inductance is set to $L = 100$ mH. The aims are to build this system in PLECS in order to analyze the converter output and verify that the converter is able to supply an average voltage per sample value which is equal to the reference average voltage per sample value. We then reduce (for the single edge PWM version) the supply voltage to 200 V and determine if the converter is still able to meet the reference average voltage per sample value U^*. The reference average voltage per sample function is taken to be of the form $U^* = 100\,\text{V}\cdot\sin\omega t$ where $\omega = 100\pi$ rad/s.

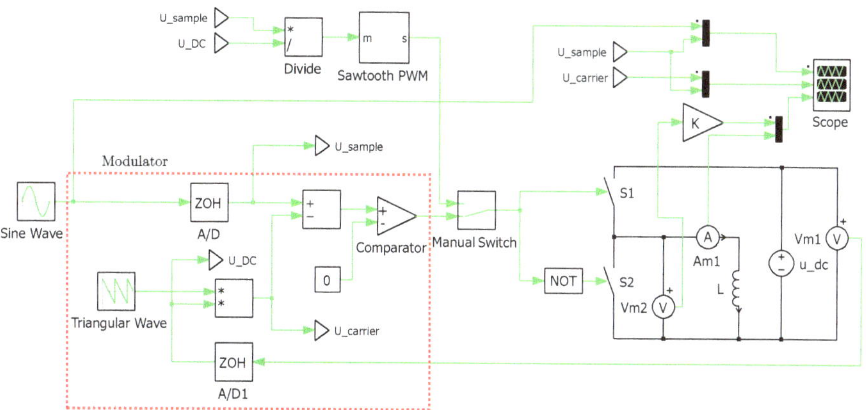

Fig. 11.17 PLECS model of uni-polar converter with "rising" edged PWM

Figure 11.17 shows a PLECS implementation of a model which addresses the requirements of this tutorial. The `Sine Wave` module produces the reference average voltage per sample value U^*, which is then fed to an A/D converter. In PLECS this module is found in the discrete library and called a "zero order hold" `ZOH`. In this (and others of the same type) module set the sampling time to 1.0 ms. The `Triangular Wave` module is found under "Sources" and within this module set in the entry "Frequency values" the value 1000 Hz and 'Maximum signal value' to 1.0 V. In addition, set the "Duty cycle" value to 0. This in effect sets up the modulator saw-tooth function as shown in Fig. 11.8 for "rising" edge PWM. The output of this module is multiplied by u_{DC}. An A/D unit is used to sample the u_{DC} value from the converter (output `u_DC`) in order to obtain $u_{DC}(t_k)$. A summation unit takes the reference and modulator average voltage per sample values and produces an input ϵ for the comparator. In PLECS a `Comparator` (found in the "Discontinuous" library) is used. The output of the comparator module is connected via a `Manual Switch` to the two switches `S1` and `S2` of the converter. Attached to the converter is an ideal inductance `L` together with a set of voltage/current measurement probes `Vm1`, `Vm2`, and `Am1`. Also shown in Fig. 11.17 is a PLECS module `Sawtooth PWM` which is a "standard library module" that is configured to generate the same control input for the converter as the set of modules shown in the "Modulator" window. At a later state use will be made of the library modulators hence at that stage the user can verify (with the aid of the manual switch) that the output of the library and discrete modulator produce identical results.

Set the "simulation parameters" to an initial run time of 5 ms (five samples) and test your PLECS model step by step. Consider first the reference average voltage per sample value before and after the A/D unit by adding a "Multiplexer" and "scope" and plot the result. An example as to the waveform you should see is given in Fig. 11.18a. The "blue" line represents the reference average voltage per sample variable U^*, the "red" line is the sampled version of this function.

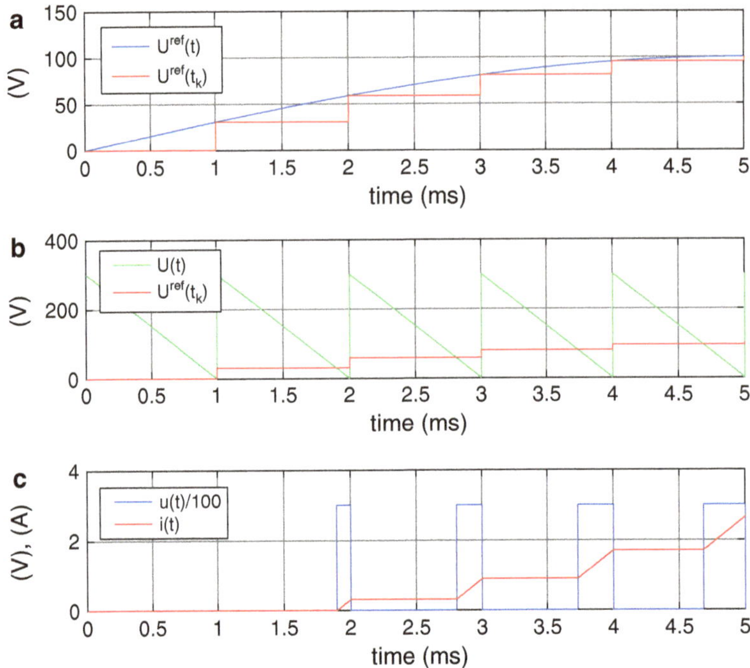

Fig. 11.18 PLECS results: uni-polar converter, $u_{\mathrm{DC}} = 300$

Next observe the modulator average voltage per sample `U_carrier` and sampled reference average voltage per sample `U_sample` an example of which is given in Fig. 11.18b. The "green" plot is the modulator output $U(t)$ and the "red" waveform is the sampled reference average voltage per sample. Check (by calculating the voltage time area within the sample interval) for one sample if the average voltage per sample produced by the converter is equal to the reference value.

Add the electrical converter components, measurement probes, and $L = 100\,\mathrm{mH}$ inductance to the PLECS model and rerun your simulation. Observe the converter voltage (divided by 100, to view the result with the other outputs) and current waveforms together by adding a "multiplexer" and Scope. An example of the converter voltage and current waveforms (processed via Matlab) is given in Fig. 11.18c. The current change will occur as a result of a converter output pulse, the area of which (being the average voltage per sample times T_s) will (for a given value of the load inductance) determine the amplitude of the variation as shown in Fig. 2.3. Note that the results shown in Fig. 11.18 could also have been generated using the PLECS library modulator `Sawtooth PWM` module (see Fig. 11.17) and using the `manual switch` to connect its output to the converter.

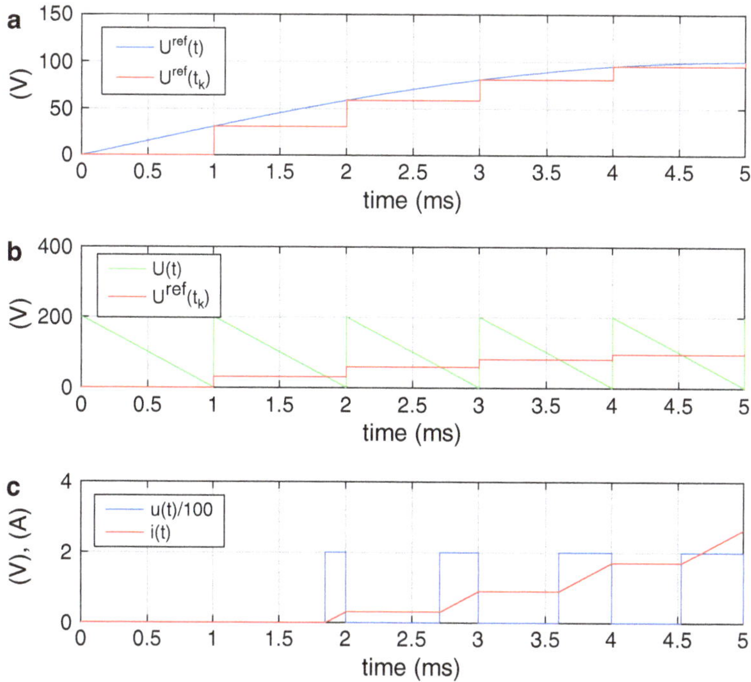

Fig. 11.19 PLECS results: uni-polar converter, $u_{DC} = 200$ and "rising edge" PW

We now look at the effect of changing the DC supply voltage level by changing U_{DC} from 300 to 200 V. Make this change in the PLECS model (using DC source `u_dc`) and rerun the simulation. Observe and plot the inductance current together with the converter voltage $u/100$. An example of the new waveforms is given in Fig. 11.19c. A comparison between Figs. 11.18c and 11.19c shows that the current waveform (shown in "red") is unaffected (in terms of the current step which takes place for each sample) by the change, which is precisely what should happen as shown in Fig. 11.9 (where use was made of "falling edge" PWM). The converter output voltage amplitude is reduced but the modulator carrier waveform `u_carrier` amplitude is also reduced as can be observed by comparing Figs. 11.18b and 11.19b. This in turn implies that the width of the output pulses will be increased given the same sampled average voltage per sample reference function is used.

It is left as an exercise to the reader to consider to reconfigure the tutorial to "falling" edge PMW. Hint: this can be realized by setting the "duty cycle" to 1 in the `Triangular Wave` module dialog box and choosing the option "rising" in the "Ramp" dialog box of the `Sawtooth PWM` module.

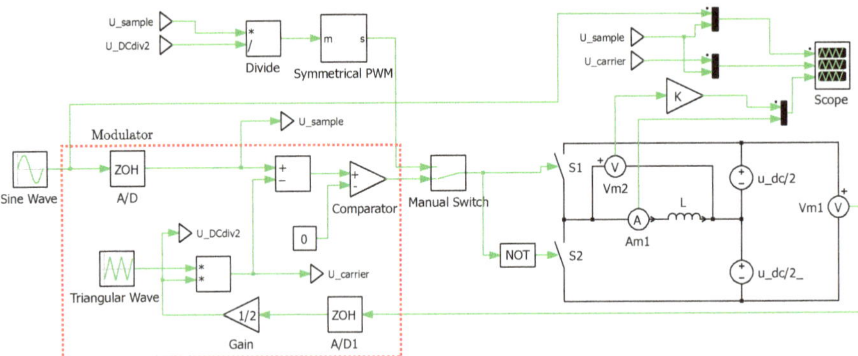

Fig. 11.20 PLECS model of bipolar half-bridge converter with double edged PWM

11.6.2 *Tutorial 2: "Double" Edged PWM with a Half-Bridge Converter*

In this tutorial we will look at building/analyzing a double edged PWM with the so-called asymmetrical sampling. The modulator will be connected to a bipolar half-bridge converter which in turn is connected to an inductive load.

The PLECS model according to Fig. 11.17 must now be modified in terms of the modulator and converter configuration. The revised simulation model shown in Fig. 11.20 contains a new `Triangular Wave` module which is located within the modulator window. In the dialog box of the `Triangular Wave` module set the "frequency" to 500 Hz, which is half the sampling frequency used in this (and previous) two tutorials. The dialog settings for "maximum" and "minimum" signals values should be set to 1 and -1, respectively. Furthermore, the measured DC bus voltage attenuated by a factor 2 (shown as variable `u_DCdiv2`) is multiplied by the output of the triangular module which leads to the carrier waveform `U_carrier`. The maximum and minimum values of this variable are set to $u_{DC}/2$ and $-u_{DC}/2$ which are the limit average voltage per sample value that can be realized with a half-bridge converter.

On the converter side a configuration change needs to be made where the inductance, arbitrarily set to $L = 200$ mH, is now connected to midpoint between the two DC supply modules as shown in Fig. 11.20. Each supply source is set to 150 V, hence the total bus voltage is equal to 300 V, as also used in the previous two cases. The sinusoidal reference average voltage per sample function (with the same frequency and amplitude used earlier) is also used for this tutorial. However change the run time for this PLECS simulation to 10 ms. Typical results which should appear after running the simulation and processing the results via Matlab are shown in Fig. 11.21. Note that the results shown in Fig. 11.21 could also have been generated by the PLECS library module `Symmetrical PWM` (see Fig. 11.20) and selecting sample option: "regular" (double edged) in the dialog box of this module. The `manual switch` (also shown in Fig. 11.20) allows the user to either use the discretely build modulator or the PLECS library module.

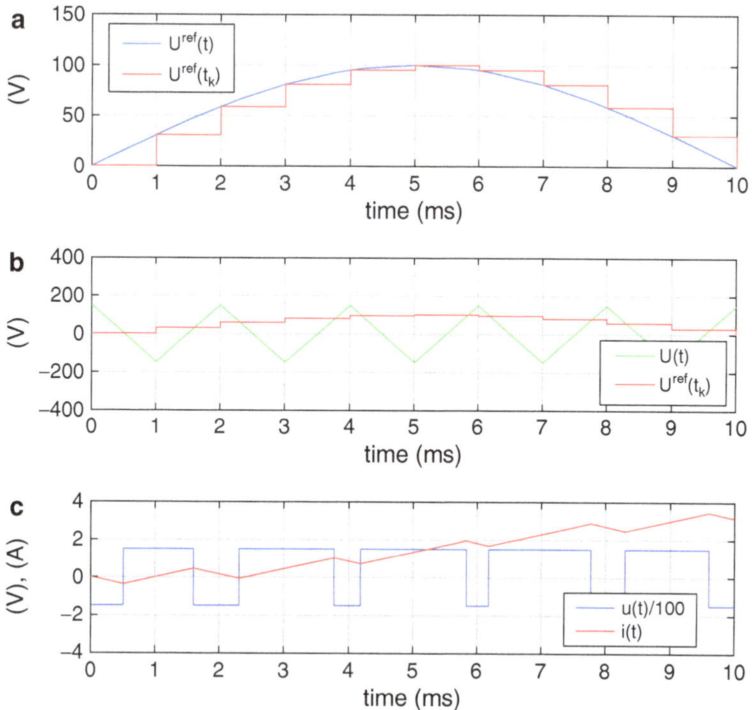

Fig. 11.21 PLECS results: bipolar half-bridge converter, $u_{DC} = 300$

11.6.3 Tutorial 3: "Double" Edged PWM with a Full-Bridge Converter

The purpose of the tutorial is to examine operation of a full-bridge converter using double edged asymmetrical pulse width modulation as introduced in the previous PLECS model (see Fig. 11.20). This implies that the PLECS module according to Fig. 11.20 must be modified in terms of the modulator and converter configuration according to the theory discussed in Sect. 11.4. For this tutorial both load and reference average voltage per sample function are assumed to be identical to those used for the half-bridge tutorial.

The revised simulation model shown in Fig. 11.22 shows a set of four switches `S1–s4` which are configured for full-bridge operation. The 200 mH inductive load `L` is now connected between the midpoints of the two half-bridge converters. Both half-bridge converters must now be controlled via two logic variables `sw_a` and `sw_b` which must be generated via the discrete modulator or the PLECS library module `Symmetrical PWM`. A `Manual Switch` is used so that the user can select either modulator. Observation of the discrete modulator shows that an additional set of modules have been introduced, which makes use of the reference

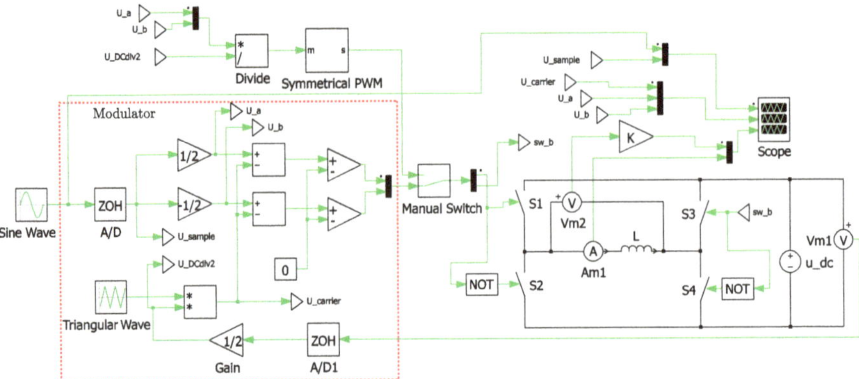

Fig. 11.22 PLECS model of full-bridge converter with double edged PWM

average voltage per sample variables U_a and U_b. These variables are in turn linked to the sampled reference average voltage per sample variable U_sample. The outputs of the two comparator modules have been combined into a single column matrix given that the PLECS library module also provides such a vector. Inputs to the latter mentioned module is a vector which is formed by the variables U_a and U_b.

The supply source u_dc should be set to 300 V, hence the bus voltage is equal to that used in the previous two tutorial examples. The sinusoidal reference average voltage per sample function (with the same frequency and amplitude used earlier) is also used for this tutorial. However change the run time for this PLECS simulation to 20 ms which implies that one full period of operation is now considered. Typical results which should appear after running this simulation and processing the results using Matlab should be according to Fig. 11.23. Subplot Fig. 11.23a shows the reference average voltage per sample waveform together with its sampled version (shown as U_sample in the PLECS diagram). Observation of subplot Fig. 11.23b reveals the carrier waveform (shown as U_carrier in the PLECS diagram) together with the reference average voltage per sample waveforms for the two half-bridge converters (shown as U_a and U_b in the PLECS diagram). Finally subplot Fig. 11.23c shows the voltage across the load (attenuated by a factor 100) and the current through the inductance. Note that said current at time $t = 10.0$ ms is precisely the same value as found with the half bridge in the previous tutorial (see Fig. 11.21). This is to be expected, as the same reference signal is used in both tutorials. Note also that the peak load voltage amplitude is now 300 V which is double the value found in the half-bridge converter. Furthermore the amplitude of the output voltage changes polarity when the reference average voltage per sample value becomes negative.

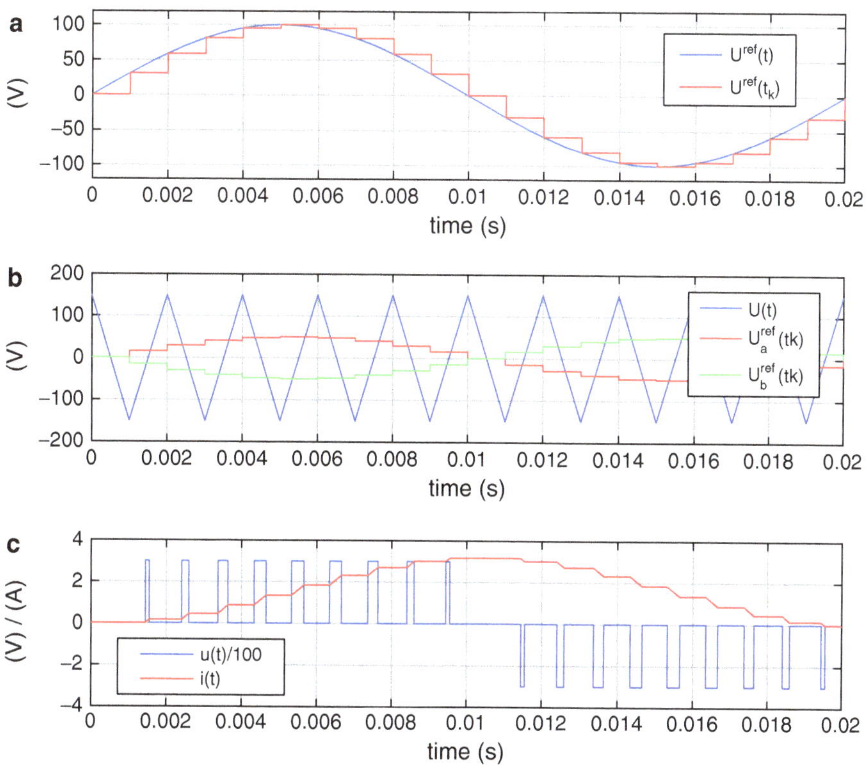

Fig. 11.23 PLECS results: bipolar full-bridge converter, $u_{\mathrm{DC}} = 300\,\mathrm{V}$

11.6.4 Tutorial 4: "Model Based" Current Control with R-L Type Load

The aim is to analyze a "model based" type current PI controller with a full-bridge converter and modulator (from the PLECS "modulator" library) discussed in the previous tutorial. A "discrete" PI controller is to be used which will deliver the required reference average voltage per sample value to the modulator needed to control the current in the load.

The load is taken to be a resistance $R = 5\,\Omega$ and inductance $L = 7.35\,\mathrm{mH}$ (series connected) which are the parameters of the "Erik" DC machine introduced in the previous chapter (see Table 10.1). This machine will also be used for the next tutorial and demonstration (proof of concept) laboratory given at the end of this chapter. A sampling frequency of 15 kHz is used, which corresponds to a sampling time of $T_s = 66.66\,\mu\mathrm{s}$. The supply voltage for the converter is set to $u_{\mathrm{DC}} = 24\,\mathrm{V}$.

An example of the PLECS model for this tutorial given in Fig. 11.24 shows the H-bridge converter configuration, modulator `Symmetrical PWM`, and current controller `PI controller`. All of the above, with exception of the current

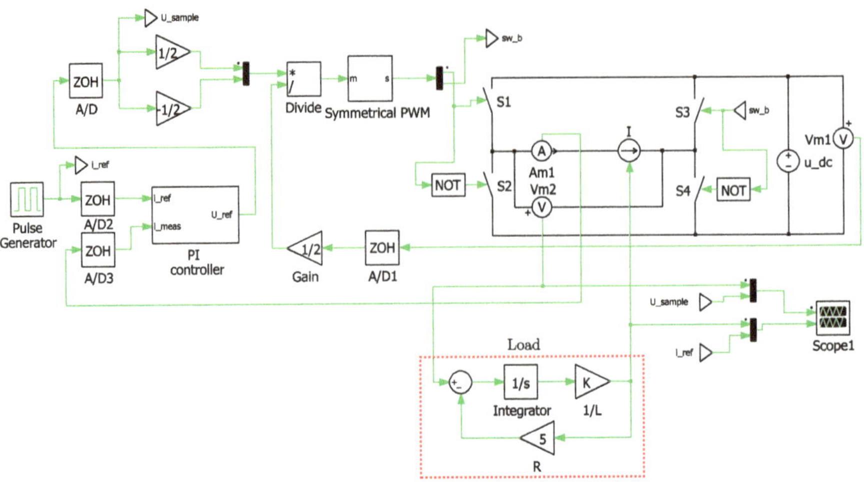

Fig. 11.24 PLECS model: modulator/converter with R/L load

controller, have been introduced in the previous tutorial (see Fig. 11.22). The load for the converter is in this tutorial represented by a set of "control block" instead of the "electrical blocks" used earlier. This implies the need to build an interface between the "electrical" converter and "control block" based load. The latter is realized by using the voltage probe `Vm2` output which now acts as an input for the load. The current output of the load is connected to a "controlled current source" `I` that is connected between the midpoints of the two half-bridge converters. A current probe `Am1` provides a control signal representation of the "measured" load current, as required by the current controller.

We will examine the performance of this system by making use of a `Pulse Generator` module which provides a ± 1 A, 2 Hz square wave reference current function shown as `i_ref` in the PLECS diagram. Set the run time of your simulation to 0.6 s.

The PLECS model of the current controller given in Fig. 11.25 contains a discrete integrator formed by a "summation block," modules `Delay1`, `Saturation`, and gain module `wixTs`. Input to the controller are the sampled reference and measured currents shown as `i_ref` and `i_meas`, respectively. The difference between these two variables [defined as ϵ in Eq. (11.20)] is multiplied by a gain factor `P`. The key dialog box entries for this current controller are the gain K_{p} and bandwidth ω_i as introduced in Eq. (11.20). Note that the feed-forward term e_{a} in said equation is zero in this example. Henceforth these variables are defined by the load inductance/resistance and sample frequency. The latter variable in the form of the sampling time must also be set in the current controller dialog box together with the limit output values set to ± 24 V in this case. The two `Saturation` modules shown in Fig. 11.25 limit the output voltage to the required value and also implement

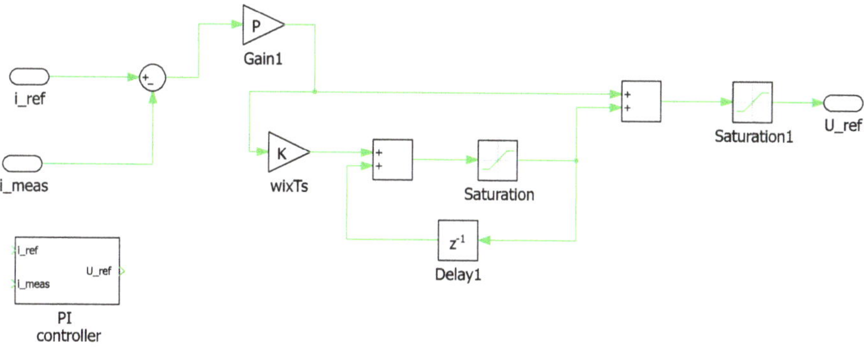

Fig. 11.25 PLECS model: PI controller

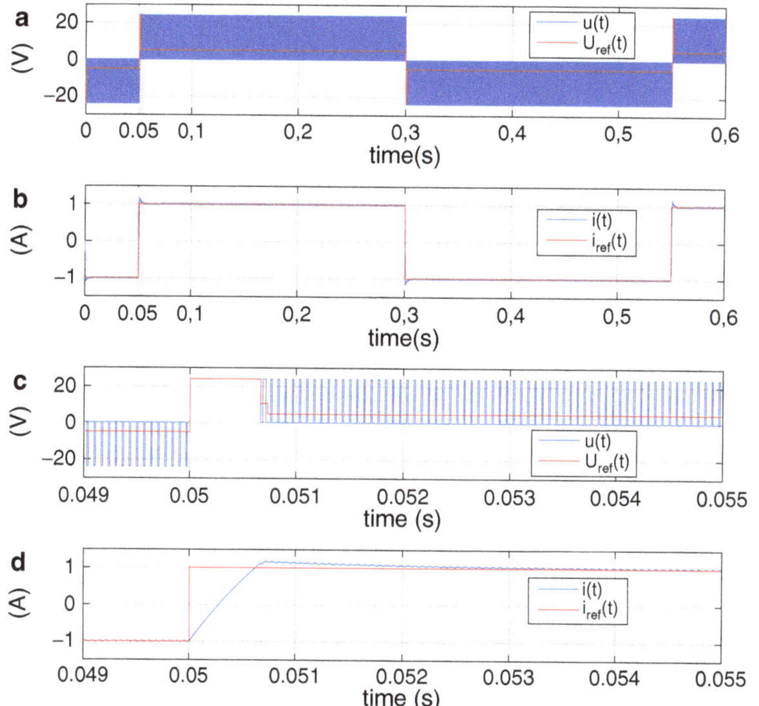

Fig. 11.26 PLECS results: modulator/full-bridge converter with *R-L* load

the integrator "anti-windup" function of the controller. A detailed discussion on discrete controller implementation is given in our book "Applied Control" [10].

An example of the results achieved with this simulation is shown in Fig. 11.26. Subplot (a) shows the converter load voltage $u(t)$ together with the current controller reference average voltage per sample output U^*. Both reference current i^* and actual load current are shown in subplot (b), which confirms that correct current control has been achieved. An expanded view of subplots (a) and (b) for the time interval 49 ms → 55 ms given in subplots (c) and (d) shows how the current controller responds to a step change in the reference current. Observe that the output voltage of the current controller is limited to 24 V, which is the maximum average voltage per sample value that the full-bridge converter can deliver. When the reference current is equal to the measured value, the current control output voltage is ±5 V, which is expected given the use of a 5 Ω load resistance and ±1 A current amplitude.

11.6.5 Tutorial 5: "Model Based" Current Control of a DC Drive Using a Full-Bridge Converter

The aim of this tutorial is to simulate a DC motor drive, with model based current control and a full-bridge converter. The DC motor model discussed in Sect. 10.6.1 on page 289 is to be used for this tutorial given that the drive setup used here will be experimentally verified in the ensuing demonstration laboratory. Hence the converter, sampling frequency, and DC bus voltage values used here must reflect those that will be used in the demonstration setup.

The previous tutorial was already configured for this purpose, hence the only change which needs to be made to the PLECS model shown in Fig. 11.24 is with respect to the converter load. A PLECS implementation example of this simulation given in Fig. 11.27 shows the DC motor `Eriks_DC`, current controller `PI controller`, and full-bridge converter. In the demonstration setup the DC motor is mechanically connected to a permanent magnet machine, hence the inertia value used in the DC model must reflect the sum inertia of both machines.

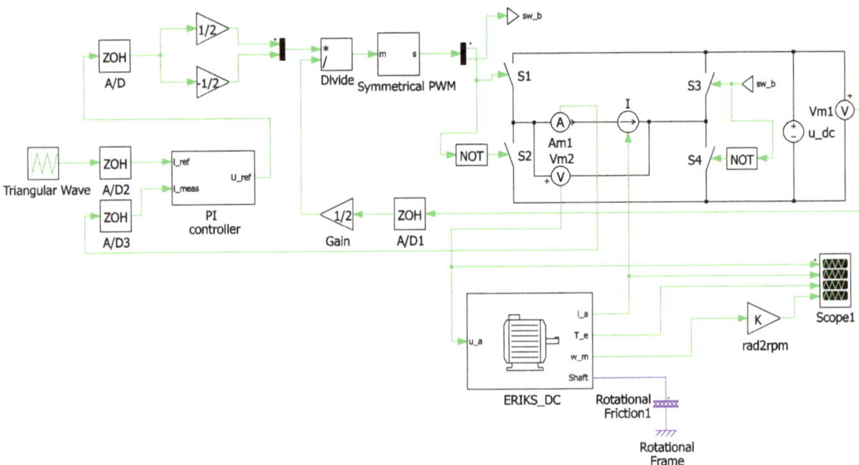

Fig. 11.27 PLECS model: DC drive using model based current control and a full-bridge converter

Furthermore, it is known from the previous demonstration example (see Sect. 10.6.4) that the combined friction of the drive is $\approx 20 \cdot 10^{-3}$ Nm, hence a mechanical block `Rotational Friction` must be added to Fig. 11.27 to accommodate this load requirement for the DC motor. We will examine the performance of this system by making use of a `Triangular Wave` module which provides a ± 1 A, 2 Hz triangular wave reference current function. Set the run time of your simulation to 2.0 s.

A typical set of results obtained after running the simulation should look like those given in Fig. 11.28. These results, which have been processed via Matlab, show operation over the time period 1.2 s $\rightarrow$ 1.8 s, which reflects steady-state drive operation. The following waveforms are shown in this diagram:

- Subplot (a): Voltage across the DC motor referred to as the "Armature voltage." This waveform consists of a set of converter pulses, as shown previously in Fig. 11.26, subplot (c).
- Subplot (b): Measured DC motor current referred to as the "Armature current." This is a triangular waveform with a 2 A peak to peak current amplitude, in

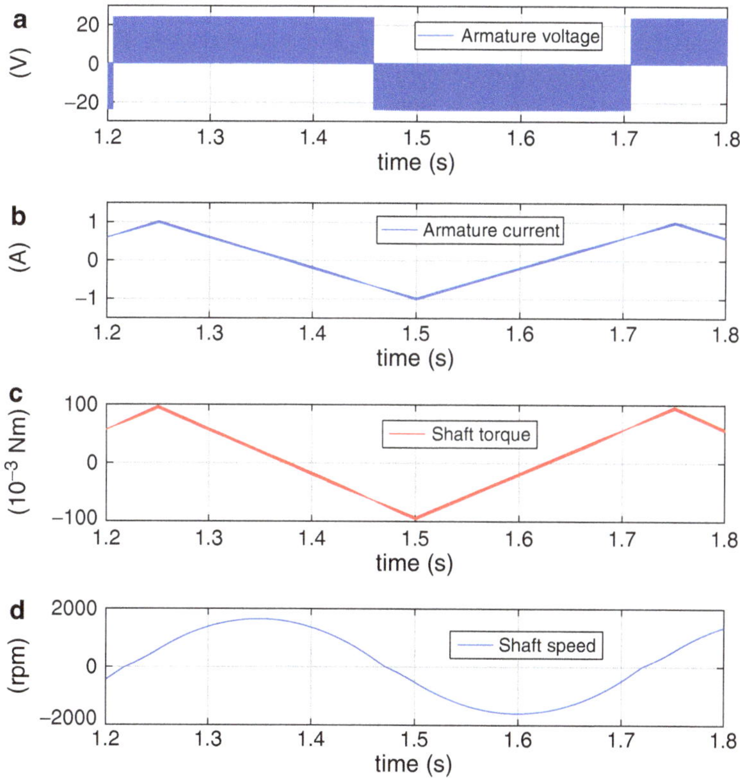

Fig. 11.28 PLECS results: DC drive using model based current control and a full-bridge converter

accordance with the 2 Hz reference waveform provided to the current controller. Note that this controller does not make use of the feed-forward term mentioned in Eq. (11.20). The reason for this is that in most practical cases adding such a term is likely to be counterproductive because this variable is derived from a measured speed signal which typically has a noise component. Hence adding a feed-forward term also implies injecting a noise source into the current controller which tends to offset any anticipated dynamic performance improvements.
- Subplot (c): the shaft torque of the machine, which will also be a triangular waveform, given that it is proportional to the armature current.
- Subplot (d): the shaft speed of the machine, which will be parabola shaped curves, due to the fact that the shaft torque is a triangular waveform.

11.6.6 Demo 4: "Model Based" Current Control of a DC Drive Using a Full-Bridge Converter

This demonstration laboratory aims to experimentally confirm the results derived with the simulation model introduced in tutorial 4 (see Sect. 11.6.5). The Texas Instruments LaunchpadXL [6] setup introduced earlier (see Fig. 8.29) is again used here, with important difference that the LVACIMTR (EMsynergy) induction machine ("gray" motor) is now replaced by the DC machine [11] used in Sect. 11.6.5. The LVSERVOMTR (Teknic) PM machine is now mechanically connected to the DC machine, as its encoder will be used to measure the shaft speed. For this demo the DC machine was connected to the aft (furthest away from the USB connector) BOOSTXL-DRV8301 module shown in Fig. 8.29. Said module has three half-bridge converters and for this purpose phases A and B are combined to form a "H-bridge" converter.

The DC drive controller is implemented using real time embedded software VisSim [14]. This software provides the user with the ability to construct any type of controller using a set of building blocks that represent the controller and also provides direct access to the attached hardware. The set of VisSim modules applicable to this demo, given in Fig. 11.29, show a "current controller" module which generates the two variables `mA_ref` and `mB_ref` that control the modulation of the two half-bridge converters. Input to the current controller is a reference current `REF`, which is generated by a 2 Hz triangular waveform generator. The amplitude of this waveform is controlled by the variable `Ampld_n` which is set by the user via a "slider." The "current controller" module itself is directly based on the PLECS model shown in Fig. 11.25. However implementation is undertaken in the so-called fixed point format [10], an approach that is applied to the complete drive controller. Furthermore, the sampling frequency for this controller is set to 15 kHz (as used in tutorial 4). The module "ADC-PWM LaunchXL-069 BOOSTXL-2" controls the PWM of the two half-bridge converters and also provides the low pass filtered converter voltages `u_A` and `u_B` and currents `i_A` and `i_B`. The latter variables are combined to form the variable `iArm`, which represents the measured "armature" current.

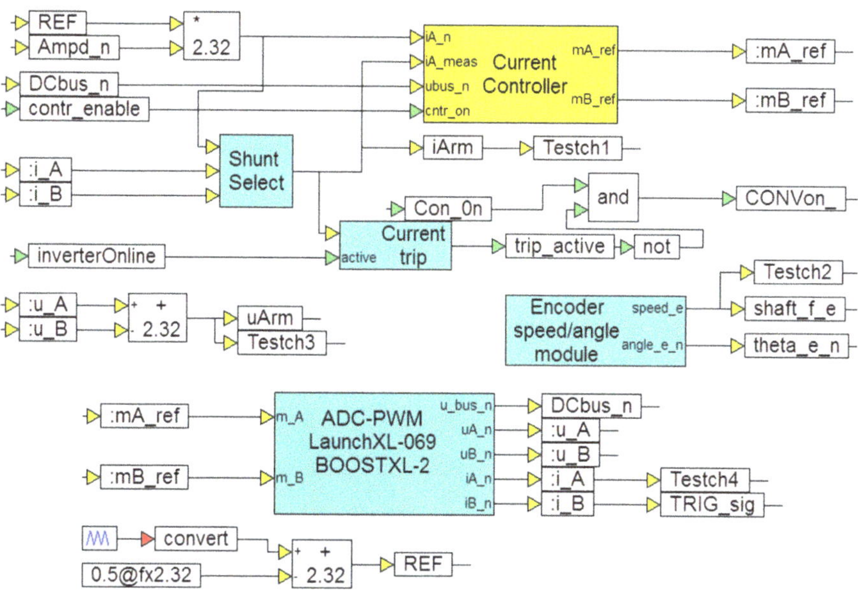

Fig. 11.29 VisSim [14] based “H-drive” with “model based” current control [10]

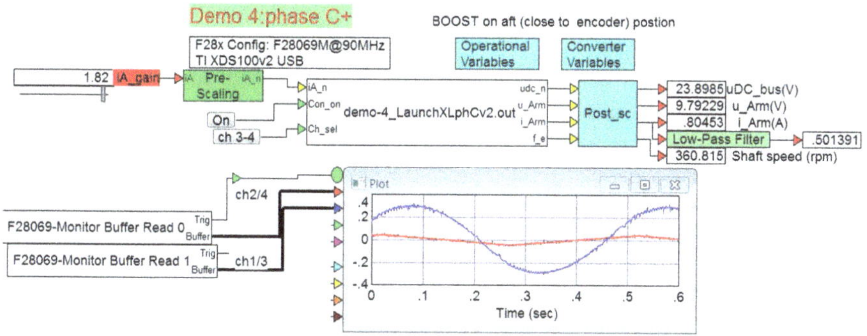

Fig. 11.30 “Run version” of VisSim [14] based “H-drive” demonstration controller with model based current control [10]

Also shown in Fig. 11.29 is the module “Encoder speed/angle,” that is used to process data from the attached shaft encoder. For this demo the scaled shaft speed variable `shaft_fe` is used and subsequently displayed on a plot module.

The set of modules given in Fig. 11.29 must be compiled by VisSim, which then generates a file `demo-4_LaunchXLphCv2.out` that is used by a VisSim target module shown in Fig. 11.30. Attached to the target module is a slider which is used to control the amplitude of the triangular reference current waveform. A scope module is also present which shows the scaled armature current and shaft speed during operation. The drive development approach briefly outlined above (and in

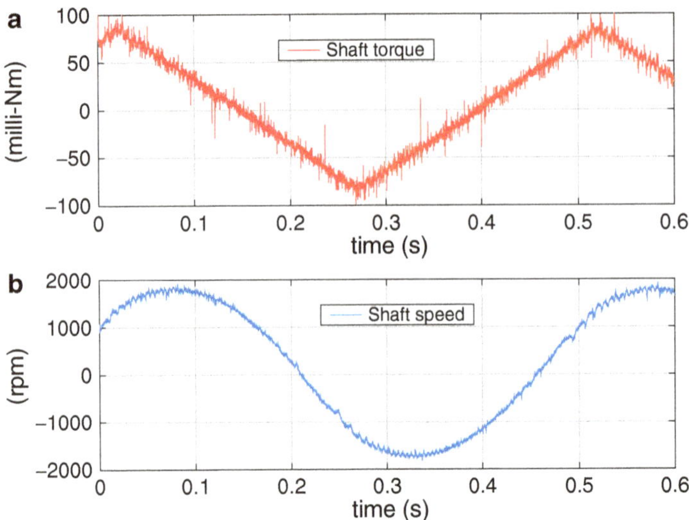

Fig. 11.31 Results: shaft torque and shaft speed versus time waveforms

the previous demonstration laboratories) is used extensively in our book "Applied Control" of Electrical Drives' [10], hence the reader is referred to this text for further guidance on this topic.

The results derived from the demonstration setup (via the Plot module) have been processed via Matlab to show the shaft torque and shaft speed for a time interval $0 \rightarrow 0.6\,\mathrm{s}$. The following comments/observations of the results according to Fig. 11.31 can be made:

- There is good agreement between the measured results and those derived via the PLECS simulation model (see subplots (c) and (d) of Fig. 11.28).
- The "measured" torque was calculated using the measured armature current which was then multiplied by the known magnet flux ψ_{m} of the machine.

An overall conclusion is that the results derived with the DEMO lab validate the theoretical and simulated results presented in this chapter.

Appendix A
Concept of Sinusoidal Distributed Windings

Electrical machines are designed in such a manner that the flux density distribution in the airgap due to a single phase winding is approximately sinusoidal. This appendix aims to make plausible the reason for this and the way in which this is realized. In this context the so-called sinusoidally distributed winding concept will be discussed.

Figure A.1 represents an ITF based transformer or IRTF based electrical machine with a finite airgap g. A two-phase representation is shown with two n_1 turn stator phase windings . The windings which carry the currents $i_{1\alpha} and\ i_{1\beta}$, respectively, are shown symbolically. This implies that the winding symbol shown on the airgap circumference represents the locations of the majority windings in each case, not the actual distribution as will be discussed shortly. If we consider the "α" winding initially, i.e., we only excite this winding with a current $i_{1\alpha}$, then the aim is to arrange the winding distribution of this phase in such a manner that the flux density in the airgap can be represented as $B_{1\alpha} = \hat{B}_{\alpha} \cos \xi$. Similarly if we only excite the "β" winding with a current i_{β}, a sinusoidal variation of the flux density should appear which is of the form $B_{1\beta} = \hat{B}_{\beta} \sin \xi$. The relationship between phase currents and peak flux density values is of the form $B_{1\alpha} = C i_{1\alpha}$, $B_{1\beta} = C i_{1\beta}$ where C is a constant to be defined shortly. In space vector terms the following relationships hold

$$\vec{i_1} = i_{1\alpha} + j i_{1\beta} \tag{A.1a}$$

$$\vec{B}_1 = \hat{B}_{1\alpha} + j \hat{B}_{1\beta} \tag{A.1b}$$

Given that the current and flux density components are linked by a constant C it is important to ensure that the following relationship holds, namely:

$$\vec{B}_1 = C \vec{i}_1 \tag{A.2}$$

© Springer International Publishing Switzerland 2016

A. Veltman et al., *Fundamentals of Electrical Drives*, Power Systems,
DOI 10.1007/978-3-319-29409-4

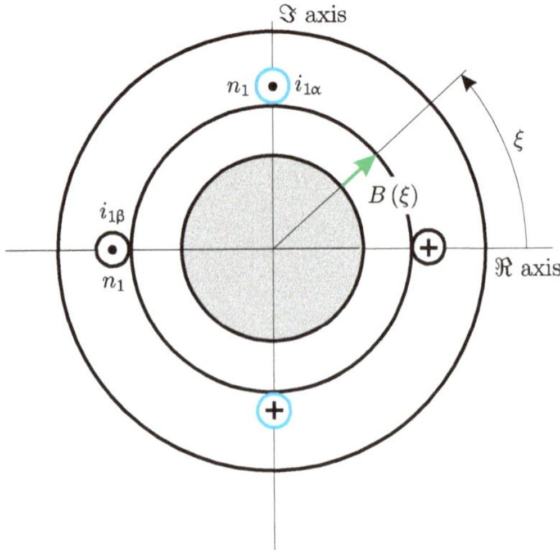

Fig. A.1 Simplified ITF model, with finite airgap, no secondary winding shown

If, for example, the current is of the form $\vec{i}_1 = \hat{i}_1 e^{j\rho}$ then the flux density should be of the form $\vec{B}_1 = C\hat{i}_1 e^{j\rho}$ for any value of ρ and values of $\hat{i}_1$ which fall within the linear operating range of the machine. The space vector components are in this case of the form $i_{1\alpha} = \hat{i}_1 \cos\rho$, $i_{1\beta} = \hat{i}_1 \sin\rho$. If we assume that the flux density distributions are indeed sinusoidal then the resultant flux density B_{res} in the airgap will be the sum of the contributions of both phases namely

$$B_{\text{res}}(\xi) = \underbrace{C\hat{i}_1 \cos\rho}_{\hat{B}_{1\alpha}} \cos\xi + \underbrace{C\hat{i}_1 \sin\rho}_{\hat{B}_{1\beta}} \sin\xi \tag{A.3}$$

Expression (A.3) can also be written as $B_{\text{res}} = C\hat{i}_1 \cos(\xi - \rho)$ which means that the resultant airgap flux density is again a sinusoidal waveform with its peak amplitude (for this example) at $\xi = \rho$, which is precisely the value which should appear in the event that expression (A.2) is used directly. It is instructive to consider the case where $\rho = \omega_s t$, which implies that the currents $i_\alpha and i_\beta$ are sinusoidal waveforms with a frequency of ω_s. Under these circumstances the location within the airgap where the resultant flux density is at its maximum is equal to $\xi = \omega_s t$. A traveling wave exists in the airgap in this case, which has a rotational speed of ω_s rad/s.

Having established the importance of realizing a sinusoidal flux distribution in the airgap for each phase we will now examine how the distribution of the windings affects this goal.

For this purpose it is instructive to consider the relationship between the flux density in the airgap at locations ξ, $\xi + \Delta\xi$ with the aid of Fig. A.2. If we consider a loop formed by the two "contour" sections and the flux density values at locations ξ, $\xi + \Delta\xi$, then it is instructive to examine the sum of the magnetic potentials along

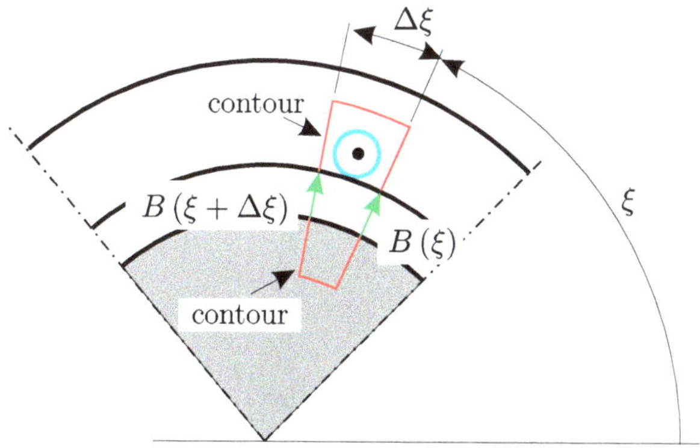

Fig. A.2 Sectional view of phase winding and enlarged airgap

the loop and the corresponding MMF enclosed by this loop. The MMF enclosed by the loop is taken to be of the form $N_\xi i$, where N_ξ represents all or part of the "α" phase winding and i the phase current. The magnetic potentials in the "red" contour part of the loop are zero because the magnetic material is assumed to be magnetically ideal (zero magnetic potential). The remaining magnetic potential contributions when we traverse the loop in the anti-clockwise direction must be equal to the enclosed MMF which leads to

$$\frac{g}{\mu_0} B(\xi) - \frac{g}{\mu_0} B(\xi + \Delta\xi) = N_\xi i \tag{A.4}$$

Expression (A.4) can also be rewritten in a more convenient form by introducing the variable $n_\xi = \frac{N_\xi}{\Delta\xi}$ which represents the phase winding distribution per radian. Use of this variable with Eq. (A.4) gives

$$\frac{B(\xi + \Delta\xi) - B(\xi)}{\Delta\xi} = -\frac{\mu_0}{g} n_\xi i \tag{A.5}$$

which can be further developed by imposing the condition $\Delta\xi \rightarrow 0$ which allows Eq. (A.5) to be written as

$$\frac{\mathrm{d}B(\xi)}{\mathrm{d}\xi} = -\frac{\mu_0}{g} n_\xi i \tag{A.6}$$

The LHS of Eq. (A.6) represents the gradient of the flux density with respect to ξ. An important observation of Eq. (A.6) is that a change in flux density in the airgap is linked to the presence of a non-zero $n_\xi i$ term, hence we are able to construct the flux density in the airgap if we know (or choose) the winding distribution n_ξ and phase current. Vise versa we can determine the required winding distribution needed to arrive at, for example, a sinusoidal flux density distribution.

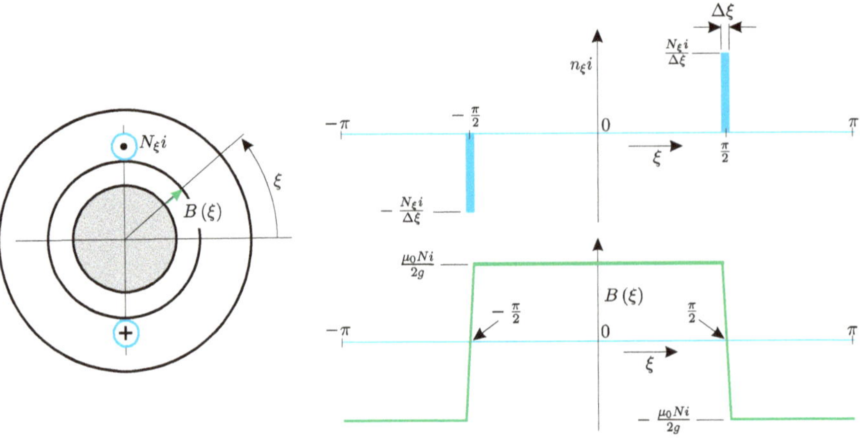

Fig. A.3 Example: concentrated winding, $N_\xi = N$

A second condition must also be considered when constructing the flux density plot around the entire airgap namely

$$\int_{-\pi}^{\pi} B(\xi)\, \mathrm{d}\xi = 0 \tag{A.7}$$

Equation (A.7) basically states that the flux density versus angle ξ distribution along the entire airgap of the machine cannot contain a non-zero average component. Two examples are considered below which demonstrates the use of Eqs. (A.6) and (A.7). The first example as shown in Fig. A.3 shows the winding distribution n_ξ which corresponds to the so-called concentrated winding. This means that the entire number of "N" turns of the phase winding is concentrated in a single slot (per winding half) with width $\Delta\xi$, hence $N_\xi = N$. The corresponding flux density distribution is in this case trapezoidal and not sinusoidal as required.

The second example given by Fig. A.4 shows a distributed phase winding as often used in practical three-phase machines. In this case the phase winding is split into three parts [and three slots (per winding half), spaced λ rad apart] hence, $N_\xi = \frac{N}{3}$. The total number of windings of the phase is again equal to N. The flux density plot which corresponds with the distributed winding is a step forward in terms of representing a sinusoidal function. The ideal case would according to Eq. (A.6) require a $n_\xi i$ representation of the form

$$n_\xi i = \frac{g}{\mu_0} \hat{B} \sin(\xi) \tag{A.8}$$

in which $\hat{B}$ represents the peak value of the desired flux density function $B(\xi) = \hat{B} \cos(\xi)$. Equation (A.8) shows that the winding distribution needs to be sinusoidal. The practical implementation of Eq. (A.8) would require a large

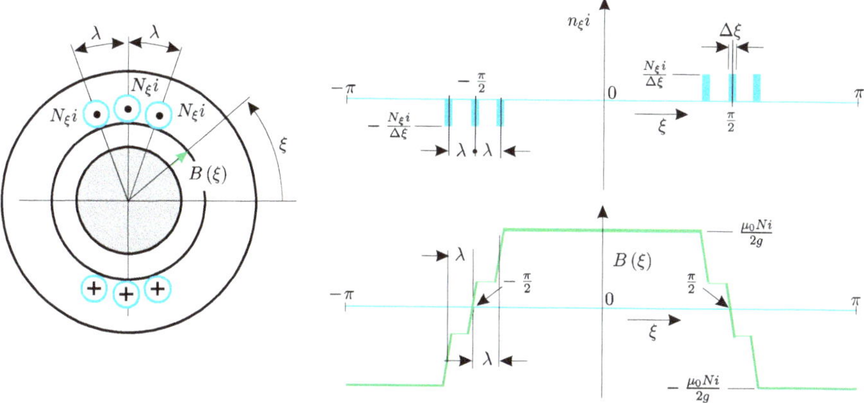

Fig. A.4 Example: distributed winding, $N_\xi = \frac{N}{3}$

number of slots with varying number of turns placed in each slot. This is not realistic given the need to typically house three-phase windings, hence in practice the three slot distribution shown in Fig. A.4 is normally used and provides a flux density versus angle distribution which is sufficiently sinusoidal.

In conclusion it is important to consider the relationship between phase flux-linkage and circuit flux values. The phase circuit flux (for the "α" phase) is of the form

$$\phi_{\mathrm{m}\alpha} = \int_{-\frac{\pi}{2}}^{\frac{\pi}{2}} B(\xi)\, \mathrm{d}\xi \tag{A.9}$$

which for a concentrated winding corresponds to a flux-linkage value $\psi_{1\alpha} = N\phi_{\mathrm{m}\alpha}$. If a distributed winding is used then not all the circuit flux is linked with all the distributed windings components in which case the flux-linkage is given as $\psi_{1\alpha} = N_{\mathrm{eff}}\phi_{\mathrm{m}\alpha}$, where N_{eff} represents the "effective" number of turns.

References

1. Boedefeld T, Sequenz H (1971) Elektrische Maschinen, 8th edn. Springer, Wien, New York. http://www.abebooks.de/buch-suchen/isbn/3211809716/
2. De Doncker RW, Pulle DWJ, Veltman A (2011) Advanced electrical drives. Springer, New York. https://books.google.de/books?id=_sEDx2lAKboC&pg=PR5&lpg=PR5&dq=978-94-007-0179-3&source=bl&ots=IMrbVms-QA&sig=YQWrZ7f_Ju1nIVdylH0Ph-C-JQk&hl=de&sa=X&ved=0ahUKEwiE5pf21JjMAhVEOhoKHdlVAgwQ6AEIMDAD#v=onepage&q=978-94-007-0179-3&f=false
3. FEMM Finite Element Software (2014). http://www.femm.info
4. Holmes D (1997) A generalised approach to modulation and control of hard switched converters. PhD thesis, Department of Electrical and Computer Engineering, Monash University, Clayton
5. Hughes A, Drury B (2013) Electric motors and drives, 4th edn. Newnes, London
6. LaunchXL-F28069M (2015) Texas instruments: InstaSPIN-MOTION. http://www.ti.com/tool/launchxl-f28069M
7. Leonhard W (2001) Control of electrical drives, 3rd edn. Springer, Berlin, Heidelberg. http://www.springer.com/us/book/9783540418207
8. Miller TJE (1989) Brushless permanent-magnet and reluctance motor drives. Monographs in electrical and electronic engineering, vol. 21. Oxford Science Publications, Oxford
9. PLEXIM® Simulation Software (2010). http://www.plexim.com
10. Pulle DWJ, Darnell P, Veltman A (2015) Applied control of electrical drives. Springer, New York
11. Schoonhoven EB (2015). http://issuu.com/eriksbv/docs/lineair-en-motion-control
12. Svensson T (1988) On modulation and control of electronic power converters. Tech. Rep. 186, Chalmers University of Technology, School of Electrical and Computer engineering
13. Veltman A (1994) The fish method: interaction between AC-machines and switching power converters. PhD thesis, Department of Electrical Engineering, Delft Technical University
14. VisSim® Embedded Software (2010). http://www.vissim.com

© Springer International Publishing Switzerland 2016

A. Veltman et al., *Fundamentals of Electrical Drives*, Power Systems,
DOI 10.1007/978-3-319-29409-4

Index

© Springer International Publishing Switzerland 2016

A. Veltman et al., *Fundamentals of Electrical Drives*, Power Systems,
DOI 10.1007/978-3-319-29409-4

GPSR Compliance
The European Union's (EU) General Product Safety Regulation (GPSR) is a set of rules that requires consumer products to be safe and our obligations to ensure this.

If you have any concerns about our products, you can contact us on

ProductSafety@springernature.com

In case Publisher is established outside the EU, the EU authorized representative is:

Springer Nature Customer Service Center GmbH
Europaplatz 3
69115 Heidelberg, Germany

www.ingramcontent.com/pod-product-compliance
Ingram Content Group UK Ltd.
Pitfield, Milton Keynes, MK11 3LW, UK
UKHW021009290726
14059UKWH00001BA/37

* 9 7 8 3 3 1 9 2 9 4 0 8 7 *